PHYSICAL CHEMISTRY FOR THE CHEMICAL AND BIOCHEMICAL SCIENCES

PHYSICAL CHEMISTRY FOR THE CHEMICAL AND BIOCHEMICAL SCIENCES

Edited by

José Luis López-Bonilla, PhD

Marat Ibragimovich Abdullin, DSc

Gennady E. Zaikov, DSc

Reviewers and Advisory Board Member

A. K. Haghi, PhD

AAP APPLE ACADEMIC PRESS

Apple Academic Press Inc. 3333 Mistwell Crescent Oakville, ON L6L 0A2 Canada	Apple Academic Press Inc. 9 Spinnaker Way Waretown, NJ 08758 USA

Exclusive worldwide distribution by CRC Press, a member of Taylor & Francis Group

Printed in the United States of America on acid-free paper

International Standard Book Number-13: 978-1-77188-149-4 (Hardcover)

International Standard Book Number-13: 978-1-77188-289-7 (eBook)

Library and Archives Canada Cataloguing in Publication

Physical chemistry for the chemical and biochemical sciences / edited by José Luis López-Bonilla, PhD, Marat Ibragimovich Abdullin, DSc, Gennady E. Zaikov, DSc ; reviewers and advisory board member A.K. Haghi, PhD.

Includes bibliographical references and index.
Issued in print and electronic formats.
ISBN 978-1-77188-149-4 (bound).--ISBN 978-1-77188-289-7 (pdf)
1. Chemistry, Physical and theoretical--Textbooks. 2. Chemistry.
3. Biochemistry. I. Zaikov, G. E. (Gennadiĭ Efremovich), 1935-, editor
II. López-Bonilla, José Luis, editor III. Abdullin, Marat Ibragimovich, editor
QD453.3.P48 2016 541 C2016-900853-3 C2016-900854-1

CIP data on file with US Library of Congress

Apple Academic Press also publishes its books in a variety of electronic formats. Some content that appears in print may not be available in electronic format. For information about Apple Academic Press products, visit our website at **www.appleacademicpress.com** and the CRC Press website at **www.crcpress.com**

CONTENTS

LIST OF CONTRIBUTORS

Rafail A. Afanas'ev
Pryanishnikov All-Russian Scientific Research Institute of Agrochemistry, d. 31A, Pryanishnikova St., Moscow, 127550, Russia, Tel: +7-499-976-25-01, E-mail: rafail-afanasev@mail.ru

A. Afzali
University of Guilan, Rasht, Iran

Alexander Alentiev
Topchiev Institute of Petrochemical Synthesis, Leninskii Prospect 29, Moscow – 119991, Russia

Z. M. Aleschenkova
Institute of Microbiology, National Academy of Sciences, Kuprevich str. 2, 220141 Minsk, Belarus, Fax: +375-17-267-47-66; E-mail: microbio@mbio.bas-net.by

E. Bakuradze
Department of Biology Faculty of Exact and Natural Sciences, Iv. Javakhishvili Tbilisi State University, Tbilisi, Georgia

O. S. Berdyugina
FGBU "I.I. Mechnikov Research Institute for Vaccines and Sera" RAMS, Moscow 105064 Maliy Kazenniy per. 5a. Russia, Tel.: +7-495-916-11-52, Fax: +7-495-917-54-60; E-mail: labpitsred@yandex.ru

A. I. Beresnev
Institute of Microbiology, National Academy of Sciences, 220141, Kuprevich Str., 2, Minsk, Belarus, Fax: +375(17)264-47-66; E-mail: zinch@mbio.bas-net.by

L. P. Blinkova
FGBU "I.I. Mechnikov Research Institute for Vaccines and Sera" RAMS, Moscow 105064 Maliy Kazenniy per. 5a. Russia, Tel.: +7-495-916-11-52, Fax: +7-495-917-54-60; E-mail: labpitsred@yandex.ru

Maria Bruma
"Petru Poni" Institute of Macromolecular Chemistry, Iasi – 700487, Romania

O. V. Dmitrieva
FGBU "I.I. Mechnikov Research Institute for Vaccines and Sera" RAMS, Moscow 105064 Maliy Kazenniy per. 5a. Russia, Tel.: +7-495-916-11-52, Fax: +7-495-917-54-60; E-mail: labpitsred@yandex.ru

D. Dzidzigiri
Department of Biology Faculty of Exact and Natural Sciences, Iv. Javakhishvili Tbilisi State University, Tbilisi, Georgia, E-mail: d_dzidziguri@yahoo.com

I. S. Eremeev
A.V. Topchiev Institute of Petrochemical Synthesis, Russian Academy of Sciences, 29, Leninsky prospect, 119991, Moscow, Russia

A. A. Fedorenchik
Institute of Microbiology, National Academy of Sciences, Kuprevich str. 2, 220141 Minsk, Belarus, Fax: +375-17-267-47-66; E-mail: microbio@mbio.bas-net.by

Vladimir S. Feofanov
N.M. Emanuel Institute of Biochemical Physics of the Russian Academy of Sciences, Kosygin St. 4-117977 Moscow, Russia, Tel.: +8(495)9361745 (office), +8(906)7544974 (mobile); Fax: (495)1374101; E-mail: komissarova-lkh@mail.ru

N. Giorgobiani
Department of Biology Faculty of Exact and Natural Sciences, Iv. Javakhishvili Tbilisi State University, Tbilisi, Georgia

A. Ya. Gorenberg
Semenov Institute of Chemical Physics, Russian Academy of Sciences, Moscow, Russia

G. P. Karpacheva
A.V. Topchiev Institute of Petrochemical Synthesis, Russian Academy of Sciences, 29, Leninsky prospect, 119991, Moscow, Russia

R. M. Khairullin
The Institute of Biochemistry and Genetics, Ufa Scientific Centre of The Russian Academy of Sciences, 71 October Ave., 450054 Ufa, Russia, Phone/Fax: +7 (347) 235-60-88; E-mail: krm62@mail.ru

S. V. Kolesov
Institute of Organic Chemistry Ufa Research Centre of Russian Academy of Sciences, 71 Prospect Oktyabrya, 450054, Ufa, Russia, E-mail: gip@anrb.ru

Lubov Kh. Komissarova
N.M. Emanuel Institute of Biochemical Physics of the Russian Academy of Sciences, Kosygin St. 4-117977 Moscow, Russia, Tel.: +8(495)9361745 (office), +8(906)7544974 (mobile); Fax: (495)1374101; E-mail: komissarova-lkh@mail.ru

G. V. Kozlov
Kh.M. Berbekov Kabardino-Balkarian State University, Chernyshevsky St., 173, 360004 Nalchik, Russian Federation; E-mail: i_dolbin@mail.ru

O. S. Kukovinets
Bashkir State University, 100 Mingazhev str., 450074, Ufa, Russia, E-mail: ku47os@yandex.ru

Z. M. Kuramshina
Sterlitamak branch of the Bashkir State University, 49 Lenin Ave., 453109 Sterlitamak, Russia, Phone/Fax: +7 (347) 343-38-69; E-mail: kuramshina_zilya@.mail.ru

G. G. Kutlugildina
Bashkir State University, 32 Z. Validi Street, Ufa 450076, Republic of Bashkortostan, Russia, E-mail: moy_mayl@mail.ru

S. V. Kvach
Institute of Microbiology, National Academy of Sciences, 220141, Kuprevich Str., 2, Minsk, Belarus, Fax: +375(17)264-47-66; E-mail: zinch@mbio.bas-net.by

G. Levin
Research Institute for Antioxidant Therapy, 137c Invalidenstr., 10115, Berlin, Germany,
E-mail: ip@antioxidant-research.com

A. L. Margolin
Emanuel Institute of Biochemical Physics, Russian Academy of Sciences, Moscow, Russia; E-mail: monakhova@sky.chph.ras.ru

N. V. Melnikova
Institute of Microbiology, National Academy of Sciences, Kuprevich str. 2, 220141 Minsk, Belarus, Fax: +375-17-267-47-66; E-mail: microbio@mbio.bas-net.by

Genrietta E. Merzlaya
Pryanishnikov All-Russian Scientific Research Institute of Agrochemistry, d. 31A, Pryanishnikov St., Moscow, 127550, Russia, Phone: +7-499-976-25-01; E-mail: lab.organic@mail.ru, User53530@yandex.ru

A. K. Mikitaev
Kh.M. Berbekov Kabardino-Balkarian State University, Chernyshevsky St., 173, 360004 Nalchik, Russian Federation; E-mail: i_dolbin@mail.ru

I. Modebadze
Department of Biology Faculty of Exact and Natural Sciences, Iv. Javakhishvili Tbilisi State University, Tbilisi, Georgia

T. V. Monakhovaa
Emanuel Institute of Biochemical Physics, Russian Academy of Sciences, Moscow, Russia; E-mail: monakhova@sky.chph.ras.ru

G. Mosidze
Department of Biology Faculty of Exact and Natural Sciences, Iv. Javakhishvili Tbilisi State University, Tbilisi, Georgia

P. M. Nedorezova
Semenov Institute of Chemical Physics, Russian Academy of Sciences, Moscow, Russia

Anatoly Iv. Opalko
Uman National University of Horticulture, Instytutska Str., Uman, Cherkassy Region, Ukraine 20305; E-mail: opalko_a@ukr.net

Olga A. Opalko
National Dendrological Park "Sofiyivka" of NAS of Ukraine, 12-a Kyivska Str., Uman, Cherkassy Region, 20300, Ukraine

S. Zh. Ozkan
A.V. Topchiev Institute of Petrochemical Synthesis, Russian Academy of Sciences, 29, Leninsky prospect, 119991, Moscow, Russia

Yu. D. Pakhomov
FGBU "I.I. Mechnikov Research Institute for Vaccines and Sera" RAMS, Moscow 105064 Maliy Kazenniy per. 5a. Russia, Tel.: +7-495-916-11-52, Fax: +7-495-917-54-60;
E-mail: labpitsred@yandex.ru

S. V. Pol'shchikov
Semenov Institute of Chemical Physics, Russian Academy of Sciences, Moscow, Russia

I. N. Popov
Research Institute for Antioxidant Therapy, 137c Invalidenstr., 10115, Berlin, Germany,
E-mail: ip@antioxidant-research.com

A. A. Popova
Plekhanov Russian University of Economics, Moscow, Russia

Inga A. Ronova
Nesmeyanov Institute of Organoelement Compounds, Vavilov Street 28, Moscow – 119991, Russia

M. Rukhadze
Department of Biology Faculty of Exact and Natural Sciences, Iv. Javakhishvili Tbilisi State University, Tbilisi, Georgia

L. Rusishvili
Department of Biology Faculty of Exact and Natural Sciences, Iv. Javakhishvili Tbilisi State University, Tbilisi, Georgia

N. N. Sazhina
Emanuel Institute of Biochemical Physics Russian Academy of Sciences, 4 Kosygin Street, 119334 Moscow, Russia, E-mail: Natnik48s@yandex.ru

N. N. Sigaeva
Institute of Organic Chemistry Ufa Research Centre of Russian Academy of Sciences, 71 Prospect Oktyabrya, 450054, Ufa, Russia, E-mail: gip@anrb.ru

G. G. Sivets
Institute of Bioorganic Chemistry, National Academy of Sciences, 220141, Kuprevich Str., 5/2, Minsk, Belarus, Fax: +375 (17) 267-87-61; E-mail: gsivets@yahoo.com

N. N. Skorlupkina
FGBU "I.I. Mechnikov Research Institute for Vaccines and Sera" RAMS, Moscow 105064 Maliy Kazenniy per. 5a. Russia, Tel.: +7-495-916-11-52, Fax: +7-495-917-54-60; E-mail: labpitsred@yandex.ru

Michail O. Smirnov
Pryanishnikov All-Russian Scientific Research Institute of Agrochemistry, d. 31A, Pryanishnikov St., Moscow, 127550, Russia, Phone: +7-499-976-25-01; E-mail: lab.organic@mail.ru, User53530@yandex.ru

J. V. Smirnova
Sterlitamak branch of the Bashkir State University, 49 Lenin Ave., 453109 Sterlitamak, Russia, Phone/Fax: +7 (347) 343-38-69; E-mail: kuramshina_zilya@.mail.ru

L. V. Spirikhin
Institute of Organic Chemistry Ufa Research Centre of Russian Academy of Sciences, 71 Prospect Oktyabrya, 450054, Ufa, Russia, E-mail: gip@anrb.ru

E. Tavdishvili
Department of Biology Faculty of Exact and Natural Sciences, Iv. Javakhishvili Tbilisi State University, Tbilisi, Georgia

R. R. Usmanova
Ufa State Technical University of Aviation, 12 Karl Marks str., Ufa 450100, Bashkortostan, Russia, E-mail: Usmanovarr@mail.ru

R. R. Vildanova
Institute of Organic Chemistry Ufa Research Centre of Russian Academy of Sciences, 71 Prospect Oktyabrya, 450054, Ufa, Russia, E-mail: gip@anrb.ru

V. P. Volodina
Institute of Organic Chemistry Ufa Research Centre of Russian Academy of Sciences, 71 Prospect Oktyabrya, 450054, Ufa, Russia, E-mail: gip@anrb.ru

I. S. Zaidullin
Ufa Eye Research Institute of Academy of Sciences of the Republic of Bashkortostan, 90 Pushkin str., 450008, Ufa, Russia, E-mail: zaidullinsrb@mail.ru

G. E. Zaikov
N.M. Emanuel Institute of Biochemical Physics of Russian Academy of Sciences, Kosygin St., 4, 119991 Moscow, Russian Federation, E-mail: chembio@sky.chph.ras.ru

Yu. S. Zimin
Bashkir State University, 32 Z. Validi Street, Ufa 450076, Republic of Bashkortostan, Russia, E-mail: ZiminYuS@mail.ru

A. I. Zinchenko
Institute of Microbiology, National Academy of Sciences, 220141, Kuprevich Str., 2, Minsk, Belarus, Fax: +375(17)264-47-66; E-mail: zinch@mbio.bas-net by

D. K. Zinnatullina
Bashkir State University, 32 Z. Validi Street, Ufa 450076, Republic of Bashkortostan, Russia, E-mail: dina.zinnatullina2013@yandex.ru

LIST OF ABBREVIATIONS

ACW	antiradical capacity of water soluble compounds
AFM	atomic force microscopy
AMF	arbuscular mycorrhizal fungi
AOA	antioxidant activity
ARAP	anti-radical ability of proteins
ASC	ascorbic acid
ATI	absolute tolerance index
BNCT	boron neutron capture of tumor therapy
BSA	bovine serum albumin
CA	cellulose acetate
CB	conduction band
CL	chemiluminescence
CNTs	carbon nanotubes
COFs	covalent organic frameworks
CRDCSC	Canadian Research and Development Center of Sciences and Cultures
CTS	chitosan
CVD	chemical vapor deposition
DD	deacetylation degrees
DHA	dialdehyde of hyaluronic acid
DM	degree of modification
DSC	differential scanning calorimetry
ECM	extracellular matrix
ESCs	embryonic stem cells
FAV	fractional accessible volume
FCC	face-centered cubic
FGG	fine-grained graphite
GAG	glycosaminoglycan
GPS	global positioning system
HCP	hexagonal close-packed
HIC	hydrophobic interaction chromatography

HNIPU	hybrid nonisocyanate polyurethane based nanocomposites
HVAC	heating, ventilating, and air conditioning
IAA	indolyl-3-acetic acid
IPN	interpenetrated polymer network
IUPAC	International Union of Pure and Applied Chemistry
L-BPA	L-borophenilalanin
LSD	least significant difference
MCTS	modified chitosan
MDI	methylenebis(phenyl diisocyanate)
MF	microfiltration
MHA	modified HA
MM	molecular masses
MMC	mitomycin C
MMT	montmorillonite
MOFs	metal organic frameworks
MOPs	microporous organic polymers
MW	molecular weight
MWCO	molecular weight cut-off
MWNTs	multi-walled carbon nanotube
NCT	neutron capture therapy
NF	nanofiltration
NIBIB	National Institute of Biomedical Imaging and Bioengineering
PAN	polyacrylonitrile
PANCMPC	polyacrylonitriles-2-methacryloyloxyethyl phosphoryl choline
PCL	polycaprolactone
PEVA	poly[ethylene-co-(vinyl acetate)]
PGA	polyglycolic acid
PP	polypropylene
PPB	potassium-phosphate buffer
PPOA	polyphenoxazine
PSD	pore size distribution
PU	polyurethane
PUCs	polyurethane cationomer
PVA	polyvinyl alcohol

PVC	poly(vinyl chloride)
RMS	root means square
RO	reverse osmosis
RTI	relative tolerance index
SANS	small angle neutron scattering
SAXS	small angle X-ray scattering
SDP	super-deep penetration
SDS	sodium dodecyl sulphate
SEM	scanning electron microscopy
SWNTs	single-walled carbon nanotubes
TEM	transmission electron microscopy
TI	tolerance index
TIC	thermo-initiated CL
TSPC	thermostable protein complex
TUFT	tubes by fiber template
UA	uric acid
UF	ultrafiltration
VB	valence band

PREFACE

By providing an applied and modern approach, this volume will help readers to understand the value and relevance of studying case studies and reviews on chemical and biochemical sciences. Presenting a wide-ranging view of current developments in applied methodologies in chemical and biochemical physics research, the papers in this collection, all written by highly regarded experts in the field, examine various aspects of chemical and biochemical physics and experimentation.

In the Part 1 of this volume, many topics such as, trends in polymeric gas separation membranes, trends in polymer/organoclay nanocomposites, synthesis of the hybrid metal-polymer nanocomposite, oxidation of polypropylene-graphite nanocomposites, and investigation on the cleaning process of gas emissions are discussed. In Part 2, several case studies and reviews in biochemical sciences are reported.

The book:

- introduces the types of challenges and real problems that are encountered in industry and graduate research;
- provides short chapters that introduce students to the subject in more bite-sized pieces;
- presents biochemical examples and applications;
- focuses on concepts above formal experimental techniques and theoretical methods.

The book is ideal for upper-level research students in chemistry, chemical engineering, and polymers. The book assumes a working knowledge of calculus, physics, and chemistry, but no prior knowledge of polymers.

ABOUT THE EDITORS

José Luis López-Bonilla, PhD

José Luis López-Bonilla obtained his PhD in theoretical physics (1982) from the National Polytechnic Institute (IPN), Mexico City, Mexico. He is a researcher of mathematical methods applied to engineering in the Higher School of Mechanical and Electrical Engineering, IPN, and an editorial board member of the *SciTech, Journal of Science and Technology; Global Engineers and Technologist Review; Journal of Interpolation and Approximation in Scientific Computing; International Journal of Chemoinformatics and Chemical Engineering; IUG Journal of Natural and Engineering Studies; International Journal of Applied Mathematics and Machine Learning;* and *Management of Sustainable Development*.

Marat Ibragimovich Abdullin, DSc

Marat Ibragimovich Abdullin, DSc, is currently dean of technological faculty and head of laboratory at Bashkir State University in Ufa, Russia. He is an expert in the fields of chemical physics, physical chemistry, chemistry and physics of high molecular compounds, as well as synthesis and modification of polymers (including degradation and stabilization of polymers and composites–nanocomposites). He has published about 750 original papers and reviews as well as several monographs.

Gennady E. Zaikov, DSc

Gennady E. Zaikov, DSc, is head of the Polymer Division at the N. M. Emanuel Institute of Biochemical Physics, Russian Academy of Sciences, Moscow, Russia, and professor at Moscow State Academy of Fine Chemical Technology, Russia, as well as professor at Kazan National Research Technological University, Kazan, Russia. He is also a prolific author, researcher, and lecturer. He has received several awards for his work, including the Russian Federation Scholarship for Outstanding Scientists. He has been a member of many professional organizations and on the editorial boards of many international science journals.

A. K. Haghi, PhD

A. K. Haghi, PhD, holds a BSc in urban and environmental engineering from University of North Carolina (USA); a MSc in mechanical engineering from North Carolina A&T State University (USA); a DEA in applied mechanics, acoustics and materials from Université de Technologie de Compiègne (France); and a PhD in engineering sciences from Université de Franche-Comté (France). He is the author and editor of 165 books as well as 1000 published papers in various journals and conference proceedings. Dr. Haghi has received several grants, consulted for a number of major corporations, and is a frequent speaker to national and international audiences. Since 1983, he served as a professor at several universities. He is currently editor-in-chief of the *International Journal of Chemoinformatics and Chemical Engineering* and *Polymers Research Journal* and on the editorial boards of many international journals. He is a member of the Canadian Research and Development Center of Sciences and Cultures (CRDCSC), Montreal, Quebec, Canada.

PART I

CHEMICAL SCIENCES

CHAPTER 1

KINETICS OF OXIDATION OF POLYVINYL ALCOHOL BY OZONE IN AQUEOUS SOLUTIONS

YU. S. ZIMIN, G. G. KUTLUGILDINA, and D. K. ZINNATULLINA

Bashkir State University, 32 Z. Validi Street, Ufa 450076, Republic of Bashkortostan, Russia

E-mail: ZiminYuS@mail.ru, moy_may1@mail.ru, dina zinnatullina2013@yandex.ru

CONTENTS

ABSTRACT

The kinetics of the oxidation of polyvinyl alcohol by ozone in aqueous solutions was investigated in the temperature interval 6–32°C. The activation parameters of the reaction were determined.

1.1 INTRODUCTION

It is known [1–15] that the oxidation by ozone of alcohols with different atomicity obeys the various kinetic laws. So, for example, in the reaction with mono- [1–4, 6–10, 12, 14, 15] and diatomic [11, 13–15] alcohols (the medium – water and organic solvents) the ozone is consumed according to the second order reaction law. At the same time within the range of great substrate concentrations the effective second-order rate constant of monoatomic alcohols (methanol [8], ethanol [7] and 2-propanol [8]) oxidation does not remain invariable, but it increases with the increase of the initial alcohol concentration. On the contrary, when oxidizing the alcohols of the higher atomicity (with the quantity of OH-groups more than 3) – glycerine [5, 8, 13–15], ethriole [13–15], pentaerythritol [5, 8, 13–15] and mannitol [13–15] – the effective rate constant decreases with the growth of the substrate concentration. The indicated variety of the experimental results on the ozonized oxidation of alcohols (S) in different solvents was explained in the framework of the kinetic scheme [14, 15], which includes the formation of the intermediate complex of alcohol with ozone $S \cdots O_3$:

$$S + O_3 \rightleftarrows S \cdots O_3 \tag{1}$$

$$S \cdots O_3 \rightarrow \text{Products} \tag{2}$$

$$S \cdots O_3 + S \rightarrow \text{Products} \tag{3}$$

Herewith, the difference in the kinetics laws is related to various stages, which limit the oxidation process.

In view of the aforementioned facts the kinetics of ozonized oxidation of polyvinyl alcohol (PVA) is of key interest. On the one hand, PVA for which we can expect the laws of the consumption the ozone specific for alcohols with higher atomicity ($n > 3$). On the other hand, the molecule of PVA (unlike the previously studied polyatomic alcohols – glycerine, ethriole, pentaerythritol and mannitol) has not only α-CH-bonds, but also β-CH-bonds, which can react with ozone and result to changes in the kinetic laws of the ozone (O_3) consumption, too.

Therefore, the aim of this work was the investigation the kinetics of the ozone consumption in the reaction with the synthetic polymer – PVA – in aqueous media.

1.2 EXPERIMENTAL PART

PVA "REAHIM" with an average molecular weight 31 kDa ([η] = 0.58 dL/g, water, 25 ± 1°C) was used as the study object. Ozone-oxygen mixture was obtained with the use of an ozonizer similar to that described in the work [16], which enables to obtain mixture of O_3–O_2, comprising 1÷2 vol. % of ozone. Double distilled water was used as a solvent.

The reaction kinetics was studied by the ozone consumption in the liquid phase spectrophotometrically at a wavelength of 270 nm (ε = 2700 L/mol·cm [7]). The experiments were conducted in the temperature-controlled cell similar to that described in the work [17], being in the cuvette chamber of spectrophotometer. After the preliminary thermostating during 10 minutes the ozone-oxygen mixture was supplied to the cell, comprising aqua's solution of PVA with the definite concentration. When the ozone concentration in the solution reaches a predetermined value, the supply was stopped and the consumption of ozone was considered. Let us note, that the saturation by ozone was quickly reached (less than 20 seconds), so the adjusted initial concentration of PVA was changed slightly at that time.

The statistical treatment of the experimental data was conducted in the confidence interval of 95%.

1.3 RESULTS AND DISCUSSION

In the interaction of the ozone with PVA O_3 consumes in two ways: (i) the thermal consumption of ozone in water, and (ii) the reaction of ozone with PVA. Therefore, to analyze the kinetics of the reaction considered in the present work it is necessary to have data describing the decomposition of ozone in the absence of PVA.

1.3.1 THE KINETICS OF THERMAL CONSUMPTION OF OZONE IN WATER

The decomposition of O_3 in bi-distilled water in the absence of PVA was studied in 6–32°C by the method described in the experimental part. Thus, the bi-distilled water was placed into the cell instead of the aqueous solution of PVA. The initial ozone concentrations were $(1.0 \div 5.3) \times 10^{-4}$ mol/L.

Analysis of the kinetic curves of ozone decomposition showed that the best description of them is reached in the framework of first order equation:

$$-\frac{d[O_3]}{dt} = k_0[O_3]$$

where k_0 is an effective rate constant of the ozone consumption. As the example Figure 1.1 shows the typical kinetic curve of the decomposition of O_3 and its anamorphosis in semi-logarithmic coordinates:

$$\ln\frac{[O_3]_0}{[O_3]_t} = k_0 t$$

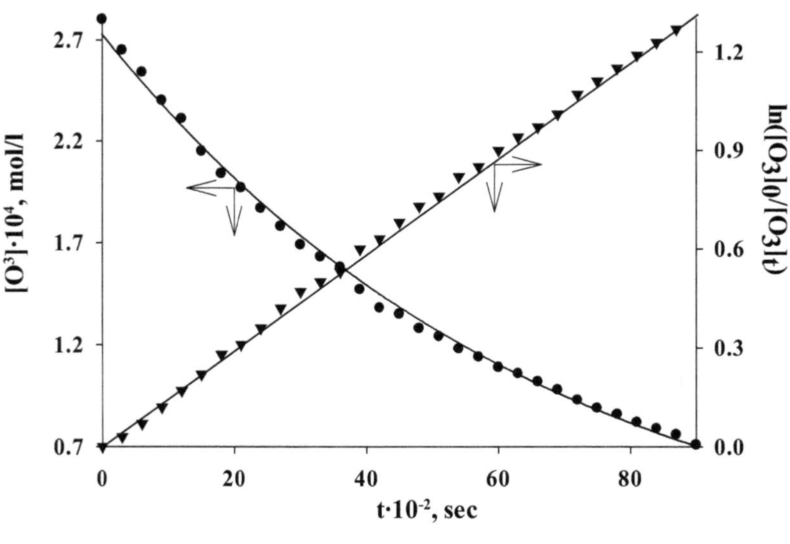

FIGURE 1.1 The kinetic curve of ozone decomposition in double distilled water and its semilogarithmic anamorphosis; 32°C.

where $[O_3]_0$, $[O_3]_t$ – initial and current concentrations of ozone (mol/L). The high value of correlation coefficient (r = 0.997) tells us about good execution of the last equation. The curve piece, which corresponds to the conversion in 50–60%, was used (on this plot the optical density of the solution was determined more reliably) for the calculation of the rate constant k_0.

The results of determination of the rate constant of ozone decomposition k_0 at different temperatures are summarized in Table 1.1, where were found the parameters of Arrhenius equation:

$$\lg k_0 = (12 \pm 3) - (63 \pm 14)/\theta,\ r = 0.996$$

where $\theta = 2.303$ RT kJ/mol.

1.3.2 THE KINETICS OF OZONE CONSUMPTION IN THE REACTION WITH POLYVINYL ALCOHOL

The reaction kinetics of ozone with PVA was studied at temperature 6÷32°C. The choice of low temperatures is determined by the necessity of studying the initial stages of oxidation of PVA. So that the concentration of products in the reaction mixture will be insignificant and their influence on the oxidation process can be neglected. The initial concentrations of PVA and ozone in the reaction mixture were varied within the ranges $(0.3 \div 2.0) \times 10^{-2}$ and $(1.0 \div 5.3) \times 10^{-4}$ mol/L, respectively; in all experiments $[S]_0 >> [O_3]_0$.

TABLE 1.1 The Temperature Dependence of the Rate Constants k_0 of Ozone Decomposition in Water

T, °C	$k_0 \times 10^4$, sec^{-1}
6	0.58 ± 0.09
12	0.75 ± 0.07
17	1.38 ± 0.10
22	2.55 ± 0.11
27	3.90 ± 0.13
32	4.72 ± 0.06

The typical kinetic curves of ozone consumption in the reaction with macromolecules of PVA and their semilogarithmic anamorphosis (r ≥ 0.995) are shown in Figure 1.2. This data explicitly indicates the first order reaction by ozone:

$$-\frac{d[O_3]}{dt} = k'[O_3]$$

where k' is an effective rate constant of the reaction. Whereas k' >> k_0 (see Section 1.1), thermal decomposition of O_3 in the conditions of our experiments can be neglected.

It was found that the effective rate constants k' are directly proportional to the initial concentrations of PVA (Figure 1.3, r ≥ 0.995):

$$k' = k\,[\Pi BC]_0$$

which indicates the first order of the reaction by the polymer.

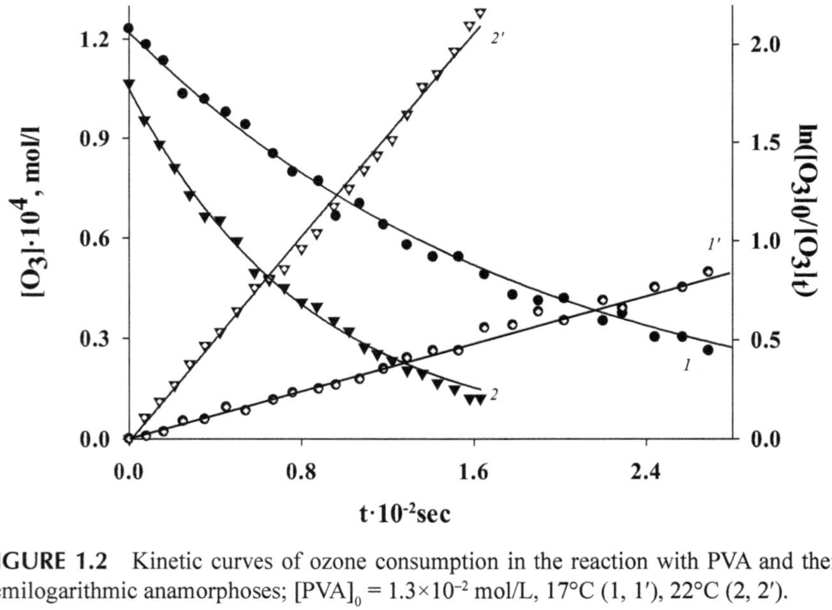

FIGURE 1.2 Kinetic curves of ozone consumption in the reaction with PVA and their semilogarithmic anamorphoses; $[PVA]_0 = 1.3\times10^{-2}$ mol/L, 17°C (1, 1'), 22°C (2, 2').

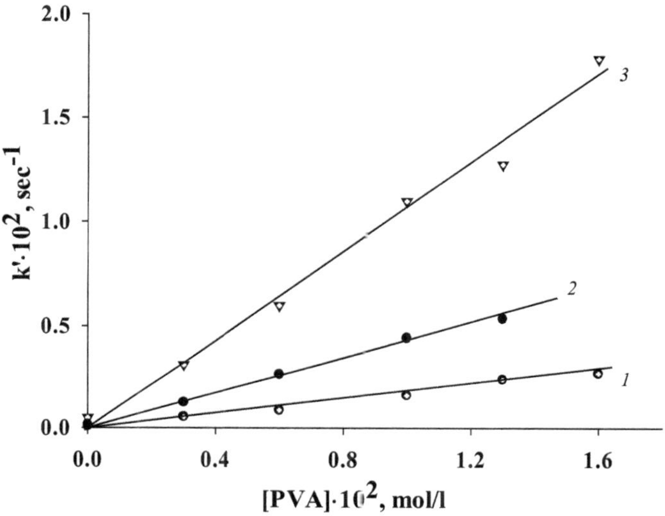

FIGURE 1.3 Dependence of k' on $[PVA]_0$ in the reaction of ozone with PVA at different temperatures: 6°C (1), 17°C (2), 22°C (3).

Therefore, the rate of ozone consumption in the reaction with the PVA is described by the following kinetic equation:

$$-\frac{d[O_3]}{dt} = k[\Pi BC][O_3]$$

where k is the rate constant of the reaction of ozone with PVA. It is significant to note that the total second order of the reaction (first for each of the reagents) is observed in the whole investigated range of process conditions. The k values were determined at different temperatures from the dependency of $k' = f([PVA]_0)$ (Table 1.2).

Processing of the obtained results (Table 1.2) in the coordinates of Arrhenius equation allowed us to determine the activation parameters of the investigated process:

$$\lg k = (14 \pm 3) - (77 \pm 2)/\theta$$

where $\theta = 2.303$ RT kJ/mol.

TABLE 1.2 The Temperature Dependence of the Rate Constant k of the Ozone in the Reaction with PVA

T, °C	$k \times 10^{-2}$, L/mol·sec
6	0.17 ± 0.05
12	0.37 ± 0.04
17	0.42 ± 0.04
22	0.83 ± 0.09
27	1.95 ± 0.08
32	2.83 ± 0.09

On the basis of the obtained results it is possible to make the following conclusion. The kinetics of ozone consumption in the reaction with the PVA described by the second order reaction law (the first – on ozone and the first – on substrate). Similar pattern, as noted above, is typical for ozone oxidation of mono- and diatomic alcohols in the field of small concentrations of the substrate and it is not typical for the oxidation of alcohols with higher atomicity (with the number of OH-groups $n \geq 3$) – glycerine, ethriole, pentaerythritol and mannitol. Obviously, here is implemented the following ratio: $k_2 + k_3[S] >> k_{-1}$ (see the scheme at the beginning of the article). Thus, all the obtained complexes $S \cdots O_3$ ($[S \cdots O_3] << [O_3]$) are transformed into products and, consequently, the rate of ozone consumption is determined by the rate of the reaction (1):

$$-\frac{d[O_3]}{dt} = k_1[S][O_3]$$

In summary, the interaction of ozone with molecules of substrate (reaction 1) is the limited step of the process of ozone oxidation of PVA as well as for mono- and diatomic alcohols.

1.4 CONCLUSION

It has been spectrophotometrically determined, that in aqueous solutions of PVA the ozone consumption explicitly obeys to the second order reaction law. The temperature dependence of the rate constant of the reaction of ozone with the PVA was studied in the range of 6÷32°C and parameters of Arrhenius equation were ascertained: $\lg k = (14 \pm 3) - (77 \pm 2)/\theta$, where $\theta = 2.303$ RT kJ/mol.

The work was supported by RFBR (project 14-03-97026 Povolgie) and project (project code: 299, 2014), executable under the project part of the state task of the Ministry of education and science of the Russian Federation in the sphere of scientific activity.

KEYWORDS

- **kinetics**
- **oxidation**
- **ozone**
- **polyvinyl alcohol**

REFERENCES

1. Williamson, D. G., Cvetanovic, R. J. J. Amer. Chem. Soc. (1970). V. 92, № 10. 2949–2952.
2. Gerchikov, A. Ya., Kuznetsova, E. P., Denisov, E. T. Kinet. Catal. (1974). V 15, № 2. 509–511 (in Russian).
3. Galimova, L. G. The mechanism of cyclohexane oxidation by ozone. Dissertation for candidate degree on chemical sciences. Ufa. IC BBAS USSR. (1975). 124 p. (in Russian).
4. Shereshovets, V. V., Shafikov, N. Ya., Komissarov, V. D. Kinet. Catal. (1980). V. 21, № 6. 1596–1598 (in Russian).
5. Shereshovets, V. V., Galieva, F. A., Tsarkov, A. V., Bikbulatov, I. K. Reakt. Kinet. Catal. Lett. (1982). V. 21, № 3. 413–418
6. Shereshovets, V. V., Galieva, F. A., Akhunov, I. R., Komissarov, V. D., Tsarkov, A. V., Bikbulatov, I. K. Russ. Chem. Bull. (1983). № 5. 1011–1015 (in Russian).
7. Shafikov, N. Ya. Kinetics, products and mechanism of ethanol oxidation by ozone. Dissertation for candidate degree on chemical sciences. Ufa. IC BBAS USSR. (1985). 166 p. (in Russian).
8. Galieva, F. A. Kinetics of gross-radical decomposition of hydrothreeoxides. Dissertation for candidate degree on chemical sciences. Ufa. IC BBAS USSR. (1986). 185 p. (in Russian).
9. Rakovski, S., Cherneva, D. Int. J. Chem. Kinet. (1990). V. 22, № 4, 321–329.

10. Siroejko, A. M., Proskuryakov, V. A. Russ. J. Appl. Chem. (1998). V. 71, № 8, 1346–1349.
11. Zimin Yu. S., Trukhanova, N. V., Shamsutdinov, R. R., Komissarov, V. D. React. Kinet. Catal. Lett. (1999). V. 68, № 2, 237–242.
12. Gerchikov, A. Ya., Zimin Yu. S., Trukhanova, N. V., Evgrafov, V. N. React. Kinet. Catal. Lett. (1999). V. 68, № 2, 257–263.
13. Zimin Yu. S., Trukhanova, N. V., Streltsova, I. V., Komissarov, V. D. Kinet. Catal. (2000). V. 41. № 6, 827–830 (in Russian).
14. Komissarov, V. D., Zimin Yu. S., Trukhanova, N. V., Zaikov, G. E. Oxid. Commun. (2005). V. 28, № 3, 559–567.
15. Zimin Yu. S. Kinetics and mechanism of ozonized oxidation of alcohols, ethers, ketones and olefins in an aqueous medium. Dissertation for doctor degree on chemical sciences. Ufa. IOC USC RAS. (2006). 302 p. (in Russian).
16. Vendillo, V. P., Emel'yanov Yu. M., Philippov Yu.V. Zavod. labor. (1959). V. 25, № 11, 1401–1402.
17. Komissarov, V. D., Gerchikov, A.Ya. Kinet. Catal. (1974). V. 15, № 4, 916–921 (in Russian).

CHAPTER 2

MODIFICATION OF CHITOSAN AND HYALURONIC ACID TO OBTAIN SUSTAINABLE HYDROGELS

R. R. VILDANOVA,[1] N. N. SIGAEVA,[1] O. S. KUKOVINETS,[2] V. P. VOLODINA,[1] L. V. SPIRIKHIN,[1] I. S. ZAIDULLIN,[3] and S. V. KOLESOV[1]

[1]*Institute of Organic Chemistry Ufa Research Centre of Russian Academy of Sciences, 71 Prospect Oktyabrya, 450054, Ufa, Russia, E-mail: gip@anrb.ru*

[2]*Bashkir State University, 100 Mingazhev str., 450074, Ufa, Russia, E-mail: ku47os@yandex.ru*

[3]*Ufa Eye Research Institute of Academy of Sciences of the Republic of Bashkortostan, 90 Pushkin str., 450008, Ufa, Russia, E-mail: zaidullinsrb@mail.ru*

CONTENTS

ABSTRACT

Hydrogels on the base of dialdehyde of hyaluronic acid (DHA) and chitosan, modified by succinic anhydride (MCTS), with entrapped mitomycin C (MMC) are obtained for antiglaucoma filtering surgery in ophthalmology. Such hydrogels contain no toxic cross-links and cross-linkers due its formation is related to cross-links –C=N– (between aminogroups of MCTS and aldehyde groups of DHA) called as Schiff base. The effect of way formation of solutions on retention of mitomycin C from hydrogels is studied.

2.1 INTRODUCTION

Biopolymers of glycosaminoglycan line – hyaluronic acid (HA) and chitosan (CTS) represent special interest due to its using as scaffolds for drug preparations because besides diffusion-controlled release of medical preparation, its address deliver to systems and organs of body, they exhibit high biocompatibility and possess broad spectrum of biological activity, they at some cases show obvious synergetic effect of action of biopolymers and pharmaceutic base [1–6]. At that same time fixing of medical products on the biopolymer have been realized so way, as to provide it's diffusion-controlled released with safety of medical properties and biocompatibility of polymer carrier. It is promoted by development of medical forms on the base of HA and CTS (hydrogels, films, complexes, different scaffolds and others) with drugs [7–9] including mitomycin C (MMC). Recently MMC is widely used in ophthalmology in glaucoma operations for retention of drainage effect for a long time. Excessive exudative-fibrotic response and hyperactive regeneration in postoperative period not only reduce the effectiveness of surgical action and also lead to severe complications. The use of cytostatics (MMC) provides long antiproliferative effect. However, there is a possibility of complications due toxic action of drug (long hypotonia, scleral staphyloma) that reduce hypotensive effect of surgical operation. Therefore, it's a challenge of designing of medical forms and medical systems, that would provide prolonged action of MMC and at the same time not lead to evolution of complications due toxic action of the drug.

Introduction different functional groups in macromolecules of HA and CTS with its modifications could allow not only preparation of hydrogels on the base of HA and CTS by their interaction but also fixing medical preparation on a polymeric matrix. Moreover, native HA in body by action of enzymes and free radicals is quickly degraded, that limits its use in cases required prolonged medical effect, and majority brands of CTS is dissolved only in acidic medium and cannot used in ophthalmology [10, 11].

Therefore the aim of this work was the modifications of HA and CTS for hydrogel preparations on their base with an extended selection of MMC.

2.2 EXPERIMENTAL PART

Hyaluronic acid from "Sigma Aldrich" and chitosan from "Bioprogress" and "Chimmed" have been used. Characteristics of used samples are given in table 1.

Modification of HA was carried out in two ways: by interaction with epichlorohydrin (EpCl) and by oxidation with sodium periodate.

On the 0.1% solution HA in water have been acted by epichlorohydrin in the mixture of NH_4OH and NaOH (10% solution NH_3 in water and 0.3% solution of NaOH) with ratios of components (mol/mol): NH_4OH:NaOH=10:10 [12]. A portion of HA containing 1 mmol disaccharide units was dissolved in 300 mL distilled water for 2 days at room temperature at constant stirring. To the obtained solution 200 mL of mixture containing 10 mmol NH_3 and 10 mmol NaOH has been added. EpCl also has been added in correlation: 2.5, 5.0, 7.0, 10.0 (mol/mol disaccharide units of HA). Reaction was carried out at constant stirring in 24 hours at 25±0.1°C. The reaction mixture was purified by dialysis against water for 3 days, replacing the water every 12 hours. After the lyophilization received a modified hyaluronic acid.

The degree of modification was determined by ^1H NMR spectra by comparison of integrated intensities of methyl protons of acetamidic fragment and triplet of CH_2Cl-protons.

Aldehyde groups have been introduced in macromolecules of HA by sodium periodate oxidation. Into the solution of HA in water (concentration 10 mg/mL) 0.5 M solution of sodium periodate has been added by drop. Reaction was carried at constant stirring in 2 (DHA-2), 4 (DHA-4),

12 and 24 (DHA-24) hours at 25±0.1°C in the darkness. Not reacted periodate was inactivated by ethylene glycol addition. After dialysis of reaction mixture polymer was recovered and dried by lyophilization [13, 14].

The degree of modification is calculated using ^{1}H NMR spectra by comparison of integrated intensities of the methyl group protons of acetamidic fragment and the sum of proton signals of hydrated form of aldehyde in DHA.

In order to modify there were used three chitosan brands, varied by deacetylation degrees (DD) and molecular masses (MM). DD was evaluated by acid-base titration with potentiometric definition of equivalence (Table 2.1), and was calculated with the formula [15]:

$$SD = \frac{203.2}{42.0 + \dfrac{1000m}{C(NaOH)(V_2 - V_1)}}$$

where m is the mass of CTS in the sample (g), C(NaOH) – concentration of sodium hydroxide solution, gone on titration of aminogroups (mol/L); 203.2 – molecular mass of acetylated monosaccharide unit of

TABLE 2.1 Characteristics of Samples of Chitosan (CTS), Modified Chitosan (MCTS), Hyaluronic Acid (HA), Modified HA (MHA and DHA)

Original samples	[η], dL/g	$M_v \times 10^{-3}$	M_z (*) or $M_{s[\eta]} \times 10^{-3}$	DD, %	Modified samples	DM, %	[η], dL/g	M_z or $M_{s[\eta]} \times 10^{-3}$
CTS-1	0.7[1]	23[1]	36*	74.0	MCTS-1	52	0.5	24*
CTS-2	1.9[1]	74[1]	87	83.6	MCTS-2	74	1.2	72
CTS-3	3.5[1]	150[1]	150*	81.4	MCTS-3	77	1.4	150*
HA	21.8[2]	1300[2]	1600	—	MHA	88	4.1	260
					DHA-2	—	5.4	190
					DHA-4	15	0.5	14*
					DHA-24	27	0.2	13*

[1]solvent: acetate buffer, pH 4.5.

[2]solvent: physiological solution (0.9% solution of NaCl in distilled water).

polysaccharide; 42.0 – the difference of molecular masses between acetylated and diacetylated of monosaccharide units; 1000 – multiplier of conversion of milliliters to liters.

Chitosan samples were modified by succinic anhydride. For this, 0.5 g of chitosan was dissolved in 40 mL 5% lactic acid with stirring for 1 day. About 1.5 g succinic anhydride was dissolved in 160 mL of methanol and was added to the chitosan solution. The reaction mixture was kept at constant stirring at room temperature for 24 hours. Modified chitosan (MCTS) was precipitated at pH 8–9 by acetone. The precipitate was filtered, redissolved in distilled water and dialyzed during 3 days. The purified product was precipitated by acetone and dried under vacuum up to constant weight [10, 16].

Characteristic viscosities of unmodified and modified samples were determined by the method of viscosimetry with viscosimeter Ubbelode at 25±0.1°C and M_v for initial samples was calculated by the Mark-Kun-Hawink equation: $[\eta]=2.9\times10^{-4}M^{0.8}$ (for HA in physiological solution) [17] and $[\eta]=1.38\times10^{-4}M^{0.85}$ (for CTS in acetate buffer, pH=4.5) [18].

For original and modified polymer samples with high MM constants sedimentation were found using ultracentrifuge MOM-3080 and molecular mass was calculated:

$$M_{S[\eta]}^{3/2} = \frac{S_0 \eta_0 N_A}{F^{1/3} P^{-1}} [\eta]^{1/3}$$

where $F^{1/3}P^{-1}$ – hydrodynamic invariant = 2.71×10^6.

Molecular characteristics of samples with low MM values before and after modification were determined by the method of sedimentation equilibrium (M_z).

$$M_z = \frac{RT}{(1-\nu\rho_0)\omega^2} tg\alpha$$

where is R = 8.31 Joule/mol·K; T – absolute temperature, K; $\omega = 2\pi n$ rp/sec; n – a number of rotor's revolutions; $(1-\nu\rho_0)$ – Archimedean multiplier, where – ν specific partial polymer volume; ρ_0 – density of solvent.

The degree of modification (DM) has been evaluated by the ninhydrin analysis.

Series of chitosan solutions in 0.5% acetic acid with concentration range of 0.05–0.2% were prepared. From each solution 3 parallel sample of 0.2 mL were selected, then 2.5 mL of phosphate buffer solution (pH=6.6) and 0.5 mL of 1% ninhydrin solution were added to each other. The mixtures were heated for 60 minutes in boiling water bath, cooled and quantitatively transferred in volumetric flasks with capacity of 50 mL, the amounts were diluted to the mark and were photometried at 570 nm. The solution obtained by the mixing of ninhydrin and phosphate buffer was used as comparison solution.

Calibration curves of the optical density of the degree of deacetylation of CTS (DD) were built. The content of aminogroups in the modified samples was determined by the calibration curve corresponding to the initial sample of chitosan by the formulas:

$$n(R-NH_2)=\frac{A(MCTS)*DD}{A(CTS)};$$

$$n(R-COOH)=DD-n(R-NH_2)$$

where A(MCTS) – the optical density of 0.2% MCTS solution at 520 nm; DD – degree of deacetylation of chitosan; A(CTS) – the optical density of 0.2% chitosan solution at 570 nm; n(R-NH_2) – the number of aminogroups in the modified chitosan, %; n(R-COOH) – the number of modified carboxyl groups in chitosan.

Interaction of polymers with MMC was estimated by UV spectroscopy in physiological solution and using methods of isomolecular series and molar ratios. The constants of complex stability (K) on the tandent of the dependence $[C_0]/A-A_0$ on 1/[MMC] were defined. Here A and A_0 are optical densities of solutions in the presence and absence of MMC; $[C_0]$ is initial concentration of the substrate; [MMC] – concentration of mitomycin C [20, 21].

For hydrogel preparations solutions of modified polymers with different concentrations and MMC solution have poured. Solutions were thermostated at room temperature. The gelling time was assessed visually.

The diffusion of MMC out of gel in physiological solution was determined by means of UV-spectroscopy for maximum absorption at 364 nm.

^{1}H NMR and ^{13}C NMR spectrum were registered with spectrometer Bruker Avance-500 in D_2O using as an internal standard –TMS. For ^{1}H NMR spectra working frequency was 500 MHz. UV-spectra were recorded on a spectrophotometer Schimadzu. IR spectra were registered on the unit Tensor-27 ("Bruker").

2.3 RESULTS AND DISCUSSION

Two ways of HA modification were used. So HA was modified by epichlorohydrin in the ammonia-alkaline solution at various initial reagents ratios. It was found that with the ratio of mol disaccharide units HA per mol of EpCl of 2.5 water-soluble products of MHA were formed. In earlier studies [12, 22, 23] have been showed that depending on the pH of reaction solution interaction of EpCl with HA may occur as primary hydroxyl group and amino groups, formed by HA deacetylation.

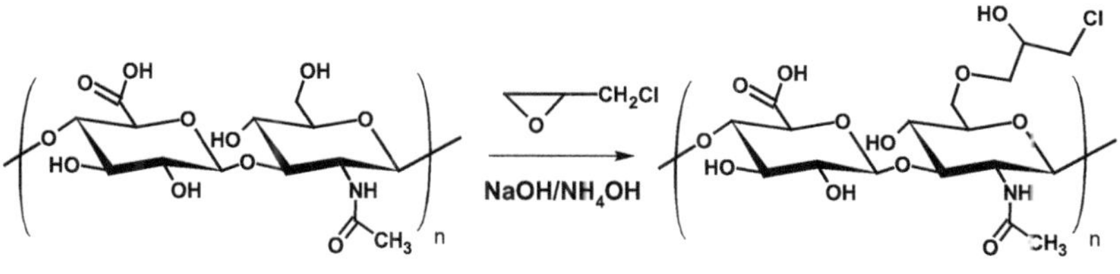

The study of the soluble sample MHA by the method of ^{13}C NMR spectroscopy showed that at spectrogram disappeared signal in the area 60.46 ppm typical for the carbon of primary CH_2OH group, but there was a coupled signal triplets in the field 65.09 ppm belonging to two –CH_2O– groups in modified fragment and triplet at 48.0 for –CH_2Cl. The band of 25 ppm of acetamide group's carbon atom is saved. Thus, under these conditions modification runs through primary hydroxyl group with formation of ether links. In the ^{1}H NMR spectrum the doublet of doublets with center at 3.65 ppm is appeared for CH_2Cl-groups with constants 7.0 and 8.0, respectively. The calculated degree of modification for MHA is 88%.

With increasing molar ratio between EpCl and HA cross-linked polymer is got that precipitates. By sodium periodate oxidation vicinal hydroxyl groups of HA are replaced with two aldehyde groups.

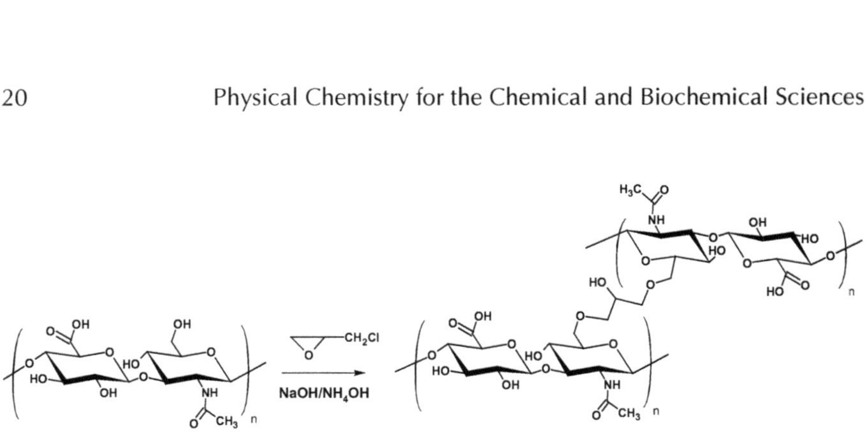

The formation of aldehyde groups in macromolecules of HA confirmed spectroscopically. In the infrared spectrum of DHA new bands in the area 2790 and 1725 cm^{-1} (vibrations of C-H and C=O groups) are received. It is typically to aldehydes. In accordance with [24] HA dialdehyde exists in solution in hydrated form, so in ^1H NMR spectrum of modified samples three signals appear in the region of 5 ppm, instead of the expected 9–10 ppm.

By comparison of integrated intensities of hydroxyl groups' signals of hydrated form of aldehyde with the signal of methyl group of acetamide fragment in ^1H NMR spectra there were degrees of DHA modifications calculated. For the sample DHA-2, modified within 2 hours, any changes in the spectrum are practically not observed. Obviously, modification of HA for 2 hours was held in a small extent. Modification of the sample DHA-4 (4 hours) was 15%, and DHA-24 (24 hours) – 27%.

Modifications of chitosan samples with different deacetylation (DD) (Table 2.2) were carried out by their interaction with succinic anhydride in acidic-ethanol solution. In CTS macromolecules polar carboxyl groups was introduced and in contrast to original CTS modified samples were dissolved in water and physiological solution.

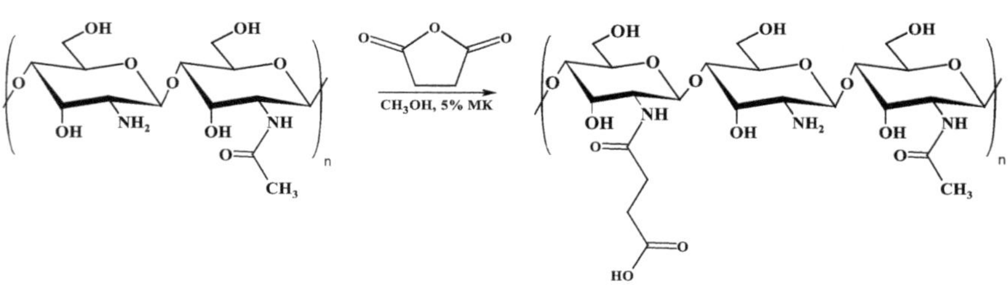

The structure of obtained samples confirmed by data of UV, IR, NMR ^{13}C and ^1H-spectroscopy. The degree of modification (DM) was estimated

TABLE 2.2 The Composition of the Modified Samples of Chitosan

Modified samples	DD (CTS), %	n(R-$COCH_3$), %	n(R-NH_2), %	n(R-COOH), %	DM, %	n(R-NH_2): n(R-COOH)
MCTS-1	74.0	26.0	36	38	52	1:1.1
MCTS-2	83.6	16.4	22	61.6	74	1:2.8
MCTS-3	81.4	18.6	15	66.4	77	1:3.6

by ninhydrin analysis, the data are given in Table 2.2. It is seen that the DM of CTS samples correlated with the degree of their deacetylation to some extent but probably also influenced by molecular characteristics of samples.

As follows from the data presented in Table 2.1, the [η] values of MCTS samples in some were decreased. Significant decline of [η] was observed after modification of HA. The decrease of the [η] can be caused by both decrease MM or affinity change of polymer to solvent. So values of MM for original and modified samples were defined by absolute methods. At the case of samples with high values of MM (CTS-2, MCTS-2, HA, MHA, DHA-2) by the method of high-speed sedimentation constants of sedimentation were determined and values of viscosity were calculated. For samples with low MM the method of sedimentation equilibrium is used and the size of M_z was determined. A significant decrease of MM is observed in case of modification of HA by sodium periodate (to 13,000). After chitosan modification MM decreases less. Apparently under the action of reagents β-glycosidic linkages in the modifications of polymers were broken partially according to random law.

By the method of UV-spectroscopy the possibility of interaction between the modified or unmodified HA and MMC in physiological solution (0.9% NaCl) was investigated.

There are three absorption bands in the region of 364, 250 and 217 nm in the UV spectrum of MMC, which belong to two types of chromophores (Figure 2.1). There are chromophores with π–π^* transition of double carbon-carbon bonds of six-membered cycle MMC and n–π transitions of three carbonyl groups of MMC. Two-carbonyl chromophore unequal involve as with electron-deficit double carbon-carbon bonds of six-membered cycle MMC and with electrodonor groups, one of which is connected with the

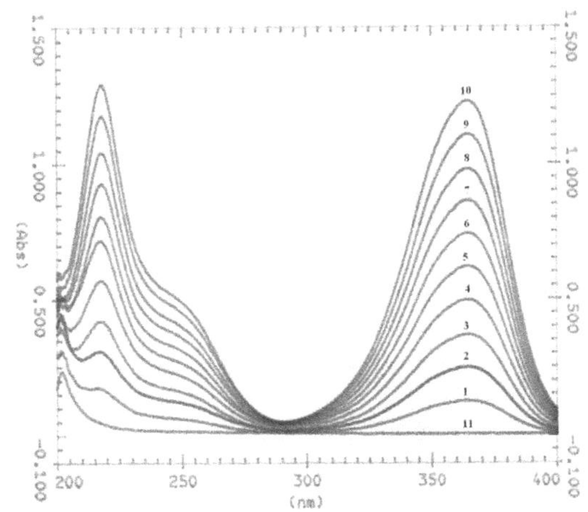

FIGURE 2.1 UV-spectra in solution: MMC (10), HA (11) and mixtures MMC:HA = 1:9 (1), 2:8 (2), 3:7 (3), 4:6 (4), 5:5 (5), 6:4 (6), 7:3 (7), 8:2 (8), 9:1 (9). Solvent: 0.9% NaCl in water, [MMC] = [HA] = 6×10^{-5} mol/L.

$-NH_2$ and the second with $-CH_3$ and N≡ fragments. The third carbonyl chromofore in MMC also carries with NH_2-group of the NH_2-COO-CH_2 depute in the five-membered heterocycle of MMC. The pairing of carbonyl chromophores in MMC with π-electron of double bonds in the UV spectra of absorption leads to considerable bathochromic shifts of maximum n–π bands to the longer-wavelength part of the spectrum (offset from the usual for carbonyl 280 nm up to 364 nm). In turn, the absorption band of carbonyl composed of the NH_2-COO-CH_2 deputy in five-membered heterocyclic, by contrast is undergoing *hypsochromic* shift of the maximum of n-π transition since it is associated with isolated electrodonor NH_{2}-group. Absorption at the 250 nm is typical for quinoines with different degrees of substitution, their positions and types of deputies (as observed for MMC). The chromophore with absorption band due to the transition of π–π electrons of double carbon-carbon bands of six-membered cycle in MMC is observed in the electron spectrum of absorption at 217 nm.

HA or MHA solutions were mixed in antibiotic relationships with solution of MMC and were kept at room temperature or at 55°C within 2 hours. In the UV spectrum for solutions of unmodified HA and MMC (Figure 2.1) as at room temperature and after warm-up there was a slight displacement (2 nm) the maximum of absorption at λ 217 nm. Research by methods of isomolecular series and molar relationships (Figure 2.2) complexation was indicated with to one disaccharide unit HA per one molecule of MMC. The constant of resistance is 3×10^5 L/mol.

After warming up of mixtures MHA and MMC in the UV spectrum absorption at 310 nm was appeared (Figure 2.3). Probably it is connected

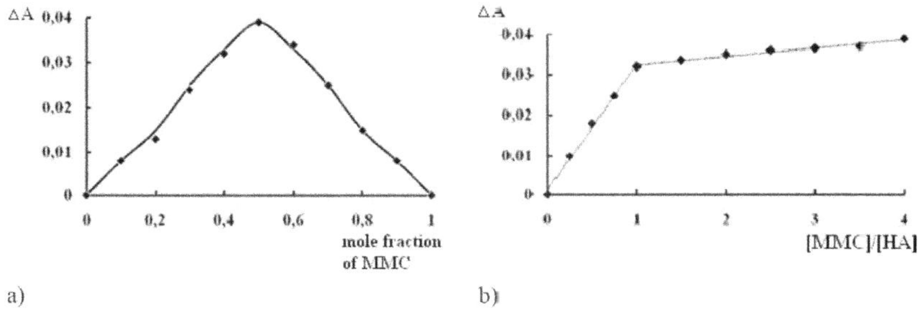

FIGURE 2.2 (a) Isomolecular chart of complex HA-MMC (method of isomolecular series); (b) Saturation curve of complex HA-MMC (method of molar relationship).

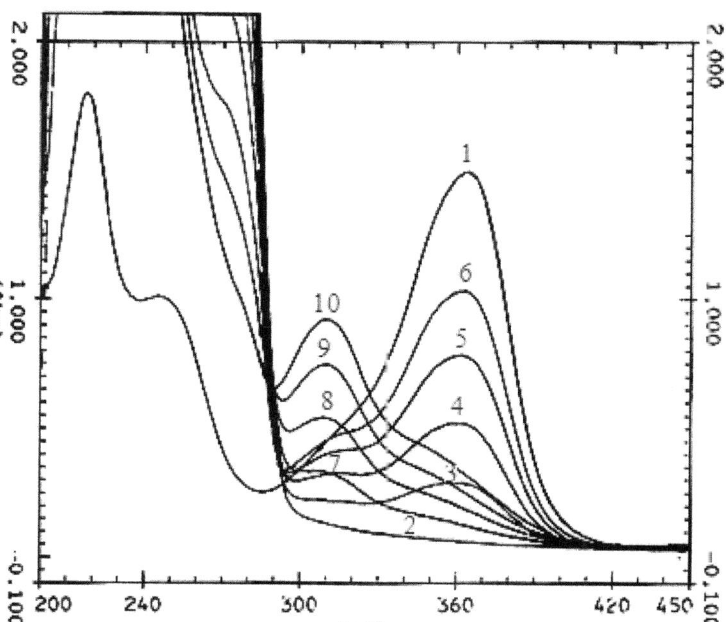

FIGURE 2.3 UV-spectra in physiological solution: MMC(1), MHA(2), mixtures of MMC:MHA, composition 1:4 (3, 7), 2:3 (4, 8), 3:2 (5, 9), 4:1 (6, 10) and after waiting for 2 hours at the temperature 25°C (1–6) and at the 55°C (7–10).

with disclosing of the most reactive asiridine cycle in the molecular of MMC due its attack by hydroxyl group of modified by epichlorohydrin fragment of HA.

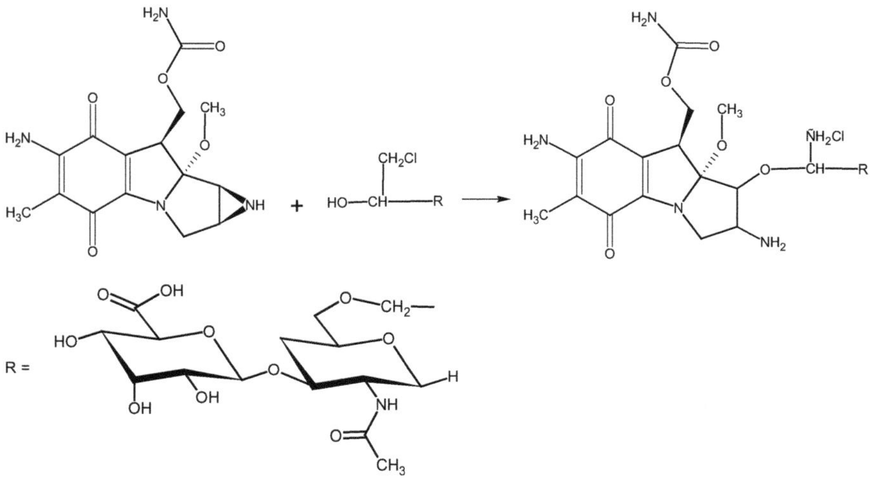

Reaction on carboxyl group not proceeds because no maximum of absorption in solution of mixtures unmodified HA and MMC at room temperature and after warming up in the area of 310 nm were observed. About disclosing of a cycle under the action of nucleophilic agents was shown in Refs. [25, 26]. Modified by epichlorohydrin HA with MMC at room temperature does not form a complex probably it is due to the formation of intramolecular hydrogen bonds between the hydroxyl group at the C6 position and nearest carboxyl group. It is blocked the formation of intermolecular complex due to steric and electronic factors. After heating modified by epichlorohydrin hyaluronic acid and mitomycin C intramolecular hydrogen bond weakens and as a result chemical interaction between MMC and MHA has been. The formation of intramolecular hydrogen bonds is confirmed by the shift of the absorption bond of carboxyl group for MHA in low field at 2.5 ppm (^{13}C NMR).

In the same way the possibility of complexation between DHA-24 and MMC in physiological solution was investigated. Using the method of isomolecular series and molar relationship the structure of complexes and constants of stability were defined (Table 2.3).

TABLE 2.3 The Constants of Stability of Complexes of HA, DHA, CTS, MCTS with MMC

Sample	$n(R\text{-}NH_2)$, %	n(R-COOH), %	$K\times10^{-5}$, L/mol (maximum at $\lambda = 364$ nm)	$K\times10^{-5}$, L/mol (maximum at $\lambda = 217$ nm)	Composition of complex
HA	0	100	–	3.3	1:1
DHA-24	0	100	–	3.0	1:1
CTS-1	74	0	1.3	0.4	1:1
MCTS-2	22	61.6	–	1.6	1:1

The possibility of complexation between CTS and MMC in physiological solution at room temperature was shown. It was found that on the one monosaccharide unit of CTS has one molecular of MMC.

Moreover as seen from the Figure 2.4 for the sample of CTS-1 more stable complexes are formed at interaction of the carbonyl group of MMC and aminogroup of CTS (λ=364 nm) than in the interaction of aminogroups of MMC with hydroxyl groups of chitosan (λ=217 nm).

Modified chitosan also forms a stable complex with other functional groups (aminogroups of MMC and carboxyl groups of MCTS).

Because of polyelectrolyte complex formation when water solutions of unmodified HA and soluble in water sample of chitosan (CTS-1) were

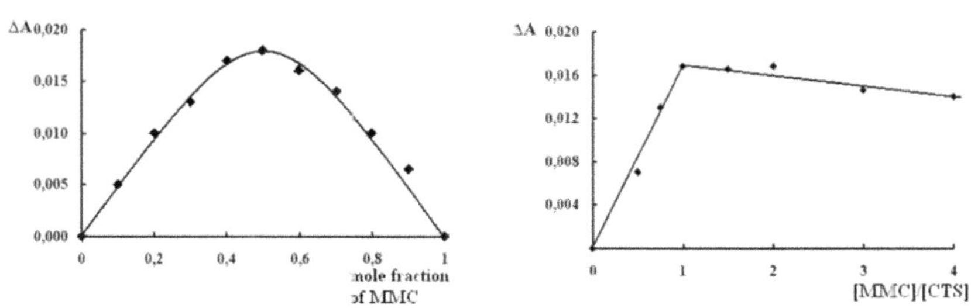

a) b)

FIGURE 2.4 (a) Isomolar chat of complex (CTS-1)-MMC in physiological solution (method of isomolecular series); (b) Curve of saturation for the complexes of (CTS-1)-MMC (method of molar relationships).

mixed sediment was formed (due the interaction –COOH groups in HA and $-NH_2$ in CTS). Hydrogels were formed after the changing of pH either acidic or alkaline region. These results agree with the data of work [27]. As for use in ophthalmology you must have a medium with pH close to neutral, that's why such hydrogels we did not use. For the same reason formation of precipitate took place in the interaction of MHA and MCTS-2 in solution.

After the pouring of water solutions DHA-24 and MCTS-2 in the same proportions hydrogel is formed. It is shown that the increase of the initial solution concentration leads to reduction of gelling time up to 3 seconds (Table 2.4). Probably, in the beginning polyelectrolyte complex between MCTS and DHA macromolecules is formed, and then imine crosslinks –C=N in the result of interaction of aldehyde groups in DHA with aminogroups in MCTS. The evidence is disappearing in the IR spectrum of obtained hydrogel's band in the region of 1725 cm^{-1}, belonging to dialdehyde of HA.

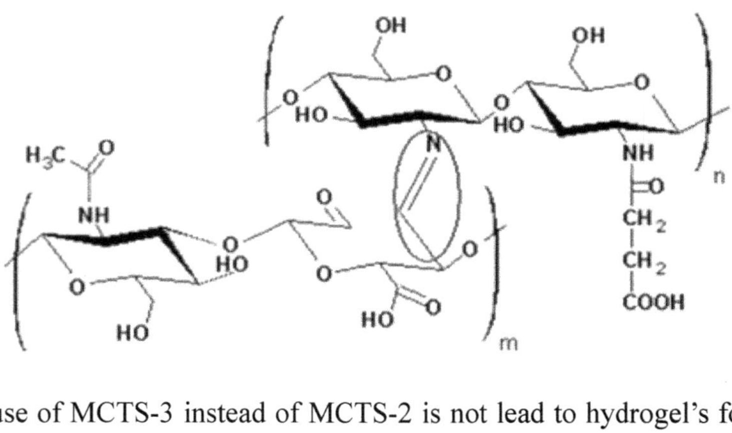

The use of MCTS-3 instead of MCTS-2 is not lead to hydrogel's formation, probably it has too high degree of modification (less aminogroups).

As follows from the results presented in Table 2.5, the possibility formation of the hydrogel is also affected by the ratio between the components in the solution. With the increasing content of DHA-24 the hydrogel is not occurred.

When MMC is introduced in the resulting hydrogels it was shown that not only the structure of hydrogel, but the order of introduction of the components have an importance for rate of its release. To study the diffusion

TABLE 2.4 Influence of the Concentration of Initial Physiological Solutions DHA-24 and MCTS on Formation of Hydrogel

Mix of solutions	Concentration of DHA-24, %	Concentration of MCTS, %	The time of gelation, sec
DHA-24: MCTS-2 = 1:1	0.1	0.1	No gel
	0.5	0.5	300
	1.0	1.0	40
	2.0	2.0	3
DHA-24: MCTS-3 = 1: 2	1.0	2.0	No gel
DHA-24: MCTS-1 = 1: 1	1.0	2.0	No gel
DHA-24: MCTS-1 = 1: 2	1.0	2.0	120

TABLE 2.5 Influence of the Ratio of 1% Solution DHA-24 and 2% Solution of MCTS-2 on Gelation

Sample	Ratio of moles DHA-24:MCTS-2	Time of gelation, sec.
1	1:1	36
2	1:2	24
3	1:4	17
4	1:8	18
5	2:1	No gel
6	4:1	No gel

of MMC there were used hydrogels based on 1% solution of DHA-24 and 2% solution of MCTS-2, which were merged with the same volume ratios but in different order. It is seen that the slowest diffusion of a drug takes place in case when the solution DHA-24 was first added to the solution of MMC and then the solution DHA-24 was added to the solution of MMC and then the solution of the modified chitosan was rushed. Apparently this is because DHA forms a more stable complexes with MMC than the chitosan. Thus, for the first six hours less than 40% of MMC were deposit (Figure 2.5). A further selection of MMC occurs within a few months.

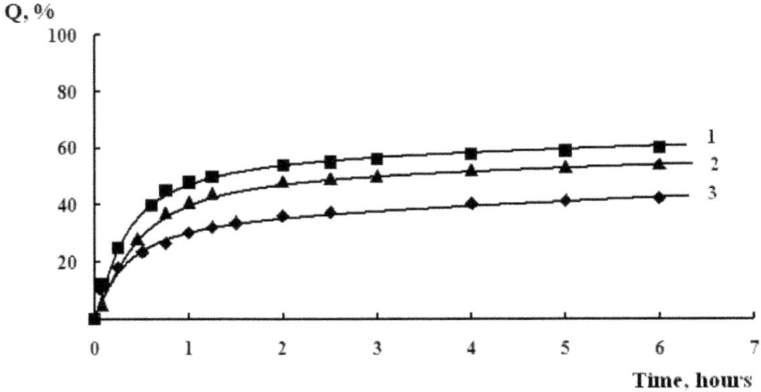

FIGURE 2.5 Diffusion MMC into the physiological solution out of hydrogels, produced by different ways: 1 – in the solution of CTS MMC was added and the DHA-24 was poured; 2 – solution of DHA mixed with a solution of MCTS and the MMC was added; 3 – in to the solution of DHA MMC was added, and then MCTS was poured.

2.4 CONCLUSION

Thus the modification of chitosan and hyaluronic acid allowed to obtain sustainable hydrogels on their basis it is permeable to water phase. Received hydrogels have biocompatibility ant it is not toxic. In addition the introduction of functional groups in the used biopolymers contributes to the solution of the task immobilization of the MMC on the polymeric matrix that provides prolongation of this release from polymer witch is necessary the medical use in ophthalmology.

KEYWORDS

- **chitosan**
- **complex**
- **diffusion**
- **hyaluronic acid**
- **hydrogel**
- **mitomycin C**
- **ophthalmology**

REFERENCES

1. Schante, C. E., Zuber, G., Herlin, C., Vandamme, T. F. Chemical modifications of hyaluronic acid for the synthesis of derivatives for a broad range of biomedical applications. Carbohydrate polymers. 2011, v. 85. Iss. 3, 469–489.
2. Zhang, L. M., Wu, C.-X., Huang, J.-Y. et al. Synthesis and characterization of a degradable composite agarose/HA hydrogel. Carbohydrate Polymers. 2012, v. 88. Iss. 4, 1445–1452.
3. Vercruysse, K. P., Prestwich, G. D. Hyauronate derivatives in drug delivery. Therapeutic Drug Carrier Systems. 1998, v. 15. Iss. 5, 513–555.
4. Lapcik, L.Jr., Lapcik, L. Hyaluronan: preparation, structure, properties, and applications. Chemical reviews.1998, v. 98. Iss. 8, 2663–2684.
5. Burdick, J. A., Prestwich, G. D. Hyaluronic Acid Hydrogels for Biomedical Applications. Adv. Mater. 2011, v. 23. Iss. 12, H41–H56.
6. Tan, H., Marra, K. G. Injectable, Biodegradable Hydrogels for Tissue Engineering. Materials. 2010, v. 3. Iss. 3, 1746–1767.
7. Khaw, P. T., Midgal, C. Current techniques in wound healing modulation in glaucoma surgery. Current Opinion in Ophthalmology. 1996, v. 7. Iss. 2, 24–33.
8. Shmyreva, V. F., Mostovoy, E. N. About application of cytostatic therapy by 5-fluorouracil in the surgery of glaucoma. Herald of Ophtalmolgy (in Rus.). 2004, Iss. 3, 7–10.
9. Kitazawa, Y., Kawaze, K., Matsushita, H., Minobe, M. Trabeculectomy with mitomycin. A comparative study with fluorouracil Arch. Ophtalmol. 1991, v. 109, 1693–1698.
10. Yan, C., Chen, D., Gu, J. et al. Preparation of N-succinyl-chitosan and their physical-chemical properties as a novel excipient. The Pharmaceutical Society of Japan. 2006, v. 126, 789–793.
11. Patent RF № 2215749, The method of preparation of chitosan water-soluble forms.
12. Simkovic C., Hricovini, M., Soltes, L. et al. Preparation of water-soluble/insoluble derivatives of hyaluronic acid by cross-linking with epichlorohydrin in aqueous NaOH/NH4OH solution. Carbohydrate Polymers. 2000, v. 41, 9–14.
13. Tan, H., Chu, C. R., Payne, K. A., Marra, K. G. Injectable in situ forming biodegradable chitosan-hyaluronic acid based hydrogels for cartilage tissue engineering. Biomaterials. 2009, v. 30, 2499–2506.
14. Ruhela, D., Riviere, K., Szoka, F. C. Efficient synthesis of an aldehyde functionalized hyaluronic acid and its application in the preparation of hyaluronan-lipid conjugates. Bioconjugate Chem. 2006, v. 17. Iss. 5, 1360–1363.
15. Kuchina, Yu. A., Dolgopyatova, N. V., Novikov, V. Yu., et al. Instrumental methods of determination of degree of deacetylation of chitin. Herald of MSTU (in Rus.). 2012, v. 15. Iss. 1, 107–113.
16. Mura, C., Manconi, M., Valenti, D., et al. In vitro study of n-succinyl chitosan for targeted delivery of 5-aminosalicylic acid to colon. Carbohydrate Polymers. 2011, v. 85. Iss. 3, 578–583.
17. Gamzazade, A. I., Šlimak, V. M., Sklar A. M., et al. Investigation of the hydrodynamic properties of chitosan solutions. Acta Polymerica. 1985, v. 36. Iss. 8, 420–424.
18. Hyaluronan. Vol. 2/Ed. by, J. F. Kennedy, G. O. Phillips, P. A. Williams, V. C. Hascall. Cambridge: Woodhead Publishing, 2002, 1152 pp.

19. Lopatin, S. A., Nemtsev, S. V., Varlamov, V. P. The new colometric method of determination of chitosan. New advances in the study of chitin and chitosan: materials of the VI international conference. M.: RSIFO, 2001, 298–299.
20. Bek M., Nad'pal, I. Investigation of complexation by new methods (in Rus.)/M.: Peace (in Rus.), 1989, 413 pp.
21. Bulatov, M. N., Kalinin, N. P. The practical handbook by photometric assay methods (in Rus.). L.: Chemistry (in Rus.), 1986, 432 pp.
22. Pat. WO/2000/046253, Process for the production of multiple cross-linked hyaluronic acid derivatives.
23. Zhao, X. B., Fraser, J. E., Alexander, C. et al. Synthesis and characterization of a novel double crosslinked hyaluronan hydrogel. J. Mat. Sci.: materials in medicine. 2002, v. 13. Iss. 1, 11–16.
24. Dahlmann, J., Krause, A., Moller, L. et al. Myocardial tissue engineering based on novel fully defined in situ cross-linkable alginate and hyaluronic acid hydrogels. Biomaterials. 2013, v. 34. Iss. 4, 940–951.
25. Plate, N. A., Vasil'ev A. E. Physiologically active polymers (in Rus.). M.: Chemistry (in Rus.), 1986, 296 pp.
26. Tomasz, M., Chowdary, D., Lipman, R. et al. Reaction of DNA with chemically or enzymatically activated mitomycin C: isolation and structure of the major covalent adduct. Biochemistry. 1986, v. 83, 6702–6706.
27. Vasile, C., Pieptu, D., Dumitriu, R. P. et al. Chitosan/hyaluronic acid polyelectrolyte complex hydrogels in the management of burn wounds. Rev. Med. Chir. Soc. Med. Nat. 2013, 117(2), 565–571.

CHAPTER 3

TRENDS IN POLYMERIC GAS SEPARATION MEMBRANES

INGA A. RONOVA,[1] ALEXANDER ALENTIEV,[2] and MARIA BRUMA[3]

[1]*Nesmeyanov Institute of Organoelement Compounds, Vavilov Street 28, Moscow – 119991, Russia*

[2]*Topchiev Institute of Petrochemical Synthesis, Leninskii Prospect 29, Moscow – 119991, Russia*

[3]*"Petru Poni" Institute of Macromolecular Chemistry, Iasi – 700487, Romania*

CONTENTS

ABSTRACT

Here, the swelling with supercritical carbon dioxide (sc-CO_2) of thin films of polyimides having various structures was investigated. It was shown that the degree of swelling is significantly influenced by the solvent which was used for the synthesis of those polyimides, by the solvent which was used for the preparation of thin films and by the conformational rigidity of the polymers. The presence of hexafluoroisopropylidene groups in the main chain of a polymer prevents its swelling with sc-CO_2. It is shown that the degree of increase of the free volume in the polymer matrix during swelling in sc-CO_2 also determined experimental conditions (pressure, temperature and the pressure relief rate). Transport parameters such as permeability and selectivity coefficients measured before and after treatment with sc-CO_2 increased from 16% to 168% and from 5% to 49%, respectively. The greatest increase in the free volume was 257% of the initial value during rapid pressure relief. This led to the formation of foams and reduction of the dielectric constant even below 1.5.

3.1 INTRODUCTION

Polyimides represent one of the most important classes of high-performance polymers due to their high-temperature durability, good mechanical properties, and excellent chemical and thermal stabilities [1–4]. It is known that most of polyheteroarylenes, particularly polyimides, have high thermal and mechanical properties due to their conformational rigidity with Kuhn segment being in a wide range of values from 10 Å to 15 Å, which is characteristic for aliphatic polymers, to a thousand of Angstroms [5]. Polyimides are used in applications demanding service at enhanced temperatures while maintaining their structural integrity and a good combination of chemical, physical and mechanical properties.

Much of the research devoted to the development of high performance polymers for gas separation application has focused on the variation of chemical structure in order to obtain novel polymers with both high permeability and high selectivity [6]. Since a typical reverse relationship between permeability and selectivity exists, to make membrane separation more competitive, an important goal of polymeric membranes design is to develop materials, which have both high permeability and selectivity. For such a purpose glassy polymers are used. Thus, various classes of polymers have been synthesized and studied for use as gas separation membranes. Among these, polyimides are in an important position since they exhibit extraordinary high gas selectivity as well as excellent thermal and mechanical stability, and film-forming ability [7, 8]. Besides, there is a broad possibility of varying the chemical structure of the repeating unit aiming to change the physical properties of polyimides, including their transport characteristics [9–12].

Some requirements to use these polymers for interlayer and intermetal dielectrics in advanced microelectronic applications are: high thermal stability, high glass transition temperature, good mechanical properties, low dielectric constant, low coefficient of thermal expansion and low moisture absorption. The dielectric constant of polyimides depends mainly on the free volume of polymer matrix and on the polarizability and hydrophobicity of macromolecular chains. One way to increase the free volume of polymer matrix is by treating with supercritical carbon dioxide (sc-CO_2) [13, 14]. The degree of swelling with sc-CO_2 depends on conformational rigidity of macromolecular chains, glass transition temperature and on the presence of fluorinated units in the polymer structure. Previously, it was shown that a large number of polymers of various structures did not swell or swelled insignificantly in sc-CO_2. This behavior was explained by the fact that those studied polymers had been synthesized in N-methylpyrrolidone as solvent, which facilitated the formation of cross-links between the macromolecular chains and thus the penetration of CO_2 molecules was hindered [15, 16].

Here, we present an investigation of the swelling process with sc-CO_2 of thin films of some polyimides, which were synthesized in *m*-cresol or in carboxylic acid medium such as benzoic acid or salicylic acid. The synthesis and general characterization of these polyimides were described previously [17, 18]. Some correlations are shown between the conformational rigidity parameters, such as free volume and characteristic ratio, and the swelling degree or glass transition temperature. The effect of swelling on the dielectric constant and gas transport parameters is evidenced.

3.2 CALCULATION METHODS

3.2.1 CALCULATION OF CONFORMATIONAL PARAMETERS

The correlation between physical properties of polymers and conformational rigidity of their chains shows that the contribution of conformational rigidity to their properties is significant [19]. The conformational rigidity of a polymer can be estimated using different parameters, such as statistical Kuhn segment (**A**$_{fr}$) and characteristic ratio (**C**$_{\infty}$). *Kuhn segment* was calculated as under the assumption of free rotation by using the Eq. (1) [5].

$$A_{fr} = \lim_{n\to\infty}\left(\frac{<R^2>}{nl_o}\right) \tag{1}$$

where $<\boldsymbol{R}^2>$ is mean square distance between the ends of the chain calculated for all possible conformations, $\boldsymbol{n}$ is the number of repeating units, $\boldsymbol{l}_o$

is the contour length of a repeating unit, and $L = nl_o$ is the contour length of the chain, a parameter which does not depend on the chain conformation.

All the values of Kuhn segment were calculated with Monte Carlo method and the geometry of the repeating unit was assigned by using quantum-chemical method AM1 [20].

We also used another parameter of conformational rigidity named *characteristic ratio* C_∞, which shows the number of repeating units in Kuhn segment, as shown by Eq. (2).

$$C_\infty = \frac{A_{fr}}{l_0} \tag{2}$$

For some of the studied polymers, we also calculated the Kuhn segment values taking into consideration the hindrance of rotation, according to the method previously described [20, 21]. Most of these polymers did not have any hindrance of rotation, or their hindrance was too low, and it was neglected. Previously, it was shown that the values of conformational parameters calculated under the assumption of free rotation in the absence of voluminous substituents are practically equal to the values found experimentally from hydrodynamic data [22].

3.2.2 CALCULATION OF FREE, OCCUPIED, ACCESSIBLE AND FRACTIONAL ACCESSIBLE VOLUME

In order to correlate the geometry of the repeating units of polymers with transport properties, the following parameters were calculated: van der Waals volume (V_w), free volume (V_f), occupied volume (V_{occ}), accessible volume (V_{acs}), fractional accessible volume (*FAV*).

The occupied volume (V_{occ}) of a repeating unit is given by Eq. (3) as being the sum of the Van der Waals volume (V_w) of the repeating unit and the volume of space around this unit that is not accessible for a given type of molecule of gas, which is named "dead volume" (V_{dead}) (Figure 3.1). It is evident that the occupied volume of a repeating unit depends on the size of the gas molecule.

$$V_{occ} = (V_w + V_{dead}) \tag{3}$$

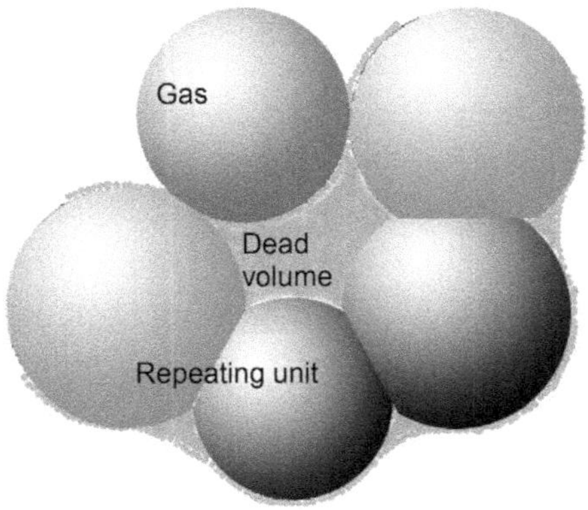

FIGURE 3.1 The definition of dead volume.

The accessible volume of a polymer (V_{acs}) is given by Eq. (4), where $N_A = 6.02\times10^{23}$ is Avogadro's number, ρ is the polymer density, and M_o is the molecular weight of the repeating unit.

$$V_{acs} = \frac{1}{\rho} - \frac{N_A \bullet V_{occ}}{M_0} \tag{4}$$

However, more often is used the so-called fractional accessible volume (***FAV***), without any dimensions, which gives a better concordance with the coefficients of diffusion and of permeability, that is given by Eq. (5) [23, 24].

$$FAV = V_{acs} \bullet \rho \tag{5}$$

To calculate the Van der Waals and the occupied volume of the repeating unit, we used the quantum chemical method AM1 to refine the structure of the monomer unit [20]. The model of the repeating unit is a set of intersecting spheres whose coordinates of centers coincide with the coordinates of atoms and the radii are equal to the Van der Waals radii of the corresponding atoms, as shown in Figure 3.2.

Van der Waals volume (V_w) of the repeating unit is the volume of the body of these overlapping spheres. The values of Van der Waals radii were taken from the reference [25]. The model of the repeating unit

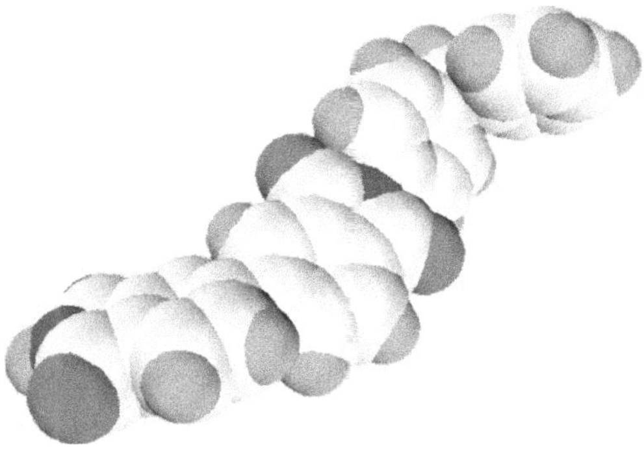

FIGURE 3.2 View of the repeating unit with Van der Waals volumes of the atoms.

was placed in a box with the parameters equal to the maximum size of repeating unit. By using the Monte Carlo method we designated the number of random points ***m*** that fall into repeating unit and the total number of tests ***M***. Their ratio is multiplied by the volume of the box, as seen in Eq. (6)

$$V_w = (m/M)V_{box} \tag{6}$$

Then we calculated the dead volume. Since the molecules of O_2, N_2 and CO_2 have ellipsoidal shape, we calculated the dead volume of the two spheres with radii corresponding to the major and to the minor axes of the ellipsoid. A number of 10^6 spheres with the radius of the gas was generated for each atom of the repeating unit. The result was a system consisting of a repeating unit, surrounded by overlapping spheres of gas. Then, the system was placed in the "box", similar to the one used in the determination of V_w, and random points were generated in the volume of the box [26, 27]. Thus, without making any assumptions about packing of the polymer chains in the glassy state, we could quickly calculate the Van der Waals volume, and the occupied and the accessible volumes.

The *free volume* (V_f) was calculated with the Eq. (7):

$$V_f = \frac{1}{\rho} - \frac{N_A \bullet V_w}{M_0} \tag{7}$$

The value V_f, thus calculated, shows the volume, which is not occupied by the macromolecules in 1 cm^3 of polymer film.

3.3 EXPERIMENTAL METHODS

3.3.1 PREPARATION OF POLYMER FILMS

The polyimides were synthesized by polycondensation reaction of an aromatic diamine with an aromatic dianhydride by traditional method using *meta*-cresol or benzoic acid and salicylic acid as solvent [17, 28], at high temperature to allow the complete imidization process and to exclude the cross-linking. The polycondensation reaction was run with equimolar quantities of diamine and dianhydride, at room temperature for 3 h, and then at 200°C for another 7 h. After cooling down to room temperature, the resulting viscous solution was poured in methanol to precipitate the polymer. The fibrous precipitate was washed with methanol and dried in vacuum oven at 100°C. These polymers showed good solubility in common solvents having low boiling point, such as chloroform and tetrahydrofuran, which are very convenient for film preparation.

The films, having the thickness usually in the range of 20–40 μm, were prepared by using solutions of polymers in chloroform, having the concentration of 15%, which were cast onto cellophane film and heated gently to evaporate the solvent. The films were carefully taken out of the substrate. To remove the residual *m*-cresol, the films were further extracted with methanol in Soxhlett apparatus, followed by heating in vacuum at 70°C for 3 days.

3.3.2 MEASUREMENT OF GLASS TRANSITION TEMPERATURE

The Glass Transition Temperature (Tg) of the polymers was measured by differential scanning calorimetry, with a DSC-822e (Mettler-Toledo) apparatus, by using samples of polymer films. The samples were heated at the rate of 10°C/min under nitrogen to above 300°C. Heat flow versus temperature scans from the second heating run was plotted and used for reporting the *Tg*. The middle point of the inflection curve resulting from the second heating run was assigned as the *Tg* of the respective polymers. The precision of this method is ±7–10°C.

3.3.3 MEASUREMENT OF DENSITY

The density of polyimide films was measured by using the hydrostatic weighing method. The study was performed with an equipment for density measurement and an electronic analytic balance Ohaus AP 250D from Ohaus Corp US, with a precision of 10^{-5}g, which was connected to a computer. With this equipment we measured the change of sample weight (density) during the experiment, with a precision of 0.001 g/cm^3 in the value of density. Ethanol and isopropanol were taken as liquids with known density. The studied polyheteroarylenes did not absorb and did not dissolve in these solvents, which for these polymers had low diffusion coefficients. The characteristic diffusion times were in the domain of 10^4–10^5 s, even for the most thin films studied here, which leads to higher times, of 1–2 order of magnitude, than that of the density measurement. This is why the sorption of solvent and the swelling of the film must have only insignificant influence on the value of the measured density. All measurements were performed at 23°C. The density was calculated with the Eq. (8):

$$\rho_s = \rho_l \times W_a/(W_a - W_l) \tag{8}$$

where ρ_s is density of the sample, W_a is the weight of the sample in air, W_l is the weight of the sample in liquid, ρ_l is the density of liquid. The error of the density measurements was 0.3–0.5%.

3.3.4 MEASUREMENT OF DIELECTRIC CONSTANT

For each polymer in this series, dielectric permittivity of polyimide films was measured by using Alpha High Resolution Dielectric Analyzer from Novocontrol-Germany, in the domain of frequencies from 10^{-3} to 10^6 Hz, and it was approximated at the frequency equal to zero to obtain the value of dielectric constant (ε_o).

3.3.5 METHOD OF TREATMENT WITH SUPERCRITICAL CARBON DIOXIDE (SC-CO_2)

The experimental set-up (Figure 3.3) and the method of treatment with sc-CO_2 were described in previous papers [29, 30]. This experimental set-up is composed of a generator, which can provide CO_2 up to 35 MPa pressure (High Pressure Equipment Company, USA). A system of valves ensures the CO_2 access to the reaction cell with the volume of 30 cm^3. The pressure generator and the reaction cell are provided with manometers to allow a control of the pressure. The temperature control allows a precision higher than ±0.2°C. The cell is designed for experiments at pressures up to 50 MPa and temperatures up to 120°C. CO_2 desorption curves were obtained using the gravimetric technique. Sample mass was measured with an Ohaus AP 250 D electronic balance interfaced with a computer.

The following experimental technique was applied: The polymer film was weighed and placed into the cell. The films had the form of a disk with 15 mm diameter and thickness in the range from several to tens of microns. The cell was purged with CO_2 to remove the air and water vapors, and it was sealed. Then it was heated to the temperatures shown in this chapter, the pressure was increased to the values also shown in this chapter, and it was kept a certain time necessary

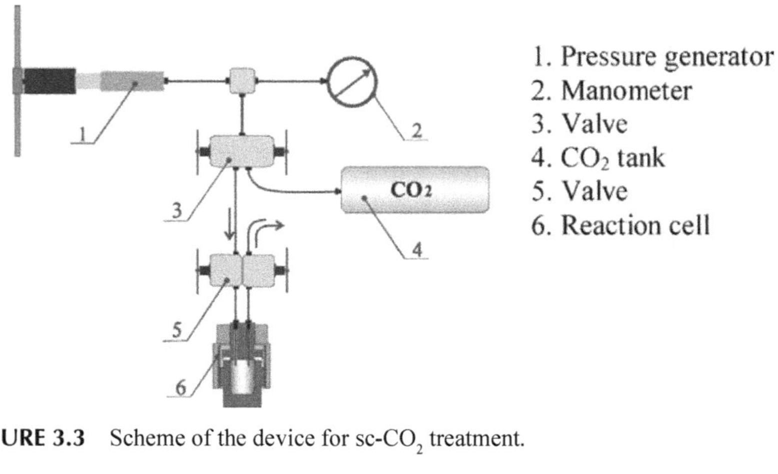

FIGURE 3.3 Scheme of the device for sc-CO_2 treatment.

to attain the equilibrium degree of swelling. Then the cell was open, the sample was taken out and it was put on Ohaus AP250D electronic balance (precision of 10^{-5} g) for less than 10 s, and the CO_2 desorption was fixed with the computer. To determine the mass degree of swelling with sc-CO_2, we recorded gravimetrically the CO_2 desorption from polymer (Figure 3.4). The scheme of gravimetric experiment is shown in Figure 3.5.

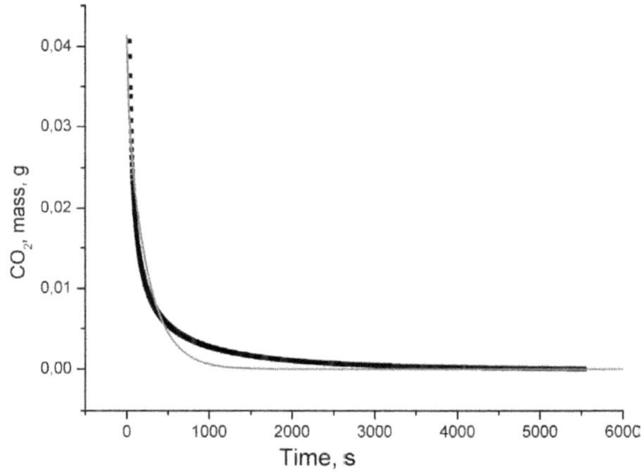

FIGURE 3.4 The curve of desorption of CO_2 from polymer film after the swelling process.

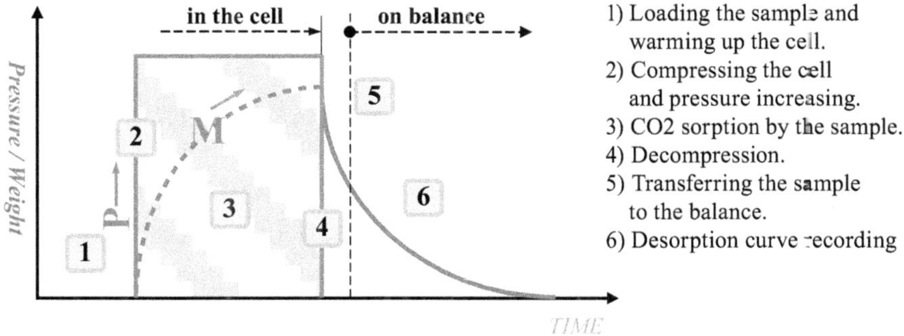

FIGURE 3.5 The scheme of gravimetric experiment. (1) Loading the sample and warming up the cell; (2) Compressing the cell and pressure increasing; (3) CO_2 sorption by the sample; (4) Decompression; (5) Transferring the sample to the balance; (6) Desorption curve recording.

3.3.6 INVESTIGATION OF THE FILMS MICROSTRUCTURE BEFORE AND AFTER SWELLING IN SC-CO_2

The dimension of pores and their distribution is very important from the point of view of mechanical and dielectric properties of porous materials. The morphology of the surface of polymer films was studied by using atomic force microscopy (AFM) in tapping mode, in air, at room temperature. The polymer films were cast on cellophane. The topographic images of the surface of polymer were obtained by using scanning probe microscope FemtoScan produced by Advanced Technologies Center, Russia. The analysis and processing of AFM images were performed with a program FemtoScan Online [31]. We used cantilevers of type fpN20 from F.V. Lukin Research Institute of Physical Problems, Russia. The medium resonance frequency of cantilever was 420 kHz. The radius of the curvature of the tip was less than 25 nm. The depth and diameter of pores were measured from the profiles of topographic images.

The investigation of the cross-sections of films before and after treatment with sc-CO_2 was performed by using transmission electron microscopy (TEM) with LEO 912AB OMEGA apparatus from Karl Zeiss, Germany. The spatial resolution was less than 0.5 nm. We prepared cross-sections of polymer films with the aid of microtome Ultracut (Reichert-Jung, Germany), having the thickness of 0.1 μm and we examined them immediately after their preparation. The cross-sections were placed on formvar coated copper grids.

3.3.7 MEASUREMENT OF TRANSPORT PARAMETERS

The transport parameters at 25 ± 3°C for He, O_2, N_2 and CO_2 were measured using a mass spectrometric technique [32, 33] and barometric techniques on a Balzers QMG 420 quadrupole mass spectrometer (Liechtenstein) MKS Barotron [34], respectively. The upstream pressure was 0.8–0.95 at, and the downstream pressure was about 10–3 mm Hg for spectrometric method, while for barometric technique that pressure was in the range of 0.1–1 mm Hg; therefore, the reverse diffusion of penetrating gas was negligible. The permeability coefficients **P** were estimated using the formula: P = Js L/Dp, where Js (cm^3 (STP)/cm^{2-}9 s) is the flux of the penetrant gas through 1 cm^2 of the film; Dp (cm Hg) is the pressure drop on the film; l (cm) is the film thickness. The diffusion coefficient **D** was determined by using the Daynes–Barrer (time lag) method: D = l2/6h, where h (s) is the time lag. The solubility coefficient **S** was estimated as the ratio: S = P/D

3.4 RESULTS AND DISCUSSION

3.4.1 THE DEPENDENCE OF DIELECTRIC CONSTANT OF POLYIMIDES ON CONFORMATIONAL RIGIDITY AND FREE VOLUME

Table 3.1 shows the chemical structure of the repeating units of the first series of the studied polyimides and some of their physical properties. Some of these polyimides contain methyl substituents on phenylene rings (**2**, **4** and **6**) and other contain hexafluoroisopropylidene groups (**5** and **6**) in the main chain [35]. Figure 3.6 presents the dependence of glass transition temperature on Kuhn segment.

It can be seen that the behavior of polymers **1–6** is as usual: with increasing of Kuhn segment, from 25.43 to 29.1 Å, the glass transition temperature increases, from 200°C to 287°C [5]. Since all the points of this dependence are situated on a straight line, described by equation "y = 324 + 20.85x" with a high correlation coefficient R = 99.17%, the glass transition temperature of polymer **3** can be calculated because it could not be measured experimentally. Thus, its calculated value is 262°C. The value of glass transition temperature of this polymer is well situated on the line showing the dependence of glass transition temperature on free volume (Figure 3.7). This dependence in Figure 3.7 shows that with the increasing of free volume, from 0.1648 to 0.2915 cm^3/g in polymers **5, 3, 1**, and from 0.1793 to 0.3056 cm^3/g in polymers **6, 4, 2**, the glass transition temperature in general decreases because the higher free volume allows the molecular fragments to change their conformation and therefore an increased mobility in polymer matrix appears which leads to the decrease of glass transition temperature.

Figure 3.8 presents the dependence of free volume on conformational rigidity. This dependence is linear, with a high correlation coefficient,

TABLE 3.1 Structure and Properties of the First Series of Polyimides

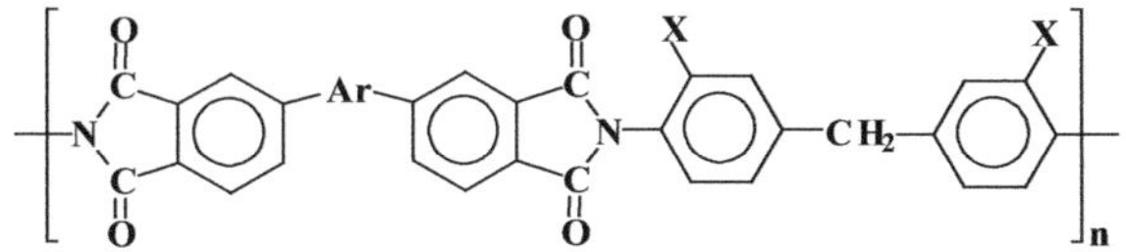

Polymer	Ar	X	ρ (g/cm³)	Tg (°C)	l_o (Å)	A_{fr} (Å)	C_∞	V_W (Å³)	V_f (cm³/g)	ε_o
1	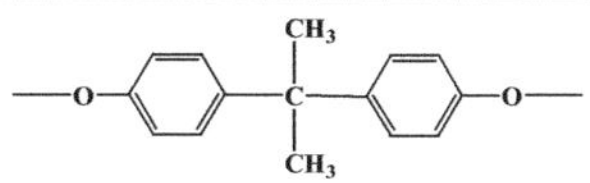	—	1.200	200	32.32	25.43	0.787	625.359	0.2915	2.54
2	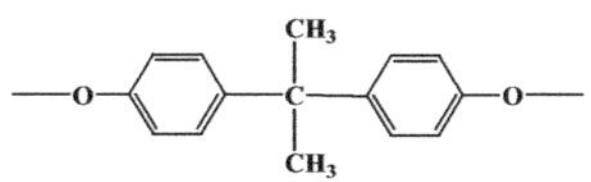	CH_3	1.169	228	32.32	26.2	0.811	656.092	0.3056	2.11
3	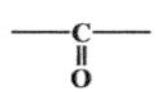	—	1.398	—	24.53	28.03	1.143	426.696	0.1889	3.28
4	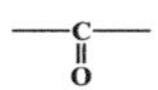	CH_3	1.337	278	24.53	28.87	1.177	460.613	0.2085	2.84
5	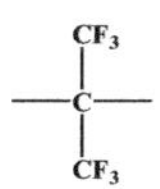	—	1.546	275	22.37	28.65	1.281	488.651	0.1648	2.31
6	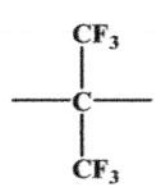	CH_3	1.488	287	22.37	29.1	1.319	522.676	0.1793	1.99

ρ = Density of polymer film; **Tg** = glass transition temperature; l_o = contour length of a repeating unit; A_{fr} = Kuhn segment; C_∞ = characteristic ratio; V_w = Van der Waals volume; V_f = free volume; ε_o = dielectric constant.

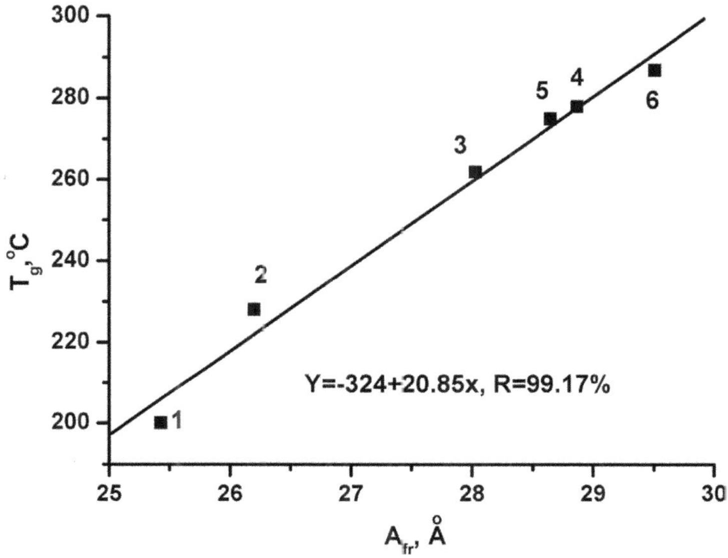

FIGURE 3.6 The dependence of glass transition temperature (T_g) on Kuhn segment (A_{fr}), for the first series of polyimides.

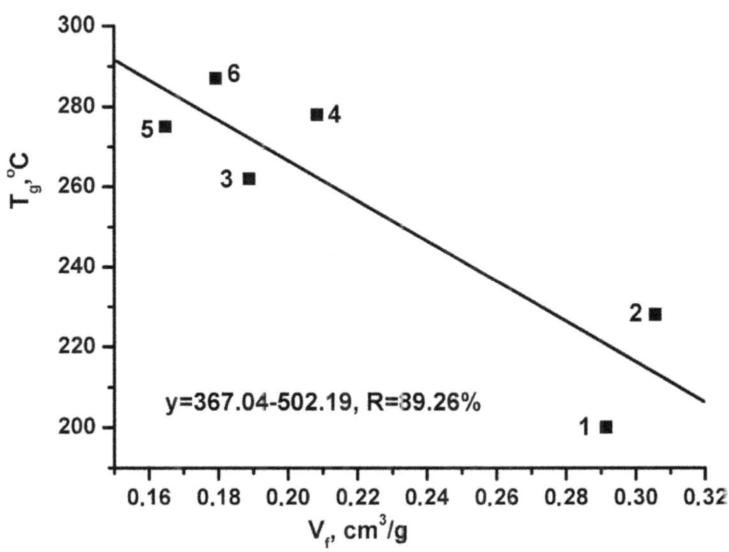

FIGURE 3.7 The dependence of glass transition temperature (T_g) on free volume (V_f), for the first series of polyimides.

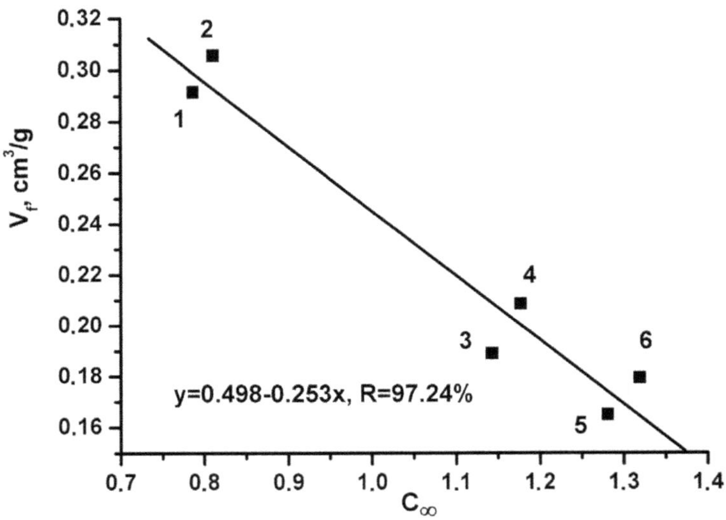

FIGURE 3.8 The dependence of free volume (V_f) on characteristic ratio C_∞, for the first series of polyimides.

97.24%. It confirms that with increasing rigidity of polymer matrix, the free volume decreases.

When comparing Figures 3.6–3.8, a contradiction appears: with increasing rigidity, the glass transition temperature should increase and the free volume should decrease. However, by going from the polymers without any substituents (**1**, **3** and **5**) to the polymers containing methyl substituents on phenylene rings (**2**, **4** and **6**, respectively) not only the glass transition temperature increases, but also the free volume increases. This can be explained in the following way: the introduction of methyl substituents on phenylene rings in ortho-position towards the bond between phenyl and imide rings increases the hindrance of rotation and consequently the rigidity of the polymer increases [22]. The amount of conformational isomers and the mobility of one repeating unit towards the others decrease, and consequently the glass transition temperature increases. On another hand, the methyl substituents do determine a loosing of packing, but the polymer cannot benefit of the increased free volume because the rotation hindrance does not allow the formation of all conformations, which should occupy that free volume when heating the samples. This is why the glass transition temperature of polymers containing methyl substituents does not decrease, but it increases in each pair of these polymers: Tg values of polymers **2, 4,** and **6**, containing methyl substituents, are 228, 278, and 287°C, respectively, being higher than those (200, 262, and 275°C, respectively) of polymers **1, 3,** and **5**, which do not have such methyl substituents. Table 3.1 shows also the values of dielectric constant ε_0. For each polymer in this series, dielectric permittivity was measured at six frequencies, from 1 Hz to 100 kHz, and it was approximated at the frequency equal to zero to obtain the value of dielectric constant, as shown in Figure 3.9. Figure 3.10 presents the dependence of dielectric constant of the studied polymers on characteristic ratio [27].

It can be seen that the points are also situated in pairs. The values of ε_0 of the polymers containing methyl substituents are lower than those of the related polymers, which do not contain methyl groups. By introduction of methyl substituents, the packing of the polymer becomes loose. However, the general dependence divides into two branches: polymers 1 and 2 are situated on the up-going part, while polymers **3–6** are situated on the down-going part of this diagram. The presence of hexafluoroisopropylidene bridge between imide rings increases the loosing effect of methyl groups in polymers **5** and **6**.

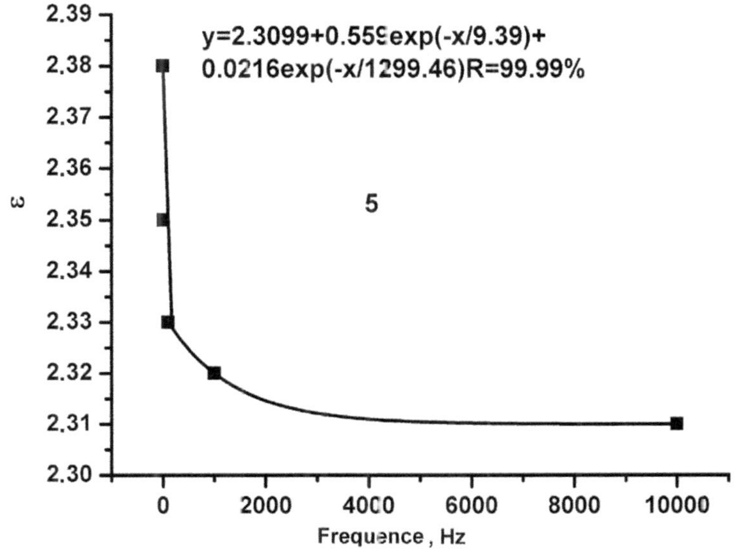

FIGURE 3.9 The dependence of dielectric permittivity (ε) on frequency, for polymer **5** from the first series of polyimides.

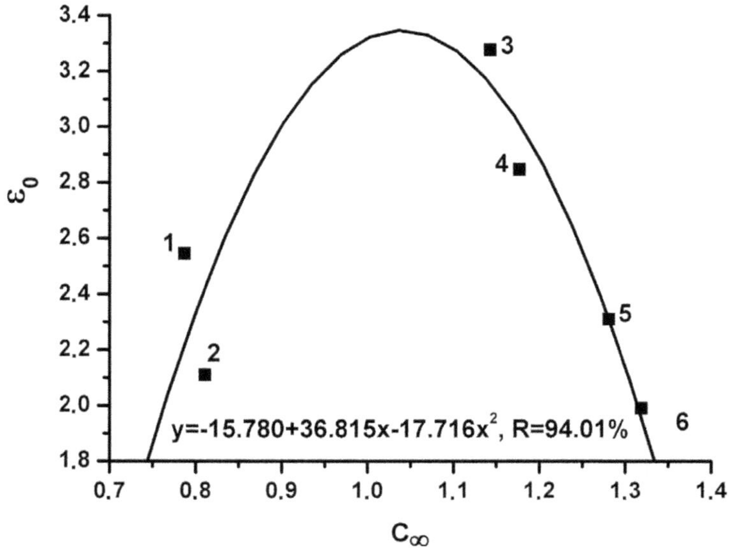

FIGURE 3.10 The dependence of dielectric constant (ε_o) on characteristic ratio C_∞, for the first series of polyimides.

It is known that fluorine atom has a highly electronegative effect [36]. Therefore, it is interesting to study the distribution of electron density in the repeating unit of macromolecules. We used the quantum-chemical method AM1 to calculate the charges on atoms, which constitute the repeating units of both polymers **5** and **6**. The charge on fluorine in hexafluoroisopropylidene segment was −0.17 e, while the charge on each hydrogen atom of isopropylidene group was only −0.08 e. Similar repulsion can appear between hexafluoroisopropylidene groups and oxygen atoms of carbonyl groups in imide rings. It follows that in glassy state repulsion appears between fluorine atoms of hexafluoroisopropylidene groups, which leads to the formation of supplemental microcavities in this polymer.

Now we examine the packing of fragments of two macromolecules: one containing hexafluoroisopropylidene and the other containing isopropylidene group. We suppose that oxygen atoms of ether groups are situated close to each other, at a distance equal to the sum of their Van der Waals radii, as shown in scheme below.

When packing, it is possible that hexafluoroisopropylidene groups of different chains are situated as shown in this scheme. Now, we make the modeling of their interaction to each other. We presume that the groups come out of the plane of this drawing, to meet each other and we compare the energy of repulsion, which appears between hydrogen or fluorine atoms during this movement as a function of the distance between these atoms. The energy of repulsion $\boldsymbol{E}$ between the three atoms of one group and the three atoms of the other group is given by Eq. (9):

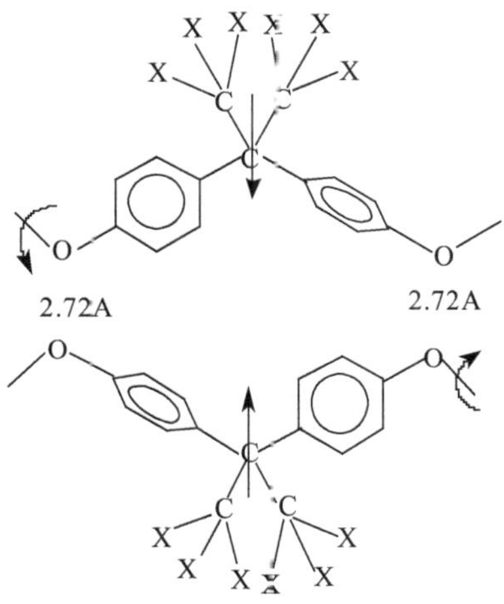

$$\boldsymbol{E} = \sum_i \frac{q_i q_j}{r_{ij}} \tag{9}$$

where q_i is the charge on hydrogen or fluorine atom, $\boldsymbol{r}$ is the distance between atoms. The sum is done at the same value of angle at which the two groups come out of the plane. Figure 3.11 presents the energy of repulsion between groups during their movement towards each other in the process when they come out of the plane.

In both cases, the groups can approach each other at a distance not shorter than the sum of Van der Waals radii of the corresponding atoms. When the distance between hydrogen or fluorine atoms is equal to the sum of Van der Waals radii of the corresponding atoms, the energy of repulsion in case of fluorine atoms (93.963 kJ/mol) is higher than in case of hydrogen atoms (25.312 kJ/mol). Similar repulsion can appear between hexafluoroisopropylidene groups and oxygen atoms of carbonyl groups in imide rings. It follows that in glassy state repulsion appears between fluorine atoms of hexafluoroisopropylidene groups, which leads to the formation of supplemental microcavities in this polymer.

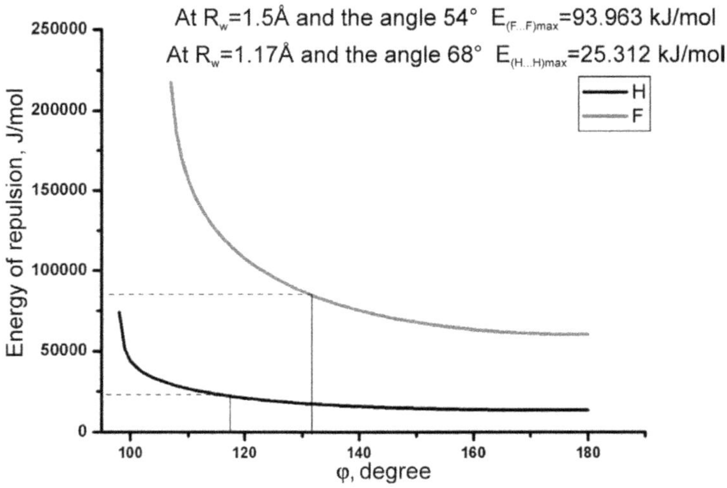

FIGURE 3.11 The dependence of repulsion energy on angle φ between hexafluoroisopropylidene groups (red line) and on the angle between isopropylidene groups (black line). Van der Waals radius of fluorine atom is 1.5 Å; Van der Waals radius of hydrogen atom is 1.17 Å.

The increase of free volume of polymers containing methyl substituents leads to the decrease of dielectric constant compared with related polymers which do not contain such substituents as seen in Table 3.1.

3.4.2 THE DEPENDENCE OF MEMBRANE PROPERTIES OF POLYIMIDES ON CONFORMATIONAL RIGIDITY AND FREE VOLUME

Table 3.2 shows the chemical structure of the repeating unit, glass transition temperature Tg, density q, and conformational parameters A_{fr} and C_∞, calculated using Equations (1) and (2), of the second series containing 12 polyimides. These polyimides are based on common dianhydride and diamine monomers and their gas transport properties toward H_2, CO, CO_2 and CH_4 have been published [37, 38]. We investigated the influence of conformational rigidity on the packing of macromolecules in glassy state and, consequently, on distribution of the free volume in polymer matrix. As seen in Table 3.2, the conformational rigidity, for example, Kuhn segment, of the studied polyimides varies in a relatively large interval, from

TABLE 3.2 Repeating Units and Properties of the Second Series of Polyimides

Polymer	Repeating unit	T_g (°C)	ρ, (g/cm³)	l_o (Å)	A_{fr} (Å)	C_∞
Pi-1	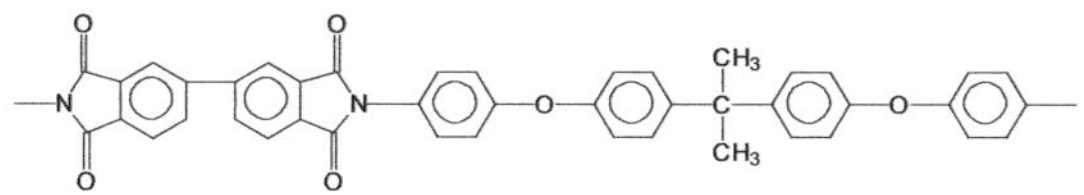	230	1.292	30.97	35.89	1.16
Pi-2	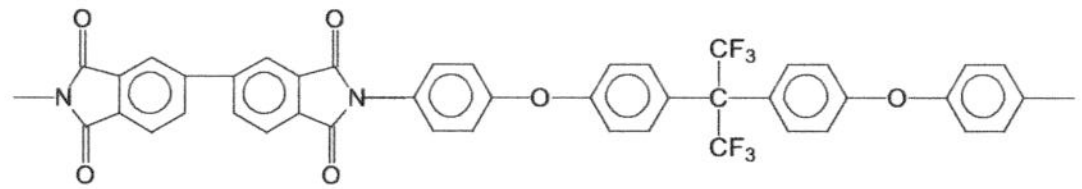	250	1.390	30.97	36.97	1.19
Pi 3	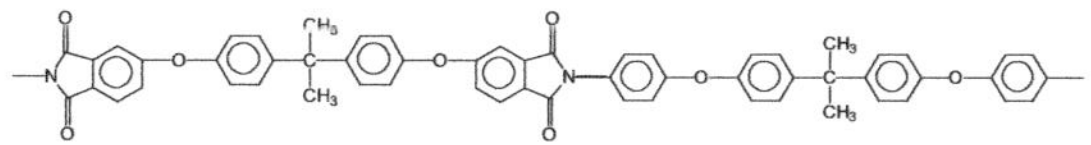	192	1.252	42.05	24.10	0.573
Pi-4	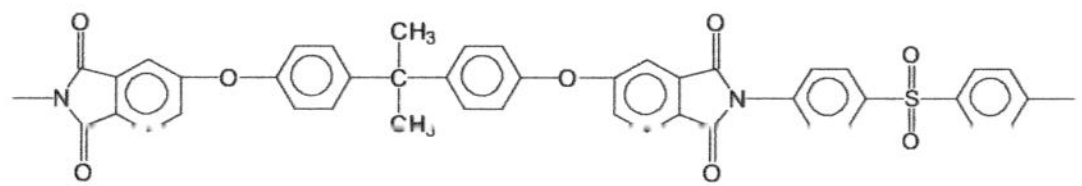	240	1.336	32.63	22.90	0.702
Pi-5	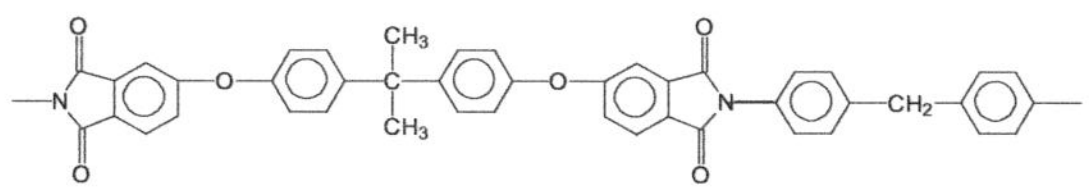	205	1.260	32.24	28.04	0.870
Pi-6 copolymer 1:1	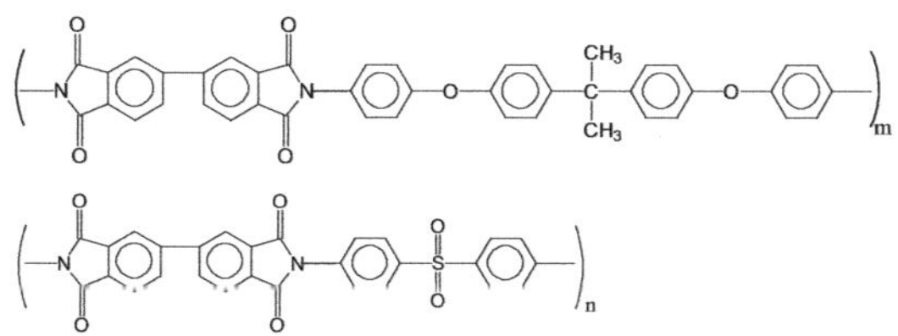	230	1.331	52.51	34.25	0.652

TABLE 3.2 Continued

Polymer	Repeating unit	T_g (°C)	ρ, (g/cm³)	l_0 (Å)	A_{fr} (Å)	C_∞
Pi-7 copolymer 1:1	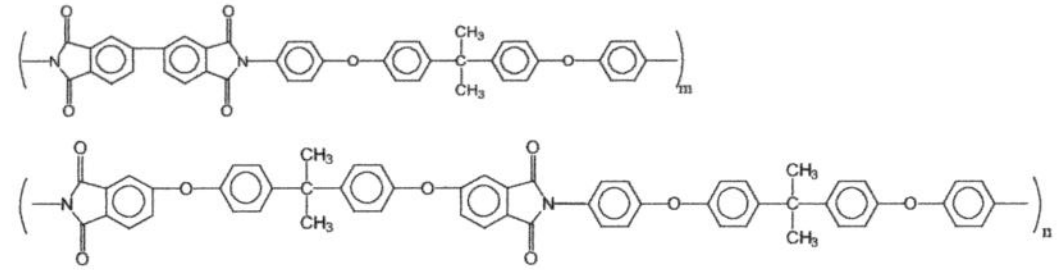	210	1.262	75.25	28.98	0.385
Pi-8	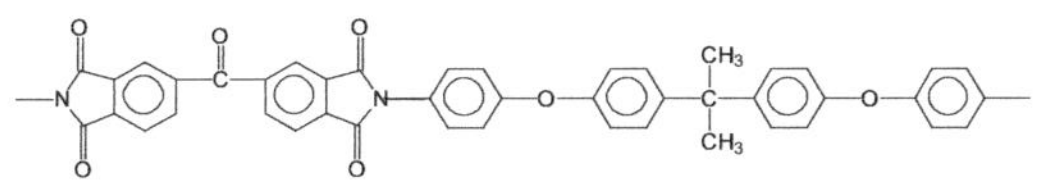	230	1.290	32.05	27.51	0.847
Pi-9	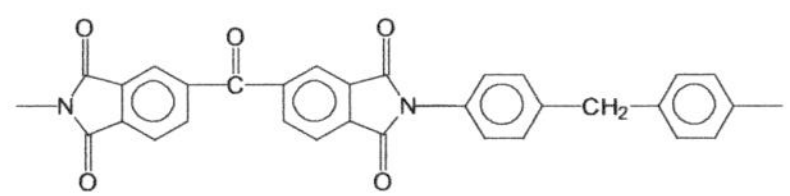	282	1.330	22.41	24.2	1.08
Pi-10	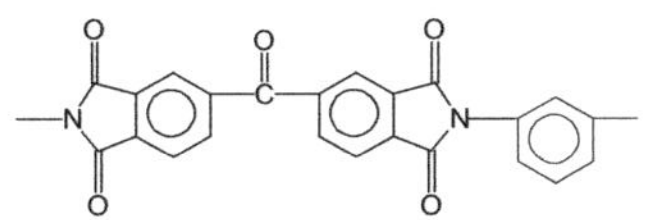	305	1.349	17.22	25.80	1.498
Pi-11	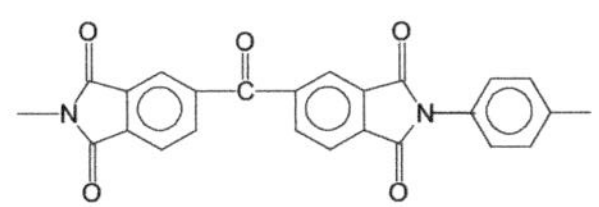	410	1.400	17.67	43.9	2,48
PAI	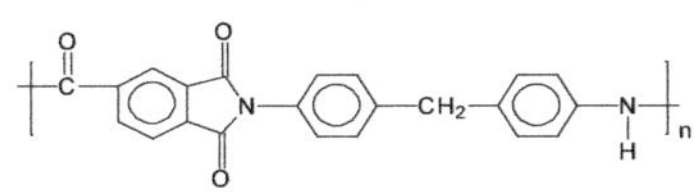	197	1.259	17.40	36.84	2.12

22 to 44 Å, and glass transition temperature from 190°C to 410°C. Thus, the selected series of polyimides is quite representative. The dependence of glass transition temperature (T_g) on rigidity (A_{fr}) divides these selected polymers in three groups (Figure 3.12) Each of them behaves in a normal way: with increasing rigidity, the glass transition temperature increases [39]. The correlation coefficients in all these three groups of polyimides are high enough, in the range of 97.06–99.99%. The dependence of glass transition temperature (Tg) on free volume (V_f) is linear and situates all these polymers on one line with the exception of polymer **PAI** (Figure 3.13). The values of V_f are given in Table 3.3. The correlation coefficient is high enough, 96.53%. With increasing free volume of the polymer, the glass transition temperature decreases.

Higher the free volume, more conformational transitions can take place in polymer matrix at heating and the highly elastic state appears faster. At the same time, the free volume of the polymer is determined by its capacity to pack in glassy state and consequently it depends on its conformational rigidity. Figure 3.14 shows the dependence of free volume (V_f) of the studied polyimides on conformational rigidity (A_{fr}). It can be seen that there are three dependences (three lines), and in each of them enter the same polymers as in Figure 3.12. The correlation coefficients of all three dependences are very high, in the range of 97.53–99.99%.

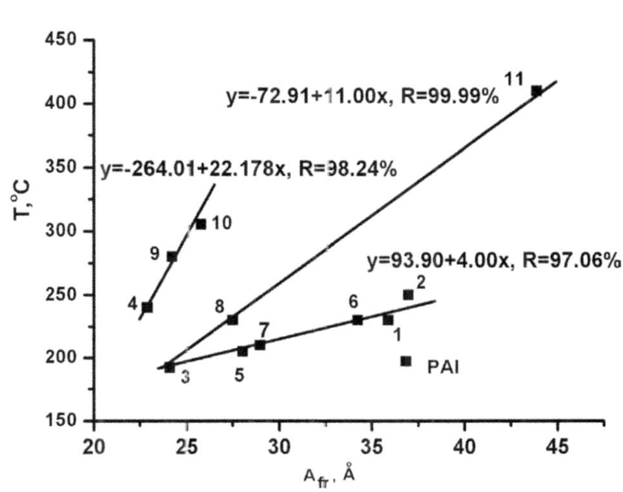

FIGURE 3.12 The dependence of glass transition temperature (T_g) on Kuhn segment (A_{fr}) for the second series of polyimides.

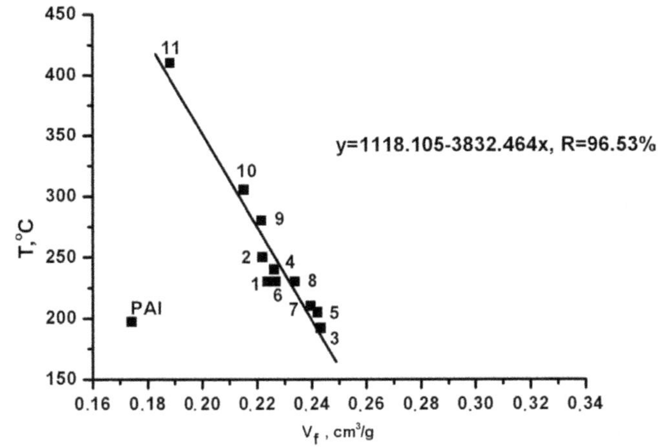

FIGURE 3.13 The dependence of glass transition temperature (T_g) on free volume (V_f) for the second series polyimides.

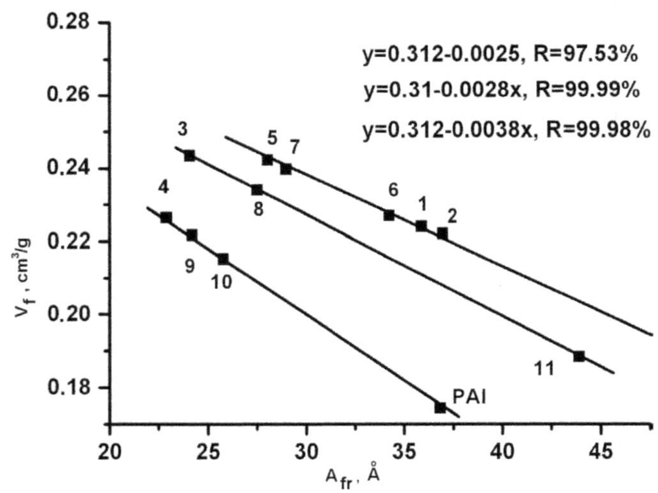

FIGURE 3.14 The dependence of free volume (V_f) on Kuhn segment (A_{fr}) for the second series polyimides.

We examine now how the conformational rigidity influences on the permeability coefficients P of the membranes made from these polymers. For these polymers we calculated the Van der Waals volume, free volume, occupied volume, and accessible volume (Table 3.3). Table 3.4 shows the permeability coefficients to four gases, H_2, CO, CO_2 and CH_4, taken from publications [37, 38].

TABLE 3.3 Van der Waals (V_w), Free (V_f), Occupied (V_{occ}) and Accessible (V_{acs}) Volumes of the Second Series of Polyimides

Polymer	V_w (Å^3)	V_f (cm^3/g)	H_2		CO		CO_2		CH_4	
			V_{occ} (Å^3)	V_{acs} (cm^3/g)	V_{occ} (Å^3)	V_{acs} (cm^3/g)	V_{occ} (Å^3)	V_{acs} (cm^3/g)	V_{occ} (Å^3)	V_{acs} (cm^3/g)
Pi-1	610.900	0.2239	646.051	0.1919	660.039	0.1794	652.406	0.1866	662.307	0.1777
Pi-2	641.686	0.2220	679.993	0.1923	695.243	0.1804	686.717	0.1817	697.961	0.1783
Pi-3	825.166	0.2434	876.832	0.2088	899.047	0.1939	886.774	0.2021	902.953	0.1913
Pi-4	635.673	0.2263	672.665	0.1958	688.284	0.1830	679.516	0.1902	691.053	0.1807
Pi-5	625.359	0.2422	661.567	0.2103	676.362	0.1972	668.369	0.2042	679.144	0.1947
Pi-6	1024.818	0.2268	1087.410	0.1948	1108.728	0.1838	1097.122	0.1898	1112.292	0.1820
Pi-7	1436.066	0.2397	1522.883	0.2060	1559.086	0.1921	1539.180	0.1997	1565.260	0.1897
Pi-8	626.346	0.2339	660.248	0.2046	675.624	0.1987	667.042	0.1987	678.383	0.1889
Pi-9	426.696	0.2216	449.094	0.1937	458.160	0.1825	453.330	0.1885	459.656	0.2402
Pi-10	333.975	0.2153	349.409	0.1910	356.025	0.1806	352.390	0.1863	357.052	0.1790
Pi-11	333.975	0.1883	349.409	0.1640	356.025	0.1520	352.390	0.1593	357.052	0.1520
PAI	325.572	0.2167	342.356	0.1870	349.811	0.1738	345.713	0.1810	351.117	0.1714

TABLE 3.4 The Permeability Coefficients (Barrer)

Polymer	H_2	CO	CO_2	CH_4
Pi-1	2.53±0.02	0.0567±0.0015	0.644±0.012	0.0329±0.0007
Pi-2	5.91±0.04	0.145±0.005	1.64±0.02	0.0504±0.0033
Pi-3	3.20±0.03	0.0678±0.0012	0.762±0.011	0.0305±0.0027
Pi-4	3.56±0.01	0.0836±0.0016	0.891±0.039	0.0418±0.0021
Pi-5	5.07±0.03	0.136±0.001	1.62±0.09	0.080±0.007
Pi-6	1.79±0.01	0.0395±0.0016	0.428±0.002	0.0099±0.0002
Pi-7	4.43±0.04	0.250±0.01	1.57±0.04	0.205±0.008
Pi-8	5.59±0.01	0.102±0.004	1.47±0.08	0.089±0.002
Pi-9	3.14±0.04	0.0367±0.0017	0.594±0.035	0.016±0.001
Pi-10	1.25±0.01	0.00397±0.00019	0.0627±0.0029	0.0011±0.0001
Pi-11	1.14±0.02	0.0176±0.0007	0.1971±0.0033	0.00795±0.00028
PAI	1.63±0.01	0.0342±0.0005	0.435±0.004	0.0163±0.0003

Since the accessible volume $\mathbf{V_{acs}}$ is considered as the volume of the space inside the polymer which can be occupied by the centers of gas molecules so that the Van der Waals spheres of the gas should not overlap on the Van der Waals spheres of the atoms of polymer chain [40], one can approximate the free volume using the Eq. (10).

$$P = A_{exp}(-B/V_{acc}) \tag{10}$$

However, more often is used the so-called fractional accessible volume (FAV), without any dimensions, which gives a better concordance with the coefficients of diffusion and of permeability [23] that is given by Eq. (5). Thus, the Eq. (10) can be written as shown in Eq. (11)

$$P = A exp(-B/FAV) \tag{11}$$

Figure 3.15 shows the dependence of permeability coefficients **P** of the studied polyimides on the fractional accessible volume **FAV** in the system one polymer–different gases. All these dependences are linear, with high correlation coefficients, being in the range of 99.05–99.91% (Table 3.5).

TABLE 3.5 The Slopes (**A** and **B**) and Correlation Coefficients (**R**) for Curves Described by Eq. (11) for the System One Polymer – Various Gases

Polymer	A	–B	R (%)
Pi-1	24.30	5.92	99.91
Pi-2	27.33	7.08	99.54
Pi-3	22.82	5.82	99.87
Pi-4	24.07	6.14	99.89
Pi-5	24.48	6.28	99.38
Pi-6	31.70	8.12	99.05
Pi-7	16.95	4.23	99.55
Pi-8	23.92	6.11	99.15
Pi-9	32.76	8.29	99.77
Pi-10	43.92	11.31	99.71
Pi-11	27.21	6.24	99.97
PAI	22.83	5.32	99.69

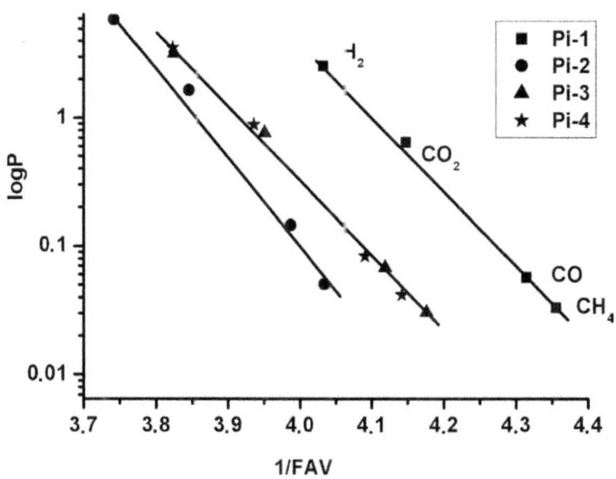

a)

FIGURE 3.15 Continued

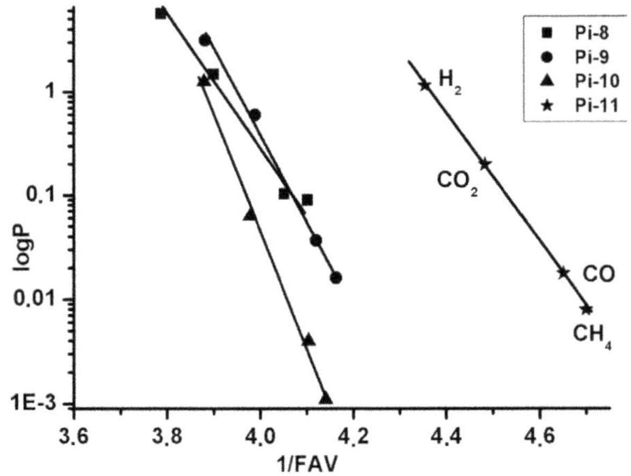

b)

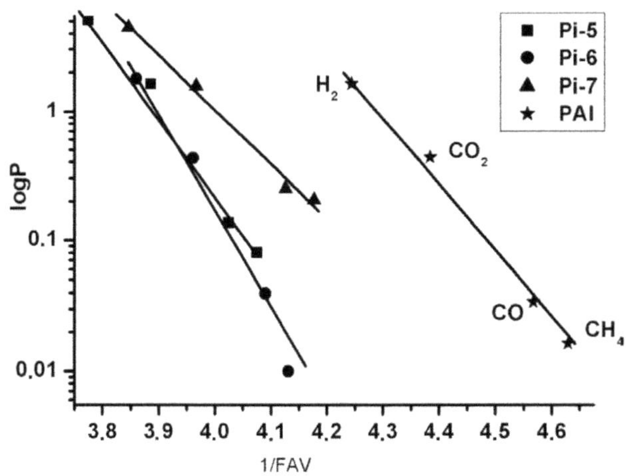

c)

FIGURE 3.15 The dependence of permeability coefficients (***P***) on fractional accessible volume (***FAV***) in the system one polymer–different gases for the second series polyimides.

Table 3.5 presents the parameters (A and B) of Eq. 11 and the correlation coefficients (R). Parameter **B** means the slope of the line representing the dependence of permeability coefficients on the fractional accessible volume. This parameter reflects the common selectivity of the respective polymer, which means that it shows how fast one gas penetrates through the polymer membrane compared with another gas. Higher the value of B, better the selectivity of the respective polymer, which means that the respective polymer will better separate one gas from another one. This behavior is determined by the packing of the polymer in glassy state that is by the distribution of microcavities in polymer matrix. The packing of polymer is significantly determined by its conformational rigidity. In Table 3.5, it can be seen that polymers **6, 9**, and **10** have the highest values of coefficient **B**, being 8.12, 8.29, and 11.31, respectively.

Figure 3.16 presents the dependence of parameter **B** on characteristic ratio $\mathbf{C}_\infty$. It can be seen that the selectivity of the polymer reaches the maximum when the characteristic ratio is 1.5, which means that Kuhn segment contains 1.5 repeating units. At such a value of $\mathbf{C}_\infty$ the packing of polymer in glassy state gives presumably a narrow distribution of micro-cavities and the permeability of one gas is significantly higher than that of the other, which leads to high selectivity. On the other hand, as seen in the dependence of free volume on conformational rigidity, with increasing rigidity the free volume decreases, and in case of polymer **PAI** and **Pi-11** it is the lowest, which means that the packing of macromolecules in these two polymers is more dense.

When we examine the dependence of permeability coefficients on fractional accessible volume for one gas and different polymers (Figure 3.17), we can see that the points corresponding to polymers **6, 9**, and **10** fall down visibly from the general dependences, for all four studied gases: CO_2, CO, H_2, and CH_4.

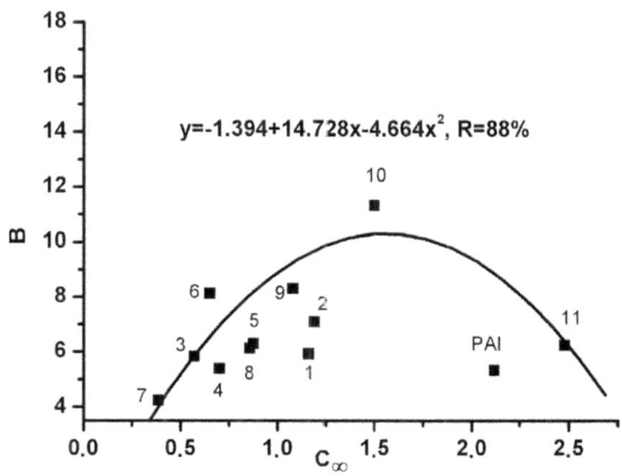

FIGURE 3.16 The dependence of slope ***B*** of Eq. (5) on characteristic ratio C_∞ for the second series of polyimides.

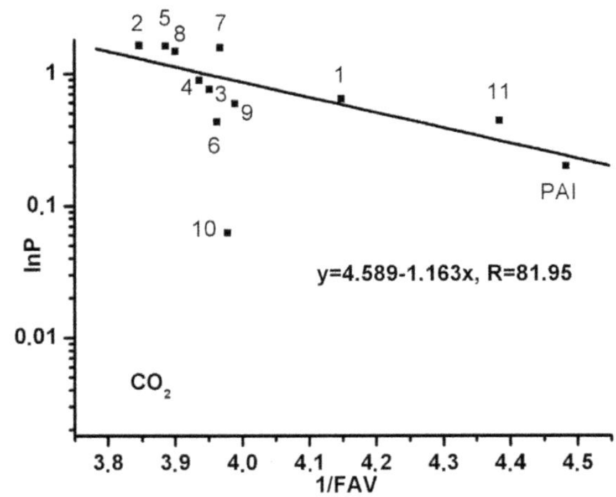

a)

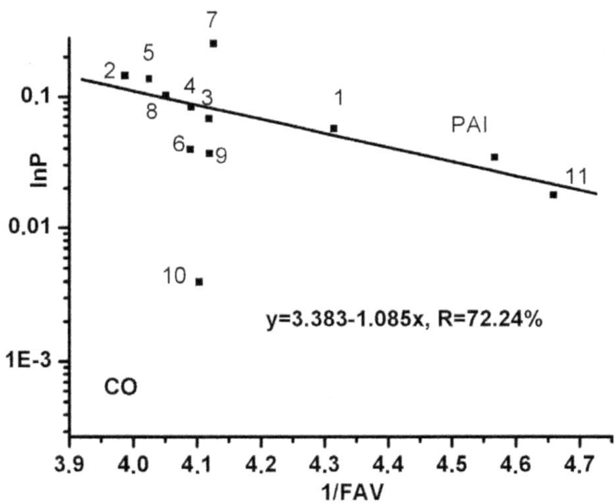

b)

FIGURE 3.17 Continued

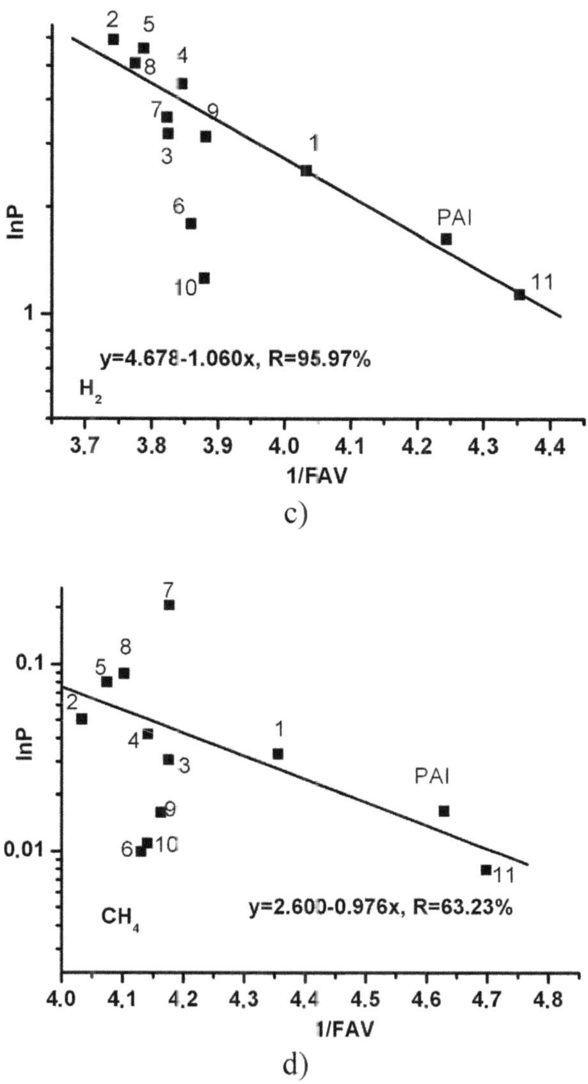

FIGURE 3.17 The dependence of permeability coefficients (***P***) on fractional accessible volume (***FAV***), for the system one gas–various polymers for the second series of polyimides.

It shows that even if the selectivity is high, the permeability of these polymers is low. The geometrical structure of the repeating unit determines the conformational rigidity of the polymer, its packing in glassy state, and subsequently its membrane characteristics: permeability and selectivity. Thus, in this section we have shown that the transport parameters and permittivity of polyimides depends on the conformational rigidity and free volume.

3.4.3 INFLUENCE ON THE FREE VOLUME OF THE MEDIUM OF SYNTHESIS AND OF THE SOLVENT FROM WHICH THE POLYMER FILMS WERE CAST

To explain the influence of solvent, which was used for the synthesis of polymers and for the preparation of solutions for casting thin films, we examined polymer 2 of the third series of polyimides (Table 3.6). First, we took polymer 2 which was synthesized in NMP [41] and the films were cast from three solvents: NMP, chloroform and tetrahydrofuran. The treatment with sc-CO_2 was performed at 200 bar and at temperatures of 40, 60 and 80°C. As seen in Table 3.7, the density of polymer films cast from different solvents, before sc-CO_2 treatment, differ in the second digit after the comma, being 1.222, 1.246 and 1.212 g/cm^3, respectively. This can be explained by the affinity of the solvent in relation with that polymer. The solvatation interaction between macromolecules and solvent molecules influences significantly the ability of polymer chains to undergo conformational transitions; it affects their equilibrium flexibility. In case of solvents like NMP, which are able to form enough strong solvate shells around macromolecules, the possibility of conformational transitions will decrease significantly, and therefore it causes a decrease of flexibility equilibrium. In such solvents, the dimensions of macromolecular coils increase which leads to the modification of their hydrodynamic properties, while the quantity of kinetically independent parts, which serve as the segments of macromolecules, is reduced. This leads to a change of the characteristics of polymer solutions. This is why after removal of solvent (NMP), the resulting film is loose, less dense, and therefore its measured density is lower. In case of solvents like chloroform in which the interactions between macromolecular chains prevent the solvatation, the

TABLE 3.6 Repeating Unit, Glass Transition Temperature and Conformational Parameters of the Third Series of Polyimides

Polymer	Repeating unit	T_g	F	l_o	$A_{fr}(A_h)$	C_∞
1	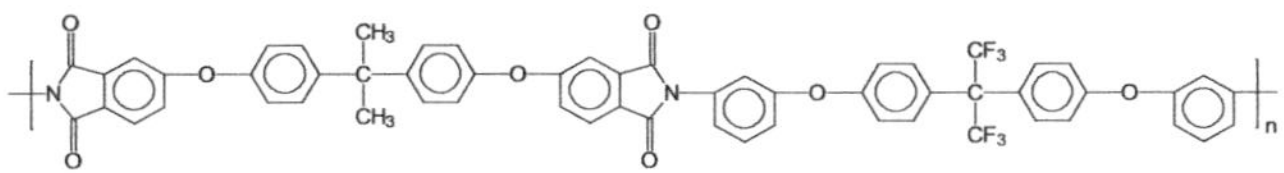 Synthesis in *meta*-cresol	180	11.36	41.86	20.28	0.484
2	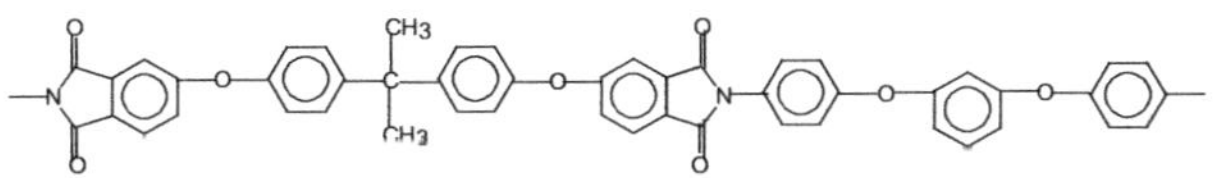 Synthesis in NMP and in salicylic acid	188	—	37.39	18.31	0.489
3	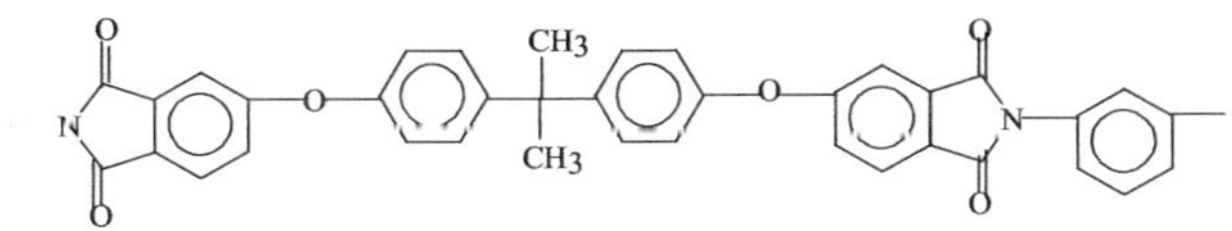 Synthesis in benzoic acid	225	—	27.13	20.93	0.771
4	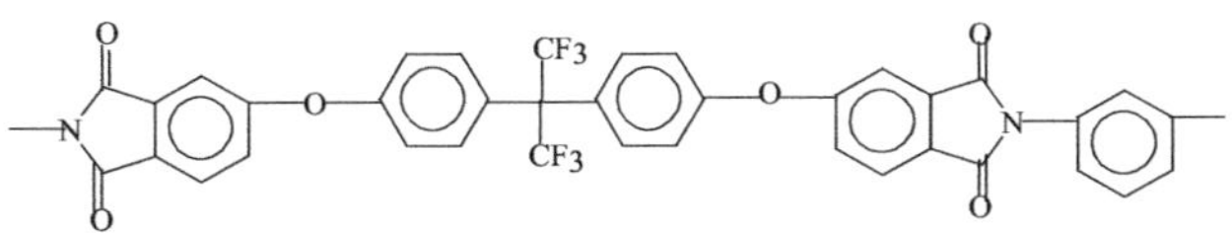 Synthesis in benzoic acid	234	16	27.13	21.71	0.800

TABLE 3.6 Continued.

Polymer	Repeating unit	T_g	F	l_o	$A_{fr}(A_h)$	C_∞
5	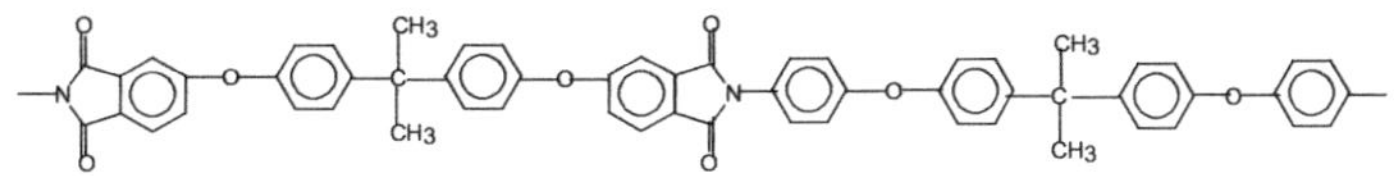 Synthesis in benzoic acid	205	—	42.05	24.81	0.590
6	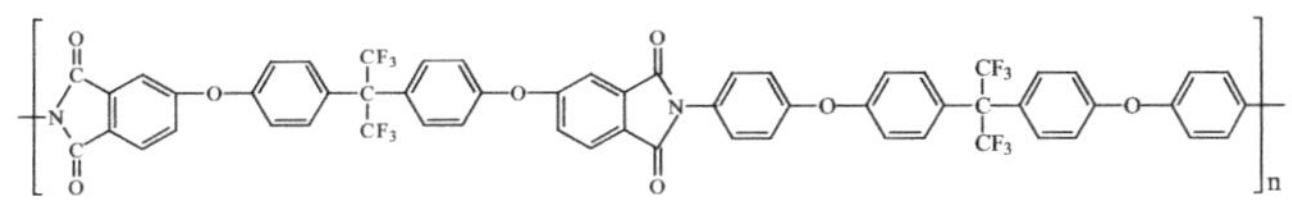 Synthesis in benzoic acid	234	25.59	32.20	29.11	0.904

F = Fluorine.

TABLE 3.7 Change of the Degree of Swelling with Increase of Temperature, at 200 bar, of Polyimide **2** of the Third Series Synthesized in NMP; the Films Were Prepared by Casting From Various Solvents: Tetrahydrofuran, Chloroform or NMP

Solvent		Tetrahydrofuran	$CHCl_3$	NMP
Before treatment	ρ_o (g/cm^3)	1.222	1.246	1.212
	V_{fo} (cm^3/g)	0.2875	0.2717	0.2942
	ρ_1 (g/cm^3)	1.238	1.266	1.234
T = 40°C	V_{f1} (cm^3/g)	0.2769	0.2590	0.2795
	ΔV_{fr}	**−0.0106**	**−0.0127**	**−0.0147**
	ρ_1 (g/cm^3)	1.211	1.215	1.214
T = 60°C	V_{f1} (cm^3/g)	0.2951	0.2922	0.2929
	ΔV_{fr}	**0.0076; 2.6%**	**0.0205; 7.5%**	**−0.0048**
	ρ_1 (g/cm^3)	1.204	1.210	1.210
T = 80°C	V_{f1} (cm^3/g)	0.2997	0.2956	0.2956
	ΔV_{fr}	**0.0122; 4.3%**	**0.0239; 8.8%**	**0.0014; 0.48%**

ΔV_f = change of free volume; V_{fo} = free volume before treatment; V_{ft} = free volume after sc-CO_2 treatment.

macromolecules will move in a relatively dense coil. The density of the films prepared from such solvents will be higher.

As seen in Table 3.7 and Figure 3.18, at 40°C, the removal of residual solvent takes place from film which was cast from chloroform solution; tetrahydrofuran solvent is removed at 40°C and 50°C, then at 60°C the film starts to swell, while NMP solvent remains in the polymer film even at 60°C, and at 80°C the degree of swelling is so low as it can be neglected. The highest degree of swelling of polymer **2** synthesized in NMP was attained when the film was cast form chloroform solution, but it did not exceed 8.8% (Table 3.8). The structure of this polyimide 2 is similar to the structure of polyimide **1** (Table 3.6) which was synthesized in m-cresol.

As seen in Table 3.8, polymer **1** swelled in sc-CO_2 quite well, which leads to the conclusion that the insignificant swelling of polymer **2** (synthesized in NMP) even in more severe conditions than for polymer **1** is most probably connected with the solvent used for its synthesis. Therefore, we synthesized polymer **2** in salicylic acid [17] and we prepared the film from chloroform solution. The treatment with sc-CO_2 was performed at 120 bar, 200 bar, and at 40°C and 60°C. As seen in Table 3.8, polymer **2** synthesized in salicylic acid swelled quite well in sc-CO_2, even if we had to use more severe

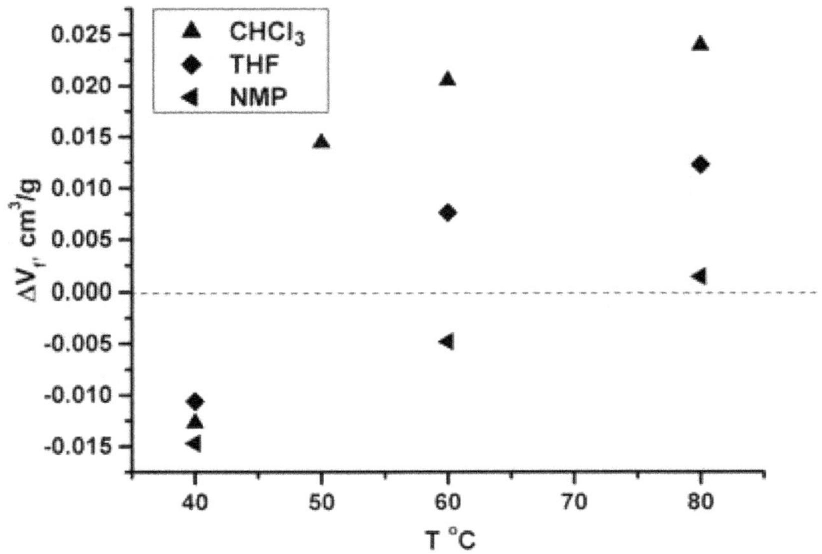

FIGURE 3.18 Relative change of free volume (ΔV_f) with temperature (T) for polyimide film **2** synthesized in NMP and cast from three solvents; the density was measured in ethanol.

conditions than for polymer **1**. Thus, we have proved that the solvent used for the synthesis of polyimides is very important when treating the film with sc-CO_2. The question is now in which way the NMP solvent can prevent the swelling of our polymers with sc-CO_2. Previously, when studying the Fourier transform infrared spectra at high temperature of polyamidic acids which are intermediary products in the synthesis of polyimides, it was shown that during the synthesis or during the preparation of thin films from polyamidic acid solutions in NMP followed by prolonged heating up to 200°C to remove that solvent, the following two concurrent processes were possible [42].

- formation of cross-links between polyamidic acids chains

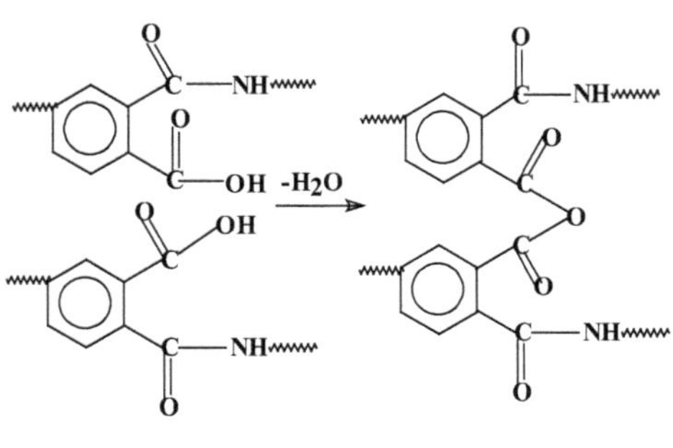

TABLE 3.8 Conditions of sc-CO_2 Treatment, Density, Change of Free Volume (*ΔVfr*) and Dielectric Constant (ε_o) of the Third Series of Polyimides

Polymer	Parameters of treatment	Density, g/cm^3		Free volume V_f (cm^3/g)		Increase of free volume (ΔV_{fr},)	ε_o
		Before treatment	After treatment	Before treatment	After treatment		
1	150 bar 40°C	1.368	1.281	0.2178	0.2681	0,0503, 23.1%	
	150 bar 65°C		1.170		0.3415	0,1237, 56.8%	
2	Initial	1.246		0.2717			
NMP	200 bar 40°C		1.266		0.2590	–0.0127	
	200 bar 50°C		1.224		0.2861	**0.0144, 5.3%**	
	200 bar 60°C		1.215		0.2922	**0.0205, 7.5%**	
	200 bar 80°C		1.210		0.2956	**0.0239, 8.8%**	
3	Initial	1.199		0.2963			
	120 bar, 40°C		1.186		0.3054	0.0091, 3.0%	
	120 bar, 60°C		1.113		0.3608	0.0644, 21.8%	3.42
	200 bar, 40°C		1.156		0.3274	0.0310, 10.5%	3.48
	200 bar, 60°C		0.995		0.4683	0.171, 58.0%	2.76
4	Initial	1.379		0.1833			3.19
	250 bar, 60°C		1.239		0.2751	0.0918, 45.9%	3.15
	250 bar, 80°C		1.155		0.3312	0.1479, 80.7%	2.78
	250 bar, 100°C		0.889		0.5906	0.4073, 209.0%	2.45

TABLE 3.8 Continued

Polymer	Parameters of treatment	Density, g/cm^3		Free volume V_f (cm^3/g)		Increase of free volume (ΔV_{fr},)	ε_o
		Before treatment	After treatment	Before treatment	After treatment		
5	Initial	1.432		0.2154			2.80
	200 bar, 80°C		1.375		0.2443	0.0289, 13.4%	2.74
	350 bar, 80°C		1.328		0.2698	0.0544, 25.3%	2.56
	500 bar, 80°C		1.318		0.2753	0.0599, 27.8%	2.46
6	Initial	1.314		0.2068			3.28
	250 bar, 60°C		1.099		0.3554	0.1486, 41.8%	3.12
	250 bar, 80°C		1.014		0.4320	0.2252, 119.2%	2.93
	250 bar, 100°C		0.967		0.4798	0.2912, 142.0%	2.87
7	Initial	1.436		0.2285			2.99
	250 bar, 80°C		1.395		0.2490	0.0205, 9.0%	2.96
	250 bar, 100°C		1.388		0.2526	0.0241, 10.5%	2.89
	350 bar, 100°C		1.137		0.4116	0.1831, 80.1%	2.77

- formation of complexes between amidic groups and NMP.

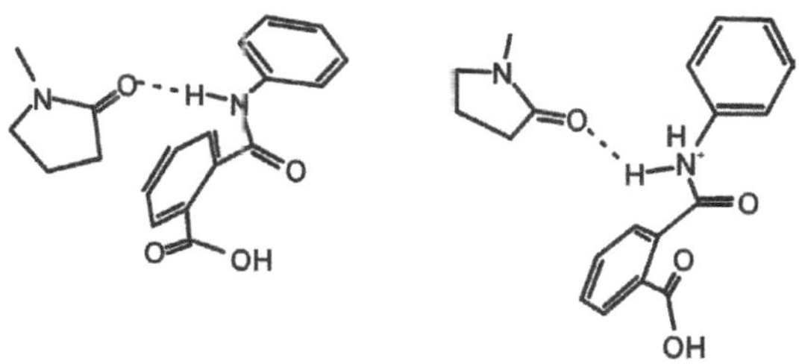

The formation of cross-links in polymers containing residual NMP prevails their swelling with sc-CO_2. The formation of complexes with amidic groups prevails the free rotation around N-phenyl bonds and leads to the decrease of the number of conformers, and therefore it facilitates the formation of inter-chain anhydride cross-links with high yield. The low degree of swelling of 3.8–7% of those polyimides is also connected with the possible formation of anhydride bridges during the synthesis of those polyimides in NMP, also leading to cross-links between chains. However, in case of polyimides, such cross-links are formed not so often as in case of polyamidic acids [43].

Another argument to the formation of cross-links is the higher density of polymer film **2** synthesized in NMP than the density of polymer film **2** synthesized in salicylic acid (Table 3.8). In both cases, the films were cast from chloroform solution of polymer. In the first case, the presence of cross-links made the polymer more rigid, and therefore its packing was more dense. Thus, the study of polymer **2** synthesized in two different solvents, NMP and salicylic acid, proved the conclusion made earlier on the basis of infrared spectra regarding the formation of cross-links during imidization of polyamidic acids in NMP.

3.4.4 EFFECT OF SOLVENT USED FOR DENSITY MEASUREMENT

The density of polymer film **2** of the third series of polyimides (Table 3.6) which was synthesized in NMP, before and after treatment with sc-CO_2, was measured in two solvents: ethanol and isopropanol. By using the value of density we calculated the free volume with Eq. (7) before and after sc-CO_2 treatment. As shown in Table 3.9, the free volume values, calculated by using the density values measured in those two solvents are different from each other. Then we calculated the change of free volume by comparison with the free volume of the sample before sc-CO_2 treatment which was measured in the same solvent: $\Delta V_f = V_{ft} - V_{fo}$, where V_{fo} is the free volume of the sample before treatment and V_{ft} is the free volume of the film after treatment with sc-CO_2 at a given temperature. Figure 3.18 presents the change of free volume with the temperature of sc-CO_2 treatment, for the three solvents. Before treatment, the film sample contains some amount of residual NMP remaining in polymer after its synthesis in this solvent.

During sc-CO_2 treatment, this solvent was removed from polymer matrix. This is why in Figure 3.19 the first value of the change of free volume when measured in ethanol was negative. The NMP solvent was extracted by CO_2 from polymer matrix and the density of polymer film increased.

When measuring the density in isopropanol the free volume of polymer included the volume of pores on the surface of film where the molecules of isopropanol could not enter due to their higher volume compared with that of ethanol. Figure 3.19 shows that the change of free volume values by swelling with sc-CO_2 at 40°C and 50°C, calculated by using the density values measured in isopropanol, is higher than in the case when density was measured in ethanol. It means that on the surface of polymer film there were some small pores in which ethanol molecules having the Van der Waals diameter of 4.16 Å could enter, while isopropanol molecules having a diameter of 6.72 Å could not enter.. The measured free volume in case of isopropanol is equal to the volume inside the polymer film plus the volume of pores on its surface. By treating the polymer film **2** of the third series of polyimides (synthesized in NMP) with sc-CO_2 at 50°C the difference between free volume values measured in ethanol and isopropanol first increases and then, at temperatures above 50°C, it decreases and it becomes negative. We presume that the pores on the film surface become smaller at low temperature of sc-CO_2 treatment, then their sizes increase at higher temperatures.

TABLE 3.9 Change of the Swelling Degree with the Increase of Temperature of sc-CO_2 Treatment of Polyimide **2** of the Third Series of Polyimides, Synthesized in NMP, Whose Film was Cast from Chloroform*

T (°C)	Ethanol			Isopropanol		
	ρ (g/cm³)	**V_{fr} (cm³/g)**	**$\_V_f = V_{ft} - V_{fo}$ (cm³/g; percentage of increase)**	**ρ (g/cm³)**	**V_{fr} (cm³/g)**	**$\_V_f = V_{ft} - V_{fo}$ (cm³/g; percentage of increase)**
Before sc-CO_2 treatment	1.246	0.2717		1.267	0.2584	
40	1.266	0.2590	–0.0127	1.250	0.2692	**0.0108; 4.2%**
50	1.224	0.2861	**0.0144; 5.3%**	1.238	0.2769	**0.018; 2.5%**
60	1.215	0.2922	**0.0205; 7.5%**			
80	1.210	0.2956	**0.0239; 8.8%**	1.235	0.2788	**0.0204; 7.9%**

$\_V_f$ = change of free volume; V_{fo} = free volume before treatment; V_{ft} = free volume after sc-CO_2 treatment.

*The treatment with sc-CO_2 was run at a pressure of 200 bar. The densities were measured in ethanol and in isopropanol.

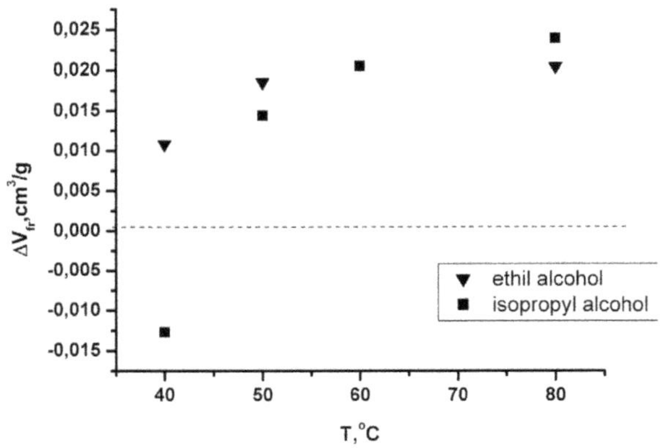

FIGURE 3.19 Variation of free volume (ΔV_f) of polymer film **1** (Table 3.6) (synthesized in NMP) by swelling with sc-CO_2 at various temperatures, when the density was measured in isopropanol or in ethyl alcohol.

3.4.5 STUDY OF THE SURFACE OF POLYIMIDE FILMS

We examined the change of pore size on the surface of the polymer films (polymer 2, Table 3.6). The pores were formed in the films during their preparation. For each sample about 100 pores were measured. The data were summarized in Table 3.10. Figure 3.20 shows the topographic images of the films before and after sc-CO_2 treatment. Information regarding the roughness of the surface can be found on a color scale: black color corresponds to minimum, and white color corresponds to maximum.

The surfaces of polymer films have a globular structure, with diameter of globules of 10–30 nm. The root mean square roughness of the defect-free regions of the surface is 0.4 nm. On the surface of the polymer films before and after sc-CO_2 treatment we found an amount of pores of 1–2 pores/μm^2. The dimension of pores varies in a large domain. The height has changed from a few Angstroms to 100 nm. The diameter ranges from 20 nm to 600 nm. The depth can be underestimated for narrow and deep pores because the probe can only touch the walls of the pores, but not the bottom. Besides pores, some particles can be observed on these images and their formation can be explained by partial degradation of the polymer surface during treatment with sc-CO_2. By sc-CO_2 treatment at 40°C the number of pores having small diameter and depth increases, which leads to the increase of total free volume (free volume inside the film plus the

TABLE 3.10 Change of Pore Size With Temperature of Swelling of the Polymer 2 in the Third Series of Polyimides

Temperature of swelling (°C)	Average pore depth (nm)	Proportion of pores (%) with depth H < 20 nm	Proportion of pores (%) with depth H > 20 nm	Average pore diameter (nm)	Proportion of pores (%) with diameter D < 100 nm	Proportion of pores (%) with diameter in the range of 100 nm < D < 200 nm	Proportion of pores (%) with diameter D > 200 nm
Before swelling	10.3	91	9	109	57	37	6
40	5.8	97	3	94	66	32	2
60	9.1	90	10	152	21	63	16
80	12.4	84	16	161	10	73	17

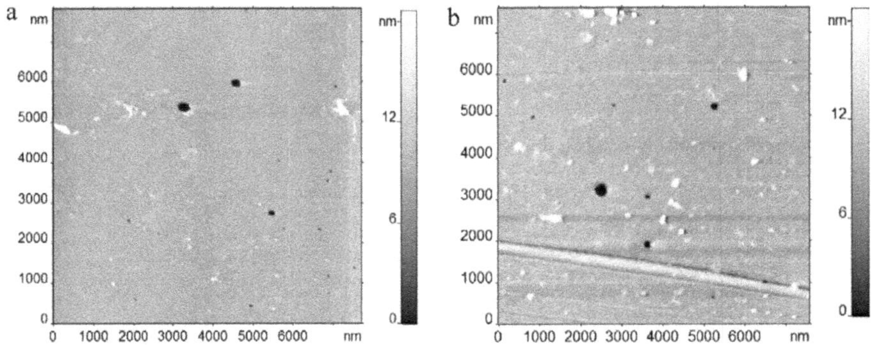

FIGURE 3.20 Topographic images (AFM) of the surface of polymer film **1** (Table 3.6) synthesized in NMP: (a) before treatment with sc-CO_2; (b) after treatment with sc-CO_2 at 40°C.

volume of the pores) of the polymer when the density was measured in isopropanol. When the polymer film was treated with sc-CO_2 above 40°C, the number of small pores decreased significantly. It is understandable why in figure 19 at the beginning, at temperatures of 40°C and 50°C, the free volume value based on measurement in isopropanol is higher than that measured in ethanol, and later at 80°C it becomes smaller. It means that the total volume of pores measured in isopropanol decreases because the depth and the diameter of pores become larger (Table 3.10) [44].

The surface structure of polymer films **5** and **6** of the third series of polyimides (Table 3.6) before and after sc-CO_2 treatment was examined by AFM (Figures 3.21 and 3.22). The films exhibited isotropic globular

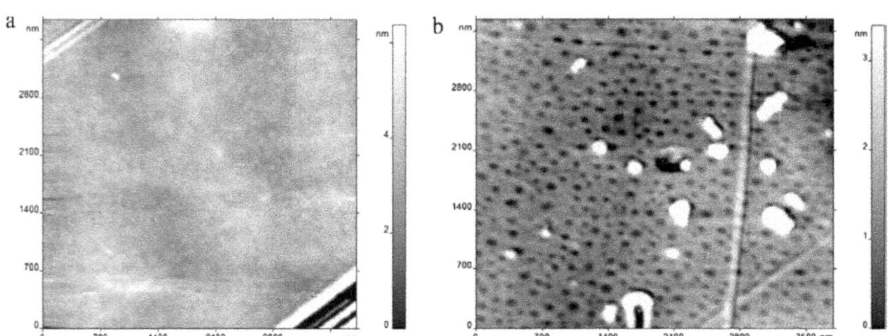

FIGURE 3.21 AFM image of the surface of polymer film **2** (Table 3.6): (a) before treatment with sc-CO_2; (b) after treatment with sc-CO_2 at 250 bar and 80°C.

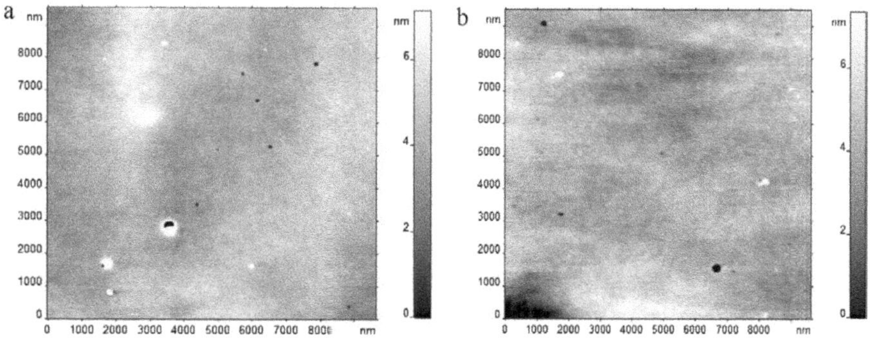

FIGURE 3.22 AFM image of the polymer film **3** (Table 3.6): (a) before treatment with sc-CO_2; (b) after treatment with sc-CO_2 at 350 bar, 100°C.

structure with globules heights of up to 1 nm relative to the surface. The polyimide films before treatment contained a small amount of surface pores (up to 15 pores per 100 μm^2) with depth of up to 30 nm.

The sc-CO_2 treatment of polymer film **5** of the third series of polyimides leads to the increase of root means square roughness (RMS) due to the appearance of some hard particles on the surface having the height of up to 350 nm. We presume that such hard particles may appear due to partial degradation of the surface during treatment with sc-CO_2 (possibly mechanical degradation). For polymer film **5** of the third series of polyimides the RMS value increases with 5 nm up to 19–33 nm after sc-CO_2 at 250 bar and at a temperature of 60–100°C. When the treatment was run at 80°C a regular pore system appeared on one surface of the film with depth of 2.5 ± 0.8 nm and diameter of 119 ± 34 nm (Figure 3.22b). When the temperature of sc-CO_2 treatment increased to 100°C, a small number of pores was observed on the surface (up to 10 pores per 100 μm^2) with depth of up to 35 nm and diameter of 100 nm up to 600 nm. The sc-CO_2 treatment of polymer film **6** of the third series of polyimides did not lead to the appearance of supplementary pores on the surface. The RMS changed insignificantly. It increased from 3 nm to 6 nm and 4 nm after sc-CO_2 treatment at 350 bar and 80°C and 100°C, respectively.

The structure inside polymer films was studied by two techniques: TEM. The TEM technique showed that polymer films **5** and **6** of the third series of polyimides became opaque after sc-CO_2 treatment. This means that there were defects in the films with dimensions comparable to the wavelength of visible light. Nano and micro-pores were formed in the core of the films. Figures 3.23a and 3.23b present cross-sections of polyimide film **3** of the third series of polyimides before and after sc-CO_2 treatment that were examined by TEM. It can be seen that pores of various diameters from 10 nm to 10 μm appear after sc-CO_2 treatment. It means that both nano-pores and micro-pores appear after sc-CO_2 treatment. In the first place, this is connected with the pressure of sc-CO_2. The formation of pores was observed only inside of the film. Near the surfaces of the films, pores were not detected. For different samples the regions without pores extended to 5–30 μm from the surface.

3.4.6 LOWERING THE DIELECTRIC CONSTANT

The influence of the increase in free volume during swelling in supercritical carbon dioxide on the dielectric constant was examined in two series of polyimides: the third series (Table 3.6) and the fourth series (Table 3.11). Two polymers of the third series (**1** and **6**) are present in the fourth series (**3** and **1**). The 3 polymer of the fourth series was swelling in sc-CO_2 in more stringent conditions. Polymer **1** of the fourth series was synthesized in another solvent. All polymers of the fourth series were synthesized in meta-cresol. All polymers of the fourth series were synthesized in *meta*-cresol [45].

At the beginning we consider the behavior of the third series of polymers. The polymer **1** of this series was synthesized in *meta*-cresol, while

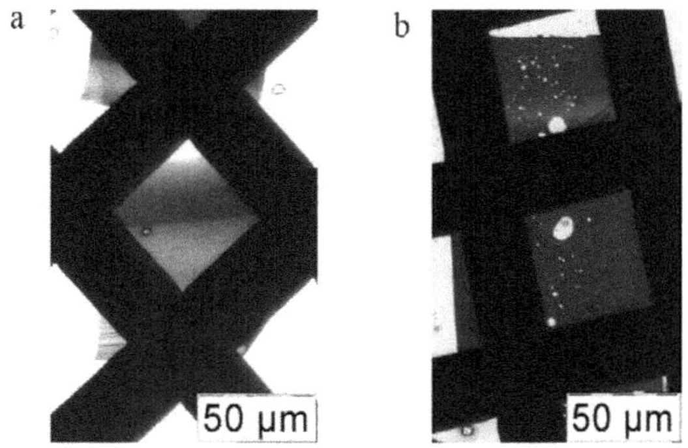

FIGURE 3.23 TEM image of the cross-sections of polymer film **3** (Table 3.6). The cross-sections are displayed on copper net having the dimension of 50μm: (a) before swelling with sc-CO_2; (b) after swelling with sc-CO_2 at 250 bar and 100°C.

TABLE 3.11 Repeating Unit and Conformational Parameters of the Fourth Series of Polyimides

Polymer	Repeating unit	l (Å)	A_h (Å)	C_∞	F (Wt %)	V_w (Å^3)	ΔV_{fr}^* (%)
1	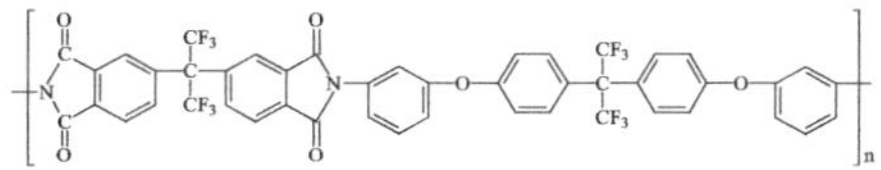	32.02	29.11	0.904	25.59	721.703	9.93
2	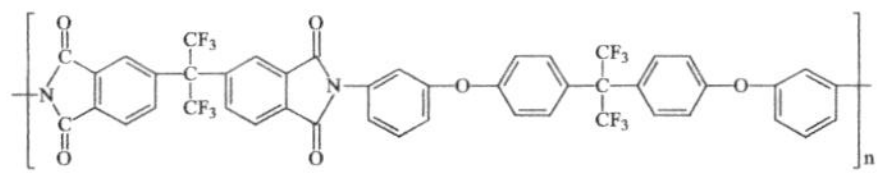	31.87	20.87	0.655	25.59	721.703	58.44
3	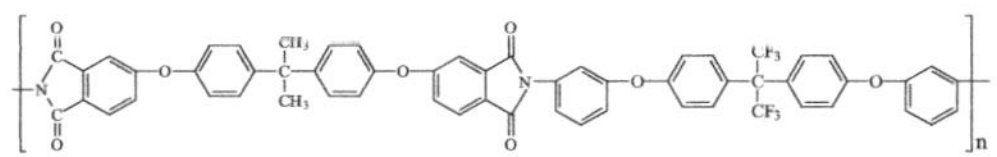	41.86	20.28	0.484	11.36	854.488	257.64
4	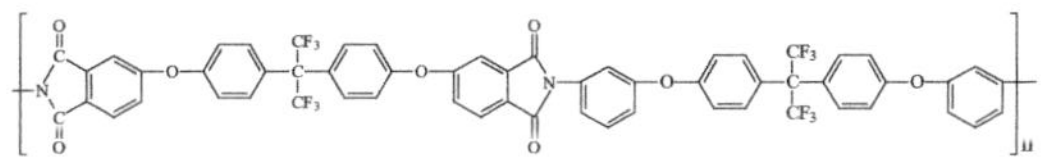	41.86	21.71	0.495	20.52	877.824	132.79
5	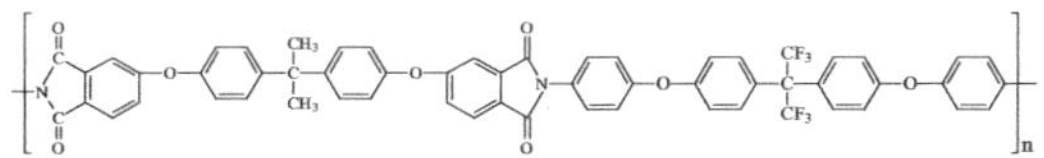	42.03	27.17	0.596	11.36	854.488	48.37
6	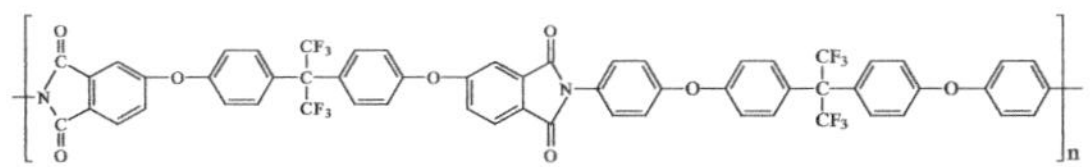	42.03	27.75	0.655	20.52	877.824	64.67

*At a pressure of 200 bar and a temperature of 60°C; F = fluorine.

the polymer **2** in NMP and salicylic acid. The other four studied polymers (Table 3.6) were synthesized in benzoic acid. To see and understand how the physico-chemical characteristics of polyimides change under the action of sc-CO_2 we analyzed them before treating with sc-CO_2. Table 3.6 presents the chemical structure of the repeating unit of the third series of polyimides, the solvent used for their synthesis, glass transition temperature, Kuhn segment value and characteristic ratio C_∞. Table 3.8 presents the values of densities before and after sc-CO_2 treatment. The dependence of T_g on characteristic ratio is linear (Figure 3.24) and it has a normal aspect: with increasing of the conformational rigidity, the probability of conformational transitions leading to the melting of polymer matrix decreases. Since all the points are situated on a line having a high correlation coefficient (99.46%), it means that all the studied polymers have enough high molecular weight and they form coils. Besides, the imidization process was complete in all those polymers and non-cyclized units were absent [46, 47]. The dependence of *Tg* on free volume confirms this explanation (Figure 3.25). With increasing of the conformational rigidity, the free volume of polymer matrix decreases (Figure 3.26), the packing of polymers becomes stronger. With increasing of the conformational rigidity, the dimensions of linear parts of the polymer chain increase, and as a consequence, the packing of chains is tighter while the free volume of polymer is smaller. On the other hand, while decreasing of free volume,

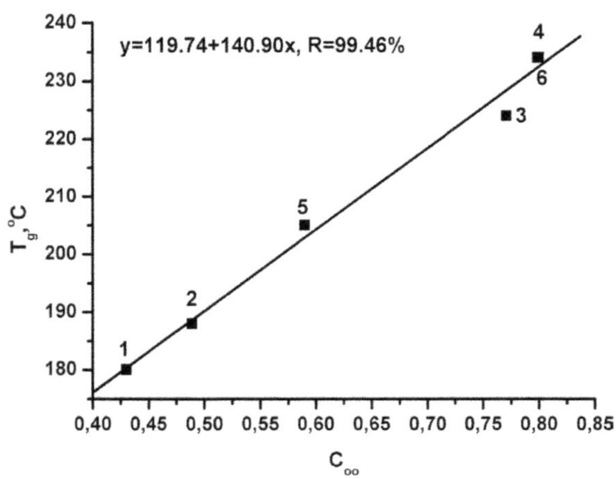

FIGURE 3.24 Dependence of glass transition temperature (***Tg***) on characteristic ratio (***C***$_\infty$) for the third series of polyimides.

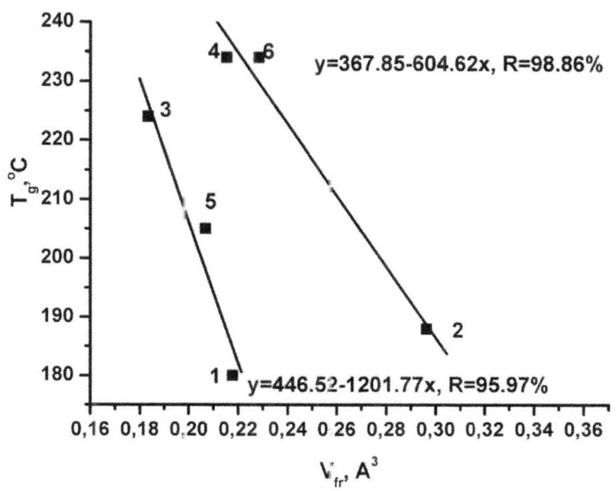

FIGURE 3.25 Dependence of glass transition temperature (***Tg***) on free volume (V_f) for the third series of polyimides.

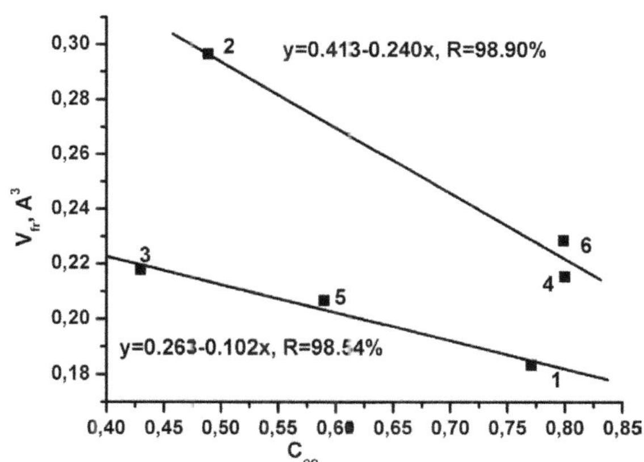

FIGURE 3.26 Dependence of free volume (V_f) on characteristic ratio (C_∞) for the third series of polyimides.

the possibility of conformational transitions decreases and thus, the glass transition temperature increases [39].

Although in Figure 3.24 all points corresponding to the studied polymers are situated on one line, in Figures 3.25 and 3.26 the corresponding points are divided in two dependences. One dependence contains polymers **1, 3** and **5** of the third series, which do not contain fluorine atoms in dianhydride segment; their *Tg* increases sharply with decreasing of free volume. The second dependence contains polymers **4** and **6** of the third series, which do contain fluorine atoms in dianhydride segment, and polymer **2**, which was synthesized in salicylic acid. As shown earlier (in Section 3.4.1), the presence of fluorine atoms in monomer units determines a loose polymer packing. The conditions of swelling in sc-CO_2 are shown in Table 3.8. The polymer films **5** and **6** of the third series of polyimides were also treated with sc-CO_2 at 250–350 bar and 60–100°C. Their densities were measured in ethanol. The free volume increased with 7% up to 140%. The dielectric constant values of those polymers are shown in Table 3.8. The values of dielectric constant are in the range of 2.45–3.48, being comparable or even lower than the value of Kapton polyimide film measured under the same conditions (3.13–3.24) [48].

As seen in Table 3.8, polymer **3** (ULTEM) of the third series (based on m-phenylene-diamine and isopropylidene-diphenoxy-bis(phthalic anhydride)) was the most sensitive to treatment with sc-CO_2. In relatively mild conditions of treatment with sc-CO_2 (250 bar, 100°C), its free volume doubled: the increase was 222.2% and its dielectric constant became significantly lower (2.45) compared with the value before swelling (3.19). The analogue polymer **4** of the third series which contains hexafluoroisopropylidene units between imide rings exhibited a lower dielectric constant (2.80) even before treating with sc-CO_2, due to the presence of fluorine substituents. This value decreased by treating with sc-CO_2 to almost the same value (2.46) as for polymer **3** of the third series. Even if polymer **3** of the third series did not contain fluorine substituents, which is very important for electronic applications, it did swell quite well with sc-CO_2 and it has the advantage of being commercially available (under the name of ULTEM).

In polymer **4** of the third series the content of fluorine is 16%. The swelling with sc-CO_2 was performed under the same temperature (80°C) and in a large domain of pressure parameters, from 200 bar to 500 bar. Regardless the severe conditions of sc-CO_2 treatment (500 bar), the increase of free volume were maximum 27.8%. It means that the presence of fluorine atoms decreases the dielectric constant, but it prevails the increase of free volume. The study of those two polymers, **3** and **4** of the third series, having identical structure except fluorine atoms in polymer **4** shows that their sc-CO_2 swelling process is more sensitive to temperature than to pressure.

Polymer **5** of the third series underwent sc-CO_2 treatment under the same conditions as polymer **3** of the third series: constant pressure of 250 bar, and various temperatures, of 60°C, 80°C, and 100°C. Its characteristic ratio is lower (0.59) than that (0.771) of polymer **3** and its glass transition temperature (205°C) is also lower than that (225°C) of polymer **3**. Its free volume increased less than that of polymer **3**, from 0.2068 to 0.4798, which means an increase of 132% compared to 222.2% increase of polymer **3**. The dielectric constant decreased but not very significantly, from 3.28 to 2.87 (Table 3.8).

In case of polymer **6** of the third series, the content of fluorine is 24% and the treatment with sc-CO_2 was performed at 80°C and 100°C, and at 250 bar and 350 bar. The increase of free volume by treatment at 250 bar and 80°C 100°C was insignificant, of only 9–10.5%. Under more severe conditions, 350 bar and 100°C, the increase of free volume was 80% and dielectric constant decreased from 2.99 to 2.77. The film of this polymer **6** of the third series having a thickness of 20 μm practically did not swell with sc-CO_2, while a thicker film, of 40 μm, did swell and the results are those shown in Table 3.8.

The low dielectric constant of polymers **4** and **6** of the third series before treatment with sc-CO_2 are due to the presence of hexafluoroisopropylidene groups, which disturb the packing of macromolecular chains in glassy state, and thus they determine the increase of free volume. Thus, the amount of polarizable groups per volume unit is lower, which also leads to the decrease of dielectric constant [30]. All these results on dielectric constant values before and after treatment with sc-CO_2 are illustrated in Figure 3.27.

As seen in Figure 3.27, the decrease of dielectric constant with increase of free volume is not identical in these polymers. The equations, which were introduced in Figure 3.27 for the linear dependence of dielectric constant on free volume, show that the correlation coefficients are high enough and the lines have different inclinations. For polymers **2** and **4** of the third series the inclination of the lines is stronger, which indicates the easy reconstruction of packing of chains in polymer matrix under the action of sc-CO_2 and rapid decrease of dielectric constant with small increase of free volume. The increase of pressure and temperature during the swelling process with sc-CO_2 in these cases could give even lower values of dielectric constant. The polymers **3**, **5** and **6** of the third series exhibit a rapid

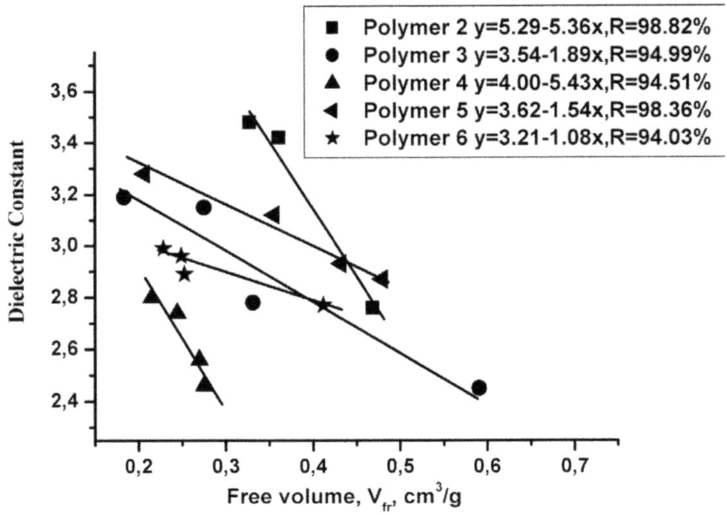

FIGURE 3.27 Variation of dielectric constant by treatment with sc-CO_2 for the third series of polyimides.

increase of free volume during swelling with sc-CO_2, but the decrease of dielectric constant with the increase of free volume takes place slower. The difference in the behavior of these two groups of polymers seems to be connected with the distribution of dipoles in the repeating unit. That corresponds well with dividing these polymers in two groups when discussing the dependence of glass transition temperature on free volume of the polymers before swelling with sc-CO_2.

To select the polymers for the fourth series we have used the results obtained when studying the previous three series of polymers. In Section 3.4.1 we have shown the dependence of the dielectric constant on the conformational rigidity and free volume. But the ability to swelling with sc-CO_2 takes place the best in case of polymers having characteristic ratio in the range of 0.4–0.8 [49]. This is why, for the fourth series, we selected the six polyimides whose structures are shown in Table 3.11.

We have calculated their Kuhn segment under the assumption of free rotation ($\mathbf{A}_{fr}$), characteristic ratio ($\mathbf{C}_{\infty}$) and Van der Waals volume ($\mathbf{V}_w$). These polymers were synthesized in *m*-cresol and the films were prepared from their solutions in chloroform. Table 3.12 shows the values of glass transition temperature of the fourth series of polyimides. These values are

TABLE 3.12 Glass Transition Temperature Tg (°C) Before and After Extraction of *m*-Cresol

Polymer	Before extraction of *m*-cresol	Tg value from literature	After extraction of *m*-cresol	Polymer	Before extraction of *m*-cresol	After extraction of *m*-cresol
1	196.9	234 [50]	251	**4**	111	181
2	131	224 [51]	211	**5**	130	207
3	102	180 [52]	164	**6**	133	204

significantly lower (in the range of 200–290°C) than those of traditional aromatic polyimides which are well above 300°C. These lower glass transition temperatures could be due to the plasticizing of solid polymers with m-cresol. In polymer films, as it was mentioned above, the residual *m*-cresol was removed by extraction with methanol and the films were further heated in vacuum at 70°C. The values of Tg measured afterwards are given in Table 3.12. For the first three polymers the values of Tg are also known from literature [50–52] and they are somewhat higher than the values measured by us after removal of *m*-cresol and methanol. It is known that Tg depends on the free volume of polymer matrix [39].

Now, we examine the dependence of Tg values measured after removal of solvents on the characteristic ratio (Figure 3.28). This dependence is linear, with a high correlation coefficient. It means that the cyclization to polyimide structure was complete, except for polymer **3** in this series. However, we can not make a clear conclusion about cyclization degree of polymer **3** since the cause of loss of points for polymer **3** (Table 3.11) could be exposed after removal of *meta*-cresol microcavities which lower its glass transition temperature Tg.

Therefore, the study was undertaken on cross-section of polymer films, by using transmission electron microscopy (TEM). A typical TEM image is shown in Figure 3.29. The thickness of the film was 80 μm before swelling with sc-CO_2 and, as can be seen in Figure 3.29, it contained pores at 10 μm under the surface having the dimensions of 18 nm – 3 μm; it shows that even before treatment with sc-CO_2 the polymer matrix was porous.

Table 3.13 presents the parameters of swelling process, increasing of free volume and of dielectric constant of the fourth series of polymers.

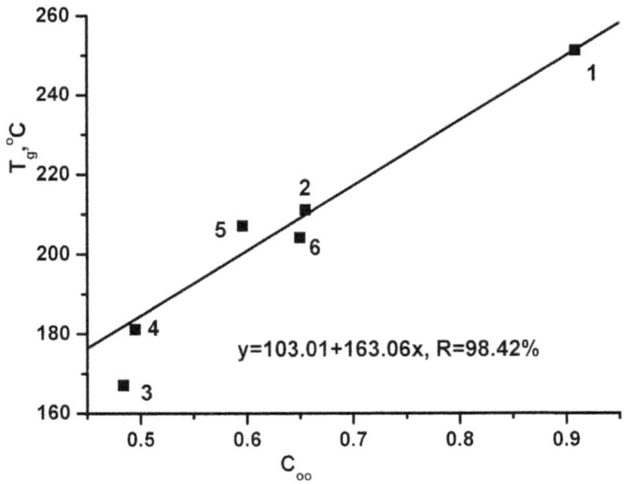

FIGURE 3.28 Dependence of glass transition temperature (***Tg***) on characteristic ratio (**C∞**) for the fourth series of polyimides.

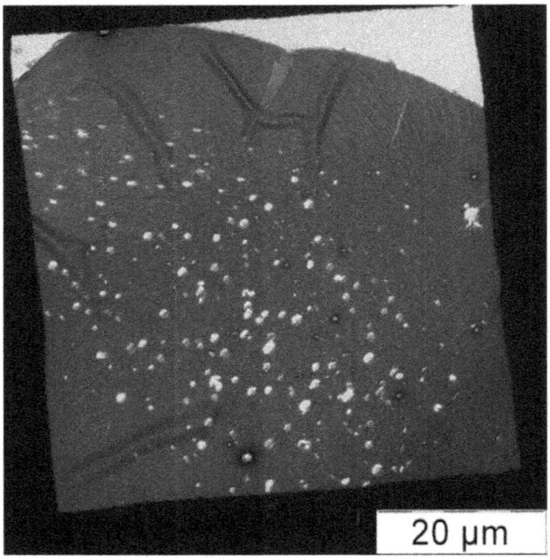

FIGURE 3.29 TEM image of a section of polymer film **3** before swelling with sc-CO_2 for the fourth series of polyimides.

The swelling was performed for three samples of each polymer under constant pressure, at three temperature values, followed by rapid decompression of CO_2 pressure. As seen in Table 3.13, the swelling at 200 bar and 60°C depends on both conformational rigidity and on the content

TABLE 3.13 Increase of Free Volume and Decrease of Dielectric Constant After Swelling in sc-CO_2 at a Pressure of 200 Bar and Various Temperatures (**T**) for the Fourth Series of Polyimides

Sample (Test)	T (°C)	Before swelling in sc-CO_2		After swelling in sc-CO_2		ΔV_f (%)	ε_o
		ρ (g/cm³)	V_f (cm³/g)	ρ (g/cm³)	V_f (cm³/g)		
1 (1)		1.458	0.1976				3.00
1 (2)	60°C	1.478	0.1883	1.439	0.2070	9.93	2.98
1 (3)	40°C	1.477	0.1891	1.401	0.2255	19.25	2.92
1 (4)	80°C	1.483	0.1864	1.421	0.2158	15.77	2.94
2 (1)		1.497	0.1799				2.54
2 (2)	60°C	1.480	0.1879	1.273	0.2977	58.44	2.04
2 (3)	40°C	1.518	0.1708	1.225	0.3286	92.38	1.96
2 (4)	80°C	1.501	0.1769	1.368	0.2430	37.37	2.16
3 (1)		1.381	0.2111				1.94
3 (2)	60°C	1.401	0.2009	0.812	0.7185	257.64	1.62
3 (3)	40°C	1.389	0.2070	0.955	0.5338	157.87	1.76
3 (4)	80°C	1.376	0.2138	0.847	0.6672	212.07	1.63
4 (1)		1.465	0.2030				2.46
4 (2)	60°C	1.467	0.2022	1.052	0.4707	132.79	1.88
4 (3)	40°C	1.476	0.1980	0.969	0.5525	179.04	1.58
4 (4)	80°C	1.384	0.2400	1.055	0.4684	95.17	2.00
5 (1)		1.268	0.2620				2.43
5 (2)	60°C	1.290	0.2485	1.116	0.3687	48.37	2.23
5 (3)	40°C	1.284	0.2517	1.174	0.3246	28.96	2.35
5 (4)	80°C	1.290	0.2480	1.182	0.3192	28.71	2.27
6 (1)		1.431	0.2193				1.89
6 (2)	60°C	1.459	0.2058	1.152	0.3389	64.67	1.58
6 (3)	40°C	1.462	0.2047	1.2240	0.3375	64.48	1.54
6 (4)	80°C	1.452	0.2094	1.201	0.3533	68.72	1.45

of fluorine in the repeating unit. For example, polymers **1** and **2** of the fourth series have the same content of fluorine, but while the characteristic ratio decreases from 0.904 to 0.655 (Table 3.11), the free volume increases almost 6 times (Table 3.13). In the case of polymers **3** and **4** of the fourth series, having almost identical conformational rigidity, the increase of fluorine content from 11.36% to 20.52% leads to the decrease of free volume almost twice. Now we compare the behavior of polymer **1** from the third series with polymer **3** from the fourth series. Both polymers were synthesized in *m*-cresol. Polymer **1** was treated with sc-CO_2 at a pressure of 150 bar, temperature of 65°C, with slow pressure-release, while polymer **3** was treated with sc-CO_2 at a pressure of 200 bar, temperature of 65°C, with rapid pressure-release. The difference in sc-CO_2 treatment conditions led to an increase of 5 times in the free volume of polymer **3** in the fourth series. This behavior may lead to the conclusion that by rapid release of CO_2 pressure, the flexible polymer (**C**∞ = 0.484) can form a foam. To confirm this conclusion, we have investigated the surface of polymer film **3** by using atom force microscopy (AFM) (Figure 3.30).

Before swelling (Figure 3.30a), the sample showed a smooth flat surface with root mean square roughness (RMS) of ~10 nm. On the surface, a low quantity of pores was seen (0.003–0.3 pores/μm^2) having the depth of 6–32 nm. After swelling with sc-CO_2 the RMS value increased significantly, to 30 nm, on the surface appeared zones with high pores concentration, up to 8 pore/ μm^2, with depth up to 270 nm, and undisclosed pores in the form of hills appeared having the length up to 24 nm and diameter of 1300 nm (Figure 3.4b). But, through-holes did not appear on all the studied films. In addition, we have investigated these samples after swelling by using transmission electron microscopy (TEM). A representative image is shown in Figure 3.31. The thickness of the sample was 130 μm. Figure 3.31 shows that the pores appeared at 5 μm below the surface and

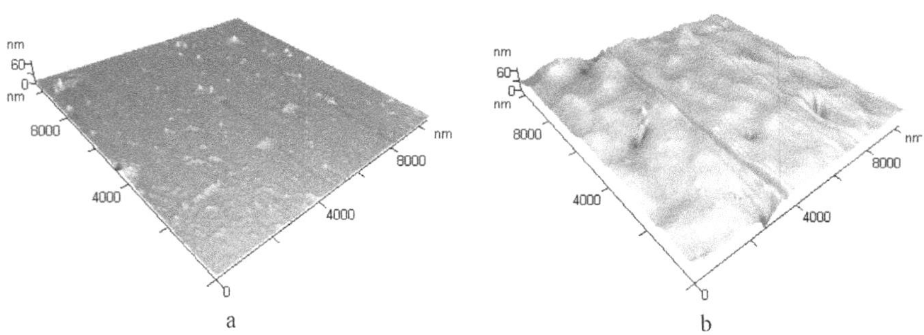

FIGURE 3.30 AFM image of a section of polymer film **3** of the fourth series of polyimides before (a) and after (b) swelling with sc-CO_2 at 200 bar and 60°C with rapid release of CO_2 pressure.

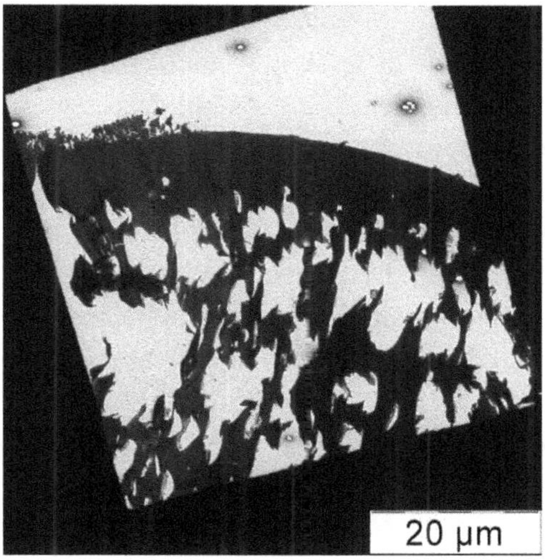

FIGURE 3.31 TEM image of a section of polymer film **3** of the fourth series of polyimides after swelling with sc-CO_2 at 200 bar and 60°C, with rapid release of CO_2 pressure.

their dimension was 100 nm to 10 μm. Thus, we have shown that when the pressure of the CO_2 is released rapidly after swelling, the flexible polymers may form foams.

For each polymer film the dielectric permittivity was measured in the domain of frequencies from 10^{-3} to 10^6 Hz and it was approximated for the frequency value of zero to obtain the dielectric constant (ε_0). Figure 3.32

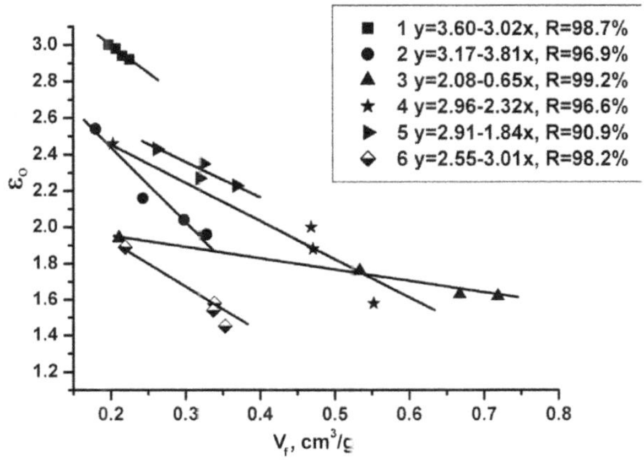

FIGURE 3.32 Dependence of dielectric constant (ε_0) on free volume (V_f) for the fourth series of polyimides.

presents the dependence of dielectric constant on free volume. As can be seen, the polymers behave differently when increasing the free volume.

For same of them the dielectric constant decreases significantly when the free volume modifies only a little; for the other polymer films the dielectric constant varies insignificantly. All the dependences are described by the equation y = A−Bx with a high correlation coefficient, R. Table 3.14 presents the values of B in these equations.

It can be seen that the inclination at rapid decrease of dielectric constant (high value of B) is strongly related with the content of fluorine in the repeating unit (polymers **1, 2** and **6** of the fourth series). Higher the fluorine content, more significantly decreases the dielectric constant with the increase of free volume. Thus, on one hand the presence of fluorine atoms and the high rigidity hinder the swelling with sc-CO_2, and on the other hand in the case of polyimides containing fluorine the dielectric constant is more sensitive to the increase of free volume.

In this section we have shown that the swelling with sc-CO_2 of polyimide films is directly connected with conformational rigidity and with fluorine content in the repeating unit of the respective polymers. The best swelling takes place when the value of characteristic ratio is 0.4–0.8. Also, higher the fluorine content of the repeating unit, more significantly decreases the dielectric constant of the polymer films. The formation of nano-foams by swelling with sc-CO_2 may lead to a dielectric constant value even below 1.5.

TABLE 3.14 Change of Free Volume (V_f) After Swelling at Pressure of 120 Bar and Temperature of 40°C of the Fourth Series of Polyimides

Polymer	Before swelling in sc-CO_2			After swelling in sc-CO_2			ΔV_f (cm³/g; %)
	ρ(g/cm³)	V_f(cm³/g)	T_g(°C)	ρ(g/cm³)	V_f(cm³/g)	Tg(°C)	
1	1.389	0.2070	167	1.328	0.2400	169	0.0330; 16%
2	1.431	0.2230	181	1.414	0.2314	180	0.0084; 3.8%
3	1.341	0.2327	207	1.321	0.2440	206	0.0113; 4.9%
4	1.428	0.2245	204	1.394	0.2416	204	0.0171; 7.6%

3.4.7 INFLUENCE OF MICROSTRUCTURE ON THE TRANSPORT PROPERTIES

To investigate the change of microstructure of polyimide thin films under the action of supercritical carbon dioxide and its influence on the transport properties we selected the polymers **3–6** of the fourth series (Table 3.11). For this study we prepared thicker films than in the investigation of dielectric properties (20–40 μm). After treatment with sc-CO_2, all these four polymers exhibited a lower density and a higher free volume compared with the untreated samples. The highest increase of free volume was observed in case of polymer **3** of the fourth series containing meta-substituted phenylene rings in the diamine segment [53, 54]. After swelling with sc-CO_2, the free volume of polymer matrix increases with 4–16% (Table 3.14). The highest increase of free volume was observed in case of polymer 3, being 16%. The glass transition temperature of these polymers did not change (in the range of the precision of DSC measurement). It means that the T_g of these polymers is predominantly determined by the chemical structure of the polymer chain itself. Tables 3.15–3.18 present the permeability and diffusion coefficients through polymer films (membranes), before and after treatment with sc-CO_2. The dependences of permeability (a) and diffusion (b) on the fractional accessible volume (FAV) for each polymer in the system one polymer—different gases are given in Figures 33–36 (before sc-CO_2 treatment, the right-side lines being described by equation $y = A + Bx$). The value of the Van der Waals radius of Helium is 1.22–1.80 Å, according to literature data [55–57].

In our calculations, we used the value of 1.22 Å for the Van der Waals radius of Helium. It came out that the point corresponding to the value of 1.22 Å clearly goes out of general dependences both of permeability coefficient and of diffusion coefficient for all four polymers. That point remains on the general dependence only in case when the Van der Waals radius of Helium is taken equal to 1.68 Å. Similar results were obtained for Helium previously [39]. The value of 1.68 Å obtained by us for the Van der Waals radius of Helium is close to the effective diameter (1.78 Å) calculated by other authors [58].

In Tables 3.15–3.18, it can be seen that after swelling with sc-CO_2, the permeability coefficients increased for different gases and different

TABLE 3.15 Change of Transport Parameters by Swelling in Supercritical Carbon Dioxide (sc-CO_2) of Polymer Film **1** of the Fourth Series

Gas	Before swelling in sc-CO_2					After swelling in sc-CO_2						
	V_{occ} (Å^3)	V_{acs} (cm^3/g)	*1/FAV*	*P* (Barrer)	$D \bullet 10^8$ (cm^3/s)	V_{acs} (cm^3/g)	*1/FAV*	*P* (Barrer)	Δ*P* (%)	$D \bullet 10^8$ (cm^3/s)	Δ*D* (%)
He*	896.194	0.1819	3.957	7.57	373	0.2150	3.502	20.3	168.2	320	–14.2
He	913.653	0.1714	4.200	7.57	373	0.2045	3.682	20.3	168.2	320	–14.2
O_2	924.118	0.1652	4.359	0.497	1.19	0.1982	3.799	1.2	141.4	0.87	–26.9
N_2	928.424	0.1626	4.428	0.0824	0.32	0.1956	3.849	0.14	69.9	0.17	–46.0
CO_2	917.956	0.1688	4.264	2.06	0.29	0.2019	3.729	5.4	162.1	0.27	–7.85

(He* – when the Van der Waals radius of Helium atom was taken equal to1.68 Å).

TABLE 3.16 Change of Transport Parameters by Swelling in Supercritical Carbon Dioxide (sc-CO_2) of Polymer Film **2** of the Fourth Series

Gas	Before swelling in sc-CO_2					After swelling in sc-CO_2					
	V_{occ} (Å^3)	V_{acs} (cm^3/g)	*1/FAV*	*P* (Barrer)	$D \bullet 10^8$ (cm^3/s)	V_{acs} (cm^3/g)	*1/FAV*	*P* (Barrer)	Δ*P* (%)	$D \bullet 10^8$ (cm^3/s)	Δ*D* (%)
He*	929.421	0.1951	3.582	16.8	869	0.2035	3.476	35.8	111.3	600	–31.0
He	948.302	0.1848	3.781	16.8	869	0.1932	3.660	35.8	111.3	600	–31.0
O_2	959.582	0.1805	3.910	1.03	2.61	0.1889	3.743	2.7	162.1	1.7	–34.9
N_2	965.011	0.1758	3.965	0.169	0.65	0.1842	3.840	0.4	136.7	0.40	–38.6
CO_2	952.702	0.1824	3.831	4.13	0.64	0.1908	3.706	10.0	142.1	0.43	–33.1

(He* – when the Van der Waals radius of Helium atom was taken equal to1.68 Å).

TABLE 3.17 Change of Transport Parameters by Swelling in Supercritical Carbon Dioxide (sc-CO_2) of Polymer Film **3** of the Fourth Series

Gas	Before swelling in sc-CO_2					After swelling in sc-CO_2					
	V_{occ} (Å^3)	V_{acs} (cm^3/g)	*1/FAV*	*P* (Barrer)	*D•10^8* (cm^3/s)	V_{acs} (cm^3/g)	*1/FAV*	*P* (Barrer)	Δ*P* (%)	*D•10^8* (cm^3/s)	Δ*D* (%)
He*	921.752	0.1923	3.877	13.7	740	0.2036	3.717	25.2	83.9	200	–73.0
He	940.301	0.1812	4.115	13.7	740	0.1925	3.932	25.2	83.9	200	–73.0
O_2	951.951	0.1742	4.280	1.09	2.44	0.1855	4.081	2.1	92. 7	1.6	–34.4
N_2	956.140	0.1717	4.343	O185	0.67	0.1830	4.137	0.4	116.2	0.40	–40.3
CO_2	945.712	0.1780	4.193	5.18	0.67	0.1892	4.000	8.6	66.0	0.47	–29.8

(He* – when the Van der Waals radius of Helium atom was taken equal to1.68 Å).

TABLE 3.18 Change of Transport Parameters by Swelling in Supercritical Carbon Dioxide (sc-CO_2) of Polymer Film **4** of the Fourth Series

Gas	Before swelling in sc-CO_2					After swelling in sc-CO_2					
	V_{occ} (Å^3)	V_{acs} (cm^3/g)	*1/FAV*	*P* (Barrer)	*D•10^8* (cm^3/s)	V_{acs} (cm^3/g)	*1/FAV*	*P* (Barrer)	Δ*P* (%)	*D•10^8* (cm^3/s)	ΔD (%)
He*	929.658	0.1964	3.566	13.6	521	0.2135	3.360	30.9	127.2	1008	93.5
He	948.954	0.1859	3.765	13.6	521	0.2030	3.533	30.9.	127.2	1008	93.5
O_2	960.356	0.1798	3.896	1.03	2.67	0.1968	3.644	2.4	133.0	1.8	–3.6
N_2	964.170	0.1777	3.951	0.189	0.71	0.1948	3.683	0.5	164.6	0.50	29.6
CO_2	953.666	0.1834	3.812	4.72	0.71	0.2005	3.578	11.1	135.2	0.66	–7.2

(He* – when the Van der Waals radius of Helium atom was taken equal to 1.68 Å).

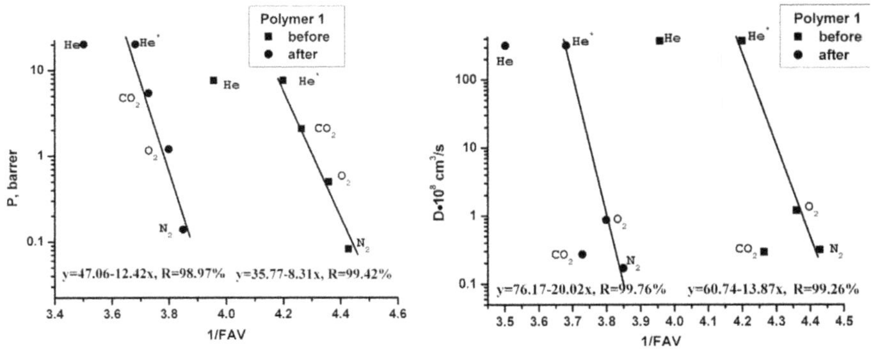

FIGURE 3.33 Dependence of permeability coefficients ***P*** (left) and diffusion coefficients ***D*** (right) on the fractional accessible volume (***FAV***) for polymer **3** of the fourth series of polyimides (He* – when the Van der Waals radius of Helium atom was taken equal to 1.68 Å).

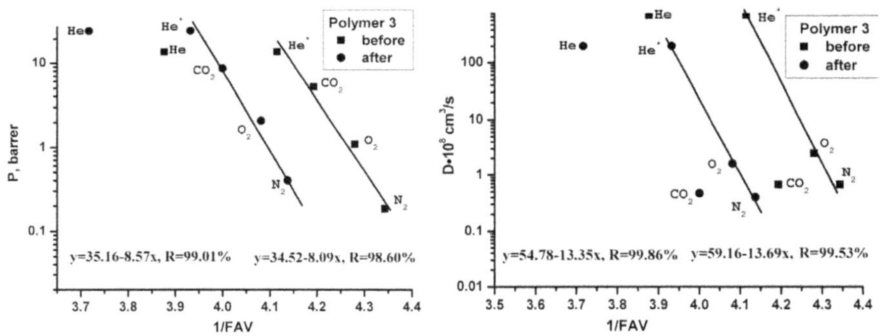

FIGURE 3.34 Dependence of permeability coefficients ***P*** (left) and diffusion coefficients ***D*** (right) on the fractional accessible volume (***FAV***) for polymer **4** of the fourth series of polyimides (He* – when the Van der Waals radius of Helium atom was taken equal to 1.68 Å).

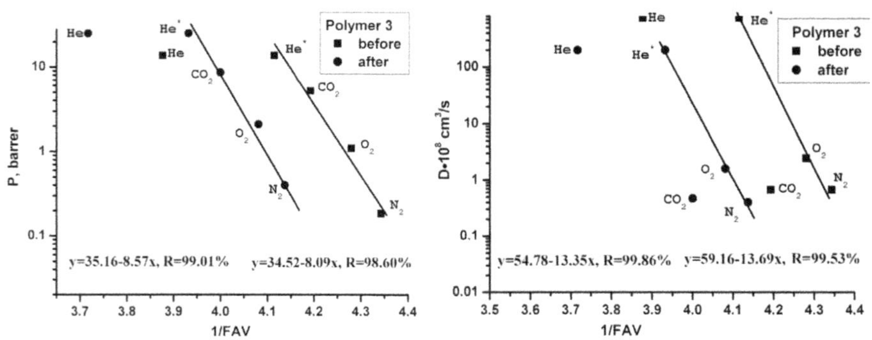

FIGURE 3.35 Dependence of permeability coefficients ***P*** (left) and diffusion coefficients ***D*** (right) on the fractional accessible volume (***FAV***) for polymer **5** of the fourth series of polyimides (He* – when the Van der Waals radius of Helium atom was taken equal to 1.68 Å).

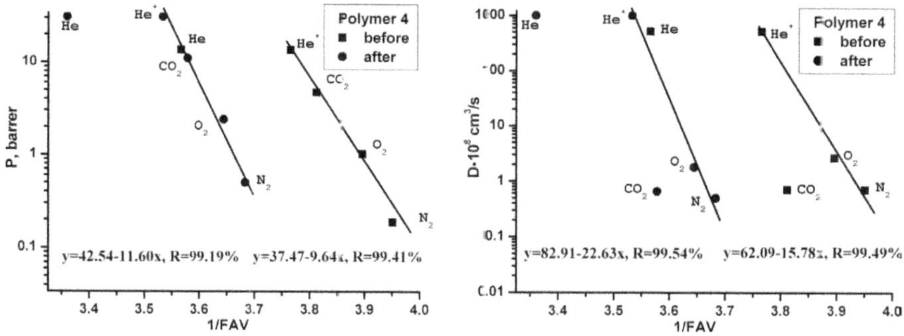

FIGURE 3.36 Dependence of permeability coefficients ***P*** (left) and diffusion coefficients ***D*** (right) on the fractional accessible volume (***FAV***) for polymer **6** of the fourth series of polyimides (He* – when the Van der Waals radius of Helium atom was taken equal to 1.68 Å).

polymers from 66 to 168%. For polymer **3**, the increase of permeability coefficients of He, O_2, and CO_2 is higher than that of N_2. The diffusion coefficients (D) decreased from 7.2 to 93% with exception He of polymer **6**. The change of D should be evidently observed for He, although the error of D(He) is quite high due to the short time lag (~0.5 s). And only for D(He) in polymer **6** increasing after treating in sc-CO_2 was observed. It is interesting that for O_2 and N_2, the diffusion decreases for all polymers in a similar way, with 27–35 and 37–46%, respectively. The minimum change of diffusion coefficient of CO_2 was observed in case of polymers **3** and **4**, by comparison with polymers **4** and **5**. For polymers **3** and **6**, the decrease of diffusion coefficient D for CO_2 is of 7%, while for polymers **4** and **5** it decreased with 30%. It can be seen that the increase of free volume of polymers **3** and **6** (Tables 3.15 and 3.18) is significantly higher than that of polymers **4** and **5** (Tables 3.16 and 3.17). This behavior when diffusion coefficient decreases as the free volume increases is not conventional. According to so-called "hole-wall" model [59], diffusion selectivity is a measure of density or ordering of chain packing in "holes." For instance, O_2/N_2 diffusion selectivity is increased for swollen polymers **3** (from 3.8 to 5.1), **4** (from 4.0 to 4.9), **5** (from 3.5 to 4.3), **6** (from 3.8 to 3.9). Thus, swelling of the polymers in sc-CO_2 results in densification of macromolecular chain packing in "walls" and the "walls" become more selective to gas transport. Similar effects were observed for the behavior of gas transport parameters in "strain aged" polymer film **3** [54]. Here, increasing of permselectivity of gas pairs was observed while free volume in the polymers and ordering of polymer **3** chain packing in "walls" increased. However, in our case (in contrary to [54]), increasing of selectivity is coupled with significant growing of free volume, and "hole" sizes, respectively, that results in growing of solubility coefficients (Table 3.19). It seems that the increase of free volume due to swelling in sc-CO_2 is not associated with disturbed packing of macromolecular chains in polymer matrix between the microcavities, and it determines that the ordering of macromolecular chains which leads to the decrease of diffusion coefficients.

Figures 3.33–3.36, left-side lines, show the dependence of permeability coefficients (a) and diffusion coefficients (b) on fractional accessible volume (FAV) in the system one polymer—different gases, after swelling with sc-CO_2. Table 3.20 presents the slope of these dependences (B) before and after swelling with sc-CO_2. This slope can be considered as general selectivity of a polymer to the studied gases. In Table 3.20, it can be seen that with regard to permeability coefficients, the selectivity increased for all polymers from 4 to 49%. With regard to diffusion coefficients, the selectivity increased for polymers **3** and **6**, but it decreased for polymers **4** and **5**. For these polymers **4** and **5**, the increase of free volume is significantly lower than in case of polymers **3** and **6** (Table 3.11).

To understand the reason of this behavior, we examine the solubility coefficients of these gases. The solubility coefficients S of each gas increases by swelling with sc-CO_2, that is by increasing of free volume (Table 3.19). For O_2, the solubility coefficients of all polymers increased with 190–285%, for CO_2 they increased with 134–254%, and for N_2 they increased with 215–323%, which shows that they increase in a similar way for these three gases through all four polymers. The increase of solubility coefficients for Helium, of 19–584%, makes an exception; the marginal values were found for polymers 5 and 6, as it was the case of diffusion coefficients of these two polymers, which can also be connected with the errors in measuring of diffusion coefficient of this gas. Since the solubility of gases in swollen polymers always increases, it can be concluded that the volume of microcavities increases which determines the increase of free volume and, therefore, of the permeability coefficients.

The significant difference in the change of transport parameters of gases after treatment with sc-CO_2 between polymer **3** and polymers **4–6** is connected with the individual selection of the swelling conditions for each

TABLE 3.19 Solubility (***S***•10^2, cm^3(STP)/cm^3cmHg) of Gases in Polymer Films of the Fourth Series Before and After Swelling in sc-CO_2

Gas	Polymer 1			Polymer 2			Polymer 3			Polymer 4		
	before	after	Δ*S* (%)	before	after	Δ*S* (%)	before	after	Δ*S* (%)	before	after	Δ*S* (%)
He	0.21	0.63	200	0.20	0.60	200	0.19	1.3	584	0.26	0.31	19
O_2	4.2	14	233	3.9	15	285	4.5	13	189	3.9	14	259
N_2	2.6	8.2	215	2.6	11.0	323	2.8	9.7	246	2.65	11.0	315
CO_2	70.5	200	184	65.0	230	254	77.0	180	134	66.0	170	158

TABLE 3.20 Coefficients B in the Dependence P(1/FAV) and D(1/FAV) for Polymers of the Fourth Series

Polymer	*P* (Barrer)			*D*•10^8 (cm^3/s)		
	B before	*B* after	Δ*B*, %	*B* before	*B* after	Δ*B*, %
1	8.31	12.42	49.4	13.87	20.02	44.34
2	10.36	10.82	4.44	17.44	17.33	–0.63
3	8.09	8.57	5.93	13.69	13.35	–2.48
4	9.64	11.60	20.33	15.78	22.63	43.41

polymer: temperature, pressure, and speed of sc-CO_2 diffusion out of the polymer matrix. We used identical conditions of sc-CO_2 treatment for all the studied polymers.

KEYWORDS

- **dielectric constant**
- **free volume**
- **gas transport parameters**
- **polyimide films**
- **porous morphology**
- **swelling in supercritical CO_2**

REFERENCES

1. Sroog, C. E. *Prog. Polym. Sci.* 1991, *16,* 561.
2. Gosh, M. K., Mittal, K. L. *Polyimides: Fundamentals and Applications*, Dekker, New York, 1996.
3. Hergenrother, P. M. *High Perform. Polym.* 2003, *15*, 3.
4. Sava, I., Resmerita, A. M., Lisa, G., Damian, V., Hurduc, N. *Polymer* 2008, *49,* 1475.
5. Ronova, I. Structural aspects in polymers. Interconnections between conformational parameters of the polymers with their physical properties. *Struct. Chem.* 2010, *21,* 541.
6. Koros, W. J., Fleming, G. K. *J Membrane Sci* 1993, *83*, 1.
7. Hergenrother, P. M. *High Perform Polym* 2003, *15*, 3.
8. Liaw, D. J., Wang, K. L., Huang, Y. C., Lee, K. R., Lai, J. Y., Ha, C. S. *Prog Polym Sci* 2012, *37*, 907.
9. Pixton, M. R., Paul, D. R. Relationship between structure and properties for polymers with aromatic backbones. In: Paul, D. R., Yampolskii, Yu. P. (eds) Polymeric gas separation membranes. CRC Press, Boca Raton 1994, 83.
10. Ghanem, B. S., McKeown, N. B., Budd, P. M., Selbie, J. D., Fritsch, D. *Adv Mater* 2008, *20*, 2766.
11. Chen, X. Y., Nik, O. G., Rodrigue, D., Kaliaguine, S. *Polymer* 2012, *53*, 3269.
12. Yampolski, Yi. *Macromolecules* 2012, *45*, 3298.
13. Beckman, E. J. *J. Supercrit. Fluids* 2004, *2,* 121.
14. Nikitin, L. N., Nikolaev, A. Yu., Said-Galiyev, E. E., Gamzazade, A. I., Khokhlov, A. R. *Supecritical fluids. Theory and practice* 2006, *1,* 77.

15. Ronova, I. A., Nikitin, L. N., Sokolova, E. A., Bacosca, I., Sava, I., Bruma, M. Swelling of polyheteroarylenes in supercritical carbon dioxide, *J. Macromolecular Science A: Pure and Applied Chemistry* 2009, *46*, 929–936.
16. Ronova, I. A., Bruma, M., Sava, I., Nikitin, L. N., Sokolova, E. A. *High Perform. Polym.* 2009, *21*, 562.
17. Kuznetsov, A. A. *High Perform. Polym.* 2000, *12*, 445.
18. Bruma, M., Hamciuc, E., Sava, I., Yampolskii, Yu. P., Alentiev, A. Yu., Ronova, I. A., Rozhkov, E. M. *Chem. Bull. "Politehnica" Univ. (Timisoara)* 2003, *48,* 110.
19. Ronova, I. A., Pavlova, S. S. A. *High Perform. Polym.* 1998, *10*, 309.
20. Dewar, M. J. S., Zoebisch, E. F., Healy, E. F., Stewart, J. J., *J. Am. Chem. Soc.* 1985, *107*, 3902.
21. Hamciuc, C., Ronova, I. A., Hamciuc, E., Bruma, M. *Angew. Makromol. Chem.* 1998, *254*, 67.
22. Pavlova, S. S. A., Ronova, I. A., Timofeeva, G. I., Dubrovina, L. V. *J. Polym. Sci. Part B: Polym. Phys.* 1993, *31*, 1725.
23. Plate, N. A., Yampolskii, Yu. P. Relationship between structure and transport properties for high free volume polymeric materials. In: Paul, D. R., Yampolskii, Y. P. (Eds.) Polymeric gas separation membranes, CRC Press, Boca Raton 1994, 155.
24. Ronova, I. A., Bruma, M. *Struct Chem* 2012, *23*(1), 47.
25. Askadskii, A. A. Computational materials science of polymers. Cambridge International Science Publishing, Cambridge 2003.
26. Rozhkov, E. M., Schukin, B. V., Ronova, I. A. *Central Eur J Chem* (Central Eur Sci J) 2003, *1(4),* 402.
27. Ronova, I. A., Bruma, M., Schmidt, H. W. *Struct Chem.* 2012, *23,* 219–226.
28. Bessonov, M. I., Koton, M. M., Kudryavtsev, V. V., Laius, L. A. Polyimides: Thermally Stable Polymers, Plenum Press, N.Y., 1987.
29. Nikitin, L. N., Said-Galiyev, E. E., Vinokur, R. A., Khokhlov, A. R., Gallyamov, M. O., Schaumburg, K. *Macromolecules* 2002, *35*, 934–940.
30. Nikitin, L. N., Gallyamov, M. O., Vinokur, R. A., Nikolaev, A. Y., Said-Galiyev, E. E., Khokhlov, A. R., Jespersen, H. T., Schaumburg, K. *J. of Supercritical Fluids* 2003, *26*, 263–273.
31. http://www.femtoscanonline.com/.
32. Yampolskii, Y. P., Novitskii, E. G., Durgar'yan SG *Zavod Lab* 1980, *46*, 256.
33. Fielding, R. *Polymer* 1980, *21*, 140.
34. Belov, N. A., Zharov, A. A., Shashkin, A. V., Shaikh, M. Q., Raetzke, K., Yampolskii, Y. P. *J Membr Sci* 2011, *383*, 70.
35. Sava, I., Chisca, S., Bruma, M., Lisa, G. *Polym Bull* 2010, *65(4),* 363.
36. Pauling L General chemistry. Freeman and Co, San Francisco 1970.
37. Lazareva, Y. N., Vidyakin, M. N., Yampolskii, Y. P., Alentiev, A. Y., Yablokova, M. Y., Semenov, G. K., Kuznetsov, A. A., Likhachev, D. Y. *Vysokomol Soedin A* 2006, *48(10)*, 1818.
38. Lazareva, Y. N., Vidyakin, M. N., Alentiev, A. Y., Yablokova, M. Y., Kuznetsov, A. A., Ronova, I. A. *Polym Sci Ser A* 2009, *51(10)*, 1068.
39. Ronova, I., Sokolova, E., Bruma, M. *J Polym Sci B* 2008, *46*, 1868.
40. Hofmann, D., Fritz, L., Ulbrich, J., Schepers, C., Bohning, M. *Macromol Theory Simul* 2000, *9(6)*, 293.

41. Bruma, M., Hamciuc, E., Sava, I., Hamciuc, C., Iosip, M. D., Robison, J. *Rev. Roum. Chim.* 2003, *48*, 629.
42. Kostina, Y. V., Moskvicheva, M. V., Bondarenko, G. N., Yablokova, M. Y., Alentiev, A. Y. Proceedings of 15th Russian Conference on "The structure and Dynamic of molecular system" Yoshkar-Ola, 2008, *1*, 133.
43. Brekner, M. J., Feger, C. *J. Polym. Sci. Part A: Polym. Chem.* 1987, *25*, 2479.
44. Ronova, I. A., Nikitin, L. N., Sinitsina, O. V., Yaminskii, I. V. *The Physics and Chemistry of Materials Processing* 2008, *4*, 54–59.
45. Ronova, I. A., Bruma, M., Sinitsyna, O. V., Sava, I., Nikolaev, A. Yu., S. Chisca, N. G. Ryvkina. *Struct. Chem.*, on line DOI 10.1007/s11224-014-0443-1.
46. Ronova, I. A., Vasilyuk, A. N., Gaina, C., Gaina, V. *High Perform. Polym.* 2002, *14*, 195.
47. Hamciuc, E., Hulubei, C., Bruma, M., Ronova, I. A., Sokolova, E. A. *Rev. Roum. Chim.* 2008, *53*, 737.
48. Damaceanu, M. D., Musteata, V. E., Cristea, M., Bruma, M. *Eur. Polym.* J. 2010, *46*, 1049.
49. Ronova, I. A., Bruma, M., Nikolaev, A. Yu., Kuznetsov, A. A. *Polym Adv. Technol.*, 2013, *24*(7), 615–622.
50. Tanaka, K., Kita, H., Okano, M., Okamoto, K. *Polymer*, 1992, *33*(3), 585.
51. Zoia, G., Stern, S. A., Clair, A. St., Pratt, J. R. *J. Polym. Sci. Part B: Polym. Phys*, 1994, *32*, 53.
52. Alentiev, A. Yu., Yampolskyi, Yu. P., Rusanov, A. L., Komarova, L. G., Likhachev, D. Yu., Kazakova, G. V. Transport properties of poly(ether imides). *Polym. Sci. Ser. A* 2003, *45*(9) 933–939.
53. Yampolskii, Y., Alentiev, A., Bondarenko, G., Kostina, Y., Heuchel, M. *Ind Eng Chem Res* 2010, *49*, 12031.
54. Ronova, I. A., Alentiev, A. Yu., Chisca, S., Sava, I., Bruma, M., Nikolaev, A. Yu., Belov, N. A., Buzin, M. I. *Struct. Chem.*, 2014, *25(1)*, 301–310.
55. Bondi, A. *J Phys Chem* 1964, 68, 441.
56. Huheey, J. E. Inorganic chemistry principles of structure and reactivity. Harper & Row Publishers, New York 1983.
57. Bizerano, J. Prediction of polymer properties. Marcel Dekker, New York 1993.
58. Teplyakov, V. V., Meares, P. *Gas Sep Purif* 1990, *4*, 66.
59. Alentiev, A. Y., Yampolskii, Y. P. *J Membr Sci* 2002, 206, 291.

CHAPTER 4

INVESTIGATION ON THE CLEANING PROCESS OF GAS EMISSIONS

R. R. USMANOVA[1] and G. E. ZAIKOV[2]

[1]*Ufa State Technical University of Aviation, 12 Karl Marks Str., Ufa 450100, Bashkortostan, Russia, E-mail: Usmanovarr@mail.ru*

[2]*N.M. Emanuel Institute of Biochemical Physics, Russian Academy of Sciences, 4 Kosygin str., Moscow 119334, Russia, E-mail: chembio@sky.chph.ras.ru*

CONTENTS

ABSTRACT

The study of fluid flow and particle separation in rotoklon allowed to consider in detail all the stages of the process of hydrodynamic interaction of phases in devices shock-inertial action. Define the boundary degree of recirculation irrigation liquid, which ensures stable operation rotoklon. It is established that the reduction of fine particle separation efficiency occurs with a decrease in viscosity irrigation liquid.

4.1 INTRODUCTION

One of the problems of gas cleaning devices is to provide apparatus of intense action with high capacity for gas. This is associated with a reduction in the dimensions of the gas cleaning systems.

In these conditions, due to the high relative velocity of the liquid and gas phases, a decisive influence on the effect of dust collection have mechanisms: inertial and direct capture of particles. This process is implemented in a shock-inertial dust collector, which include investigated apparatus.

In the literature practically there are no data on effect of viscosity of a trapping liquid on dust separation process. Therefore, one of the purposes of our work was revealing of effect of viscosity of a liquid on efficiency of a dust separation.

4.2 THE URGENCY OF THE PROBLEM

The problem formulation leans on following rules. In the conditions of full circulation of a liquid, at constant geometrical sizes of a deduster it is possible to secure with a constancy of operational parameters is a relative speed of traffic of a liquid and an aerosol, concentration of a dust in gas, a superficial tension of a liquid or an angle of wetting of a dust. Concentration of a dust growing in a time in a liquid conducts to unique essential change – to increase in its viscosity. After excess of certain concentration suspension loses properties of a Newtonian fluid. Deduster working conditions at full circulation of a liquid are approached to what can be gained in the periodical regime when at maintenance fresh water is not inducted into dust-collecting plant. Collected in the apparatus, the dust detained by a liquid, compensates volume losses of the liquid necessary on moistening of passing gas and its ablation. In the literature there are no works, theoretically both experimentally presenting effect of viscosity and effect of rheological properties of slurry on efficiency of a dust separation.

As it seems to us, a motive is that fact that in the capacity of operating fluid water is usually used, and dedusters work, predominantly, at constant temperatures. Simultaneously, at use of partial circulation certain level of concentration of a dust in a liquid is secured. In turn, accessible dependences in the literature specify in insignificant growth of viscosity of slurry even at raise of its concentration for some percent.

The reasoning's proving possibility of effect of viscosity of slurry on efficiency of a dust separation, it is possible to refer to as on the analysis of the basic mechanisms influencing sedimentation of corpuscles on an interphase surface, and on conditions of formation of this developed surface of a liquid. Transition of corpuscles of a dust from gas in a liquid occurs, mainly, as a result of the inertia affecting, effect of "sticking" and diffusion. Depending on type of the wet-type collector of a corpus of a dust deposit on a surface of a liquid which can be realized in the form of drops, moving in a stream of an aerosol, the films of a liquid generated in the apparatus, a surface of the gas vials formed in the conditions of a barbotage and moistened surfaces of walls of the apparatus.

In the monography [3] effect of various mechanisms on efficiency of sedimentation of corpuscles of a dust on a liquid surface is widely presented. The description of mechanisms and their effect on efficiency of a dust separation can be found practically in all monographies, for example, [4, 13] concerning a problem wet clearings of gas emissions of gases. In the literature of less attention it is given questions of formation of surfaces of liquids and their effect on efficiency of a dust separation.

Observing the mechanism of the inertia act irrespective of a surface of the liquid entraining a dust, predominantly it is considered that for hydrophilic types of a dust collision of a part of a solid with a liquid surface to its equivalent immediate sorption by a liquid, and then immediate clearing and restoration of the surface of a liquid for following collisions. In case of a dust badly moistened, the time necessary for sorbtion of a corpuscle by a liquid, can be longer, than a time after which the corpuscle will approach to its surface. Obviously, it is at the bottom of decrease in possibility of a retardation of a dust by a liquid because of a recoil of the corpuscle going to a surface, from a corpuscle which are on it. It is possible to consider this effect real as in the conditions of a wet dust separation with a surface of each fluid element impinges more dust, than it would be enough for monolayer formation. Speed of sorbtion of corpuscles of a dust can be a limiting stage of a dust separation.

Speed of sorbtion of a corpuscle influences not only its energy necessary for overcoming of a surface tension force, but also and its traverse speed in the liquid medium, depending on its viscosity and rheological properties. Efficiency of dust separation Kabsch [5] connects with speed of ablation of a dust a liquid, having presented it as weight m_s, penetrating in unit of time through unit of a surface and in depth of a liquid as a result of a collision of grains of a dust with this surface:

$$r = \frac{m_s}{A \cdot t}$$

Giving to shovels of an impeller sinusoidal a profile allows to eliminate breakaway a stream breakaway on edges. Thus, there is a flow of an entrance section of a profile of blades with the big constant speed and increase in ricochets from a shaped part of blades in terms of which it is possible to predict insignificant increase in efficiency of clearing of gas.

Speed of linkage of a dust a liquid depends on physicochemical properties of a dust and its ability to wetting, physical and chemical properties of gas and operating fluid, and also concentration of an aerosol. Wishing to confirm a pushed hypothesis, Kabsch [5] conducted the researches concerning effect of concentration speed of linkage of a dust by a liquid. The increase in concentration of a dust in gas called some increase in speed of linkage, however to a lesser degree, than it follows from linear dependence.

The cores for technics of a wet dust separation of model Semrau, Barth'a and Calvert'a do not consider effect of viscosity of slurry on effect of a dust separation. In-process Pemberton'a [6] it is installed that in case of sedimentation of the corpuscles, which are not moistened on drops, their sorbtion in a liquid is obligatory, and their motion in a liquid submits to principle Stokes'a.

The traverse speed can characterize coefficient of resistance to corpuscle motion in a liquid, so, and a dynamic coefficient of viscosity of a liquid. Possibility of effect of viscosity of a liquid on efficiency of capture of corpuscles of a dust a drop by simultaneous Act of three mechanisms: the inertia, "capture" the semiempirical equation Slinna [6] considering the relation of viscosity of a liquid to viscosity of gas also presents.

In general, it is considered that there is a certain size of a drop [14] at which optimum conditions of sedimentation of corpuscles of a certain size are attained, and efficiency of subsidence of corpuscles of a dust on a drop sweepingly decreases with decrease of a size of these corpuscles.

Jarzkbski and Giowiak [9], analyzing work of an impact-sluggish deduster have installed that in the course of a dust separation defining role is played by the phenomenon of the inertia collision of a dust with water drops. Efficiency of allocation of corpuscles of a dust decreases together with growth of sizes of the drops oscillated in the settling space, in case of a generating of drops compressed air, their magnitude is defined by equation Nukijama and Tanasawa [10] from which follows that drops to those more than above value of viscosity of a liquid phase. Therefore, viscosity growth can call reduction of efficiency of a dust separation.

The altitude of a layer of the dynamic foam formed in dust-collecting plant at a certain relative difference of speeds of gas and liquid phases, decreases in process of growth of viscosity of a liquid [15] that calls decrease in efficiency of a dust separation, it is necessary to consider that the similar effect refers to also to a layer of an intensive barbotage and the drop layer partially strained in dust removal systems.

Summarizing it is possible to assert that in the literature practically there are no data on effect of viscosity of a trapping liquid on dust separation process. Therefore, one of the purposes of our work was extraction of effect of viscosity of a liquid on efficiency of a dust separation.

4.2.1 THE PURPOSE AND OBJECTIVES OF RESEARCH

The conducted researches had a main objective acknowledging of a hypothesis on existence of such boundary concentration of slurry at which excess the overall performance of the dust removal apparatus decreases.

The concept is devised and the installation, which is giving the chance to implementation of planned researches is mounted. Installation had systems of measurement of the general and fractional efficiency and typical systems for measurement of volume flow rates of passing gas and water resistances. The device of an exact proportioning of a dust, and also the air classifier separating coarse fractions of a dust on an entry in installation is mounted. Gauging of measuring systems has secured with respective repeatability of the gained results.

4.3 LABORATORY FACILITY AND TECHNIQUE FOR CONDUCTING EXPERIMENT

Laboratory facility basic element is the deduster of impact-sluggish act – a rotoklon c adjustable guide vanes [11] (Figure 4.1). An aerosol gained by dust introduction in the pipeline by means of the batcher. Application of the batcher with changing productivity has given the chance to gain the set concentration of a dust on an entry in the apparatus.

Have been investigated a dust, discriminated with the wettability (a talcum powder the ground, median diameter is equal δ_{50} = 25 microns, white black about δ_{50} = 15 microns solubility in water of 10^{-3}% on weight (25°C) and a chalk powder).

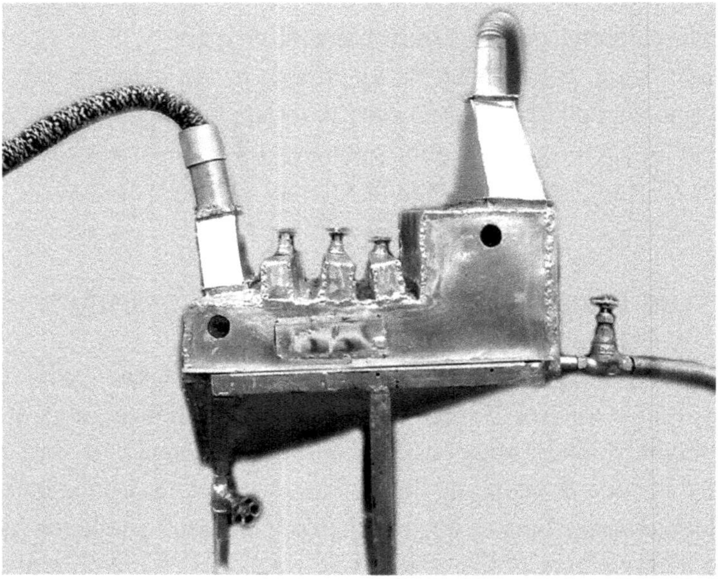

FIGURE 4.1 The laboratory facility.

The gas-dispersed stream passed shovels of an impeller 7 in a working zone of the apparatus, whence through the drip pan 8 cleared, was inferred outside. Gas was carried by means of the vacuum pump 10, and its charge measured by means of a diaphragm 1. A Gas rate, passing through installation, controlled, changing quantity of air sucked in the pipeline before installation. The composition of each system includes group of three sondes mounted on vertical sections of pipelines, on distance about 10 diameters from the proximal element changing the charge. The taken test of gas went on the measuring fine gauge strainer on which all dust containing in test separated. For this purpose used fine gauge strainer. In the accepted solution have applied system of three measuring sondes which have been had in pipelines so that in the minimum extent to change a regime of passage of gas and to select quantity of a dust necessary for the analysis. The angle between directions of deducing of sondes made 120 °, and their ends placed on such radiuses that surfaces of rings from which through a sonde gas was sucked in, were in one plane. It has allowed to scale down a time of selection of test and gave average concentration of a dust in gas pipeline cross-section.

Fractional composition of a dust on an entry and an exit from the apparatus measured by means of analogous measuring systems, Chapter 10.

For definition of structurally-mechanical properties of slurry viscosity RV-8 (Figure 4.2) has been used. The viscosity gauge consists of the internal twirled cylinder (rotor) (r = 1.6 centimeter) and the external motionless cylinder (stator) (r = 1.9 centimeter), having among themselves a positive allowance of the ring form with a size 0.3 see the Rotor is resulted in twirl by means of the system consisting of the shaft, a pulley (T_0 = 2.23 centimeter), filaments, blocks and a cargo. To the twirl termination apply a brake. The twirled cylinder has on a division surface on which control depth of its plunging in slurry.

The gained slurry in number of 30 sm^3 (in this case the rotor diving depth in sludge makes 7 sm) fill in in is carefully the washed out and dry external glass which put in into a slot of a cover and strengthen its turn from left to right. After that again remove the loaded cylinder that on a scale of the internal cylinder precisely to define depth of its plunging in sludge. Again, fix a glass and on both cups put the minimum equal cargo (on 1), fix the spigot of a pulley by means of a brake and reel up a filament,

FIGURE 4.2 Measurement of viscosity of slurry.

twirling a pulley clockwise. It is necessary to watch that convolutions laid down whenever possible in parallel each other.

Install an arrow near to any division into the limb and, having hauled down a brake, result the internal cylinder in twirl, fixing a time during which the cylinder will make 4–6 turns. After the termination of measurements fix a brake and reel up a filament. Measurement at each loading spends not less than three times. Experiences repeat at gradual increase in a cargo on 2 gr. until it is possible to fix a time of an integer of turns precisely enough. After the termination of measurements remove a glass, drain from it sludge, wash out water, from a rotor sludge drain a wet rag then both cylinders are dry wiped and leave the device in the collected aspect.

After averaging of the gained data and calculation of angular speed the schedule of dependence of speed of twirl from the enclosed loading is under construction, Viscosity is defined by formula [15]:

$$\eta = \frac{(R_2^2 - R_1^2)Gt}{8\pi^2 L R_1^2 R_2^2 L} \qquad (1)$$

4.4 RESULTS AND DISCUSSION

For each dust used in researches dependence of general efficiency of a dust separation on concentration of slurry and the generalizing schedule of dependence of fractional efficiency on a corpuscle size is presented. Other schedule grows out of addition fractional efficiency of a dust separation for various, presented on the schedule, concentration of slurry. In each case the first measurement of fractional efficiency is executed in the beginning of the first measuring series, at almost pure water in a deduster.

Analyzing the gained results of researches of general efficiency of a dust separation, it is necessary to underline that in a starting phase of work of a rotoklon at insignificant concentration of slurry for all used in dust researches components from 93.2% for black to 99.8% for a talc powder are gained high efficiency of a dust separation. Difference of general efficiency of trapping of various types of a dust originates because of their various fractional composition on an entry in the apparatus, and also because of the various form of corpuscles, their dynamic wettability and density. The gained high values of general efficiency of a dust separation testify to correct selection of constructional and operational parameters of the studied apparatus and specify in its suitability for use in technics of a wet dust separation.

The momentous summary of the spent researches was definition of boundary concentration of slurry various a dust after which excess general efficiency of a dust separation decreases. Value of magnitude of boundary concentration, as it is known, is necessary for definition of the maximum extent of recirculation of an irrigating liquid. As appears from presented in Figures 4.3–4.6 schedules, dependence of general efficiency of a dust separation on concentration of slurry, accordingly, for a powder of talc, a chalk and white black is available possibility of definition of such concentration.

Boundary concentration for a talcum powder – 36%, white black – 7%, a chalk – of 18% answer, predominantly, to concentration at which slurries lose properties of a Newtonian fluid.

The conducted researches give the grounds to draw deductions that in installations of impact-sluggish type where the inertia mechanism is the core at allocation of corpuscles of a dust from gas, general efficiency of

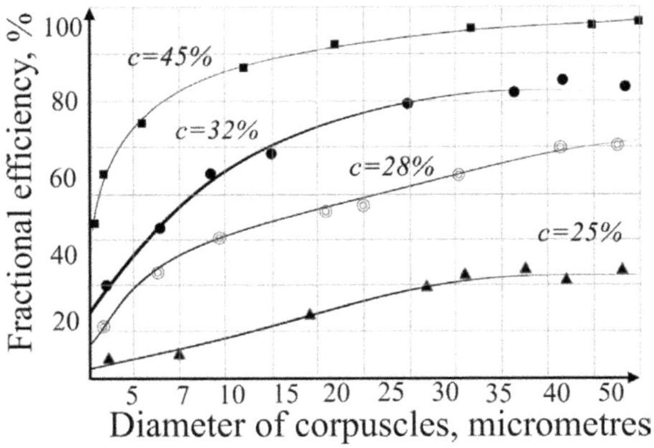

FIGURE 4.3 Dependence of fractional efficiency on diameter of corpuscles of a talcum powder and their concentration in a liquid.

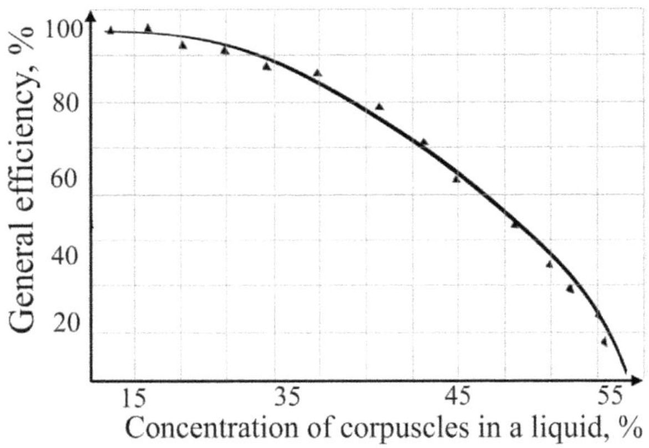

FIGURE 4.4 Dependence of general efficiency of concentration in a liquid of corpuscles of a talcum powder.

a dust separation essentially drops when concentration of slurry answers such concentration at which it loses properties of a Newtonian fluid. As appears from presented in Figures 4.3–4.6 dependences, together with growth of concentration of slurry above a boundary value, general efficiency of a dust separation decreases, and the basic contribution to this phenomenon small corpuscles with a size less bring in than 5 microns.

To comment on the dependences presented in drawings, than 5 microns operated with criterion of decrease in efficiency of a dust separation of corpuscles sizes less at the further increase in concentration of slurry at

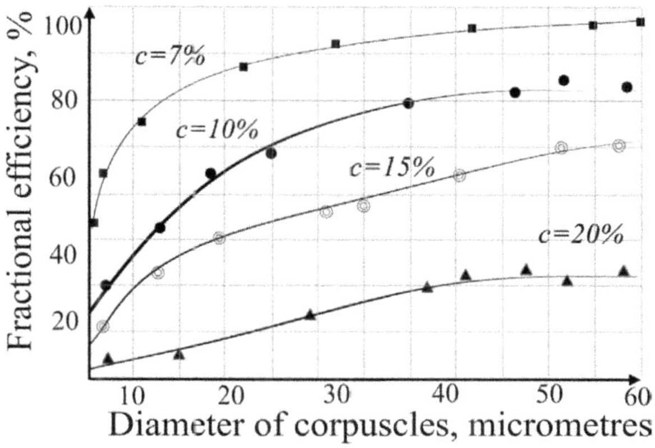

FIGURE 4.5 Dependence of fractional efficiency on diameter of corpuscles of white black and their concentration in a liquid.

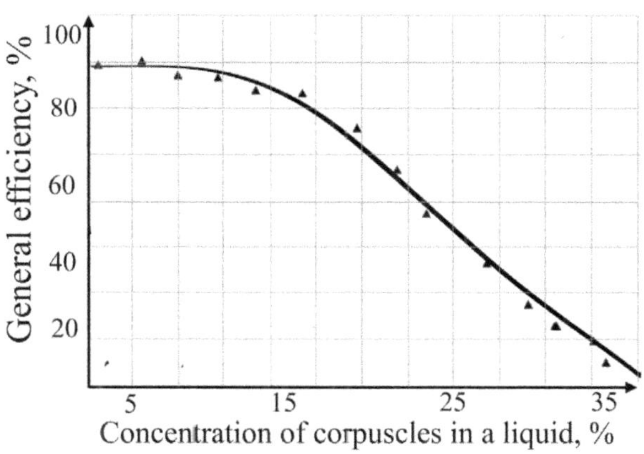

FIGURE 4.6 Dependence of general efficiency on concentration in a liquid of corpuscles of white black.

10% above the boundary. Taking it in attention, it is possible to notice that in case of allocation of a dust of talc growth of concentration of slurry from 36% to 45% calls reduction of general efficiency of a dust separation from 98% to 90% at simultaneous decrease in fractional efficiency of allocation of corpuscles, smaller 5 microns from $\eta = 93\%$ to $\eta = 65\%$.

Analogously for white black: growth of concentration from 7% to 20% calls falling of fractional efficiency from $\eta = 65\%$ to $\eta = 20\%$, for a chalk: growth of concentration from 18% to 30% calls its decrease from $\eta = 80\%$ to $\eta = 50\%$.

Most considerably decrease in fractional efficiency of a dust separation can be noted for difficultly moistened dust – white black (about 50%).

Thus, on the basis of the analysis set forth above it is possible to assert that decrease in general efficiency of a dust separation at excess of boundary concentration of slurry is connected about all by decreasing ability of system to detain small corpuscles. Especially it touches badly moistened corpuscles. It coincides with a hypothesis about updating of an interphase surface. Updating of an interphase surface can be connected also with difficulties of motion of the settled corpuscles of a dust deep into liquids, for example, with viscosity of medium.

The analysis of general efficiency of a dust separation, and, especially, a talcum powder and white black powder specifies that till the moment of achievement of boundary concentration efficiency is kept on a fixed level. In these boundary lines, simultaneously with growth of concentration of slurry dynamic viscosity of a liquid, only this growth grows is insignificant – for talc, for example, to 2.7×10^{-3} Pascal×second. At such small increase in viscosity of slurry the estimation of its effect on efficiency of a dust separation is impossible. Thus, it is possible to confirm, what not growth of viscosity of slurry (from 1 to 2.7×10^{-3} Pascal×second), and change of its rheological properties influences decrease in efficiency of a dust separation.

The method of definition of boundary extent of circulation of a liquid in impact-sluggish apparatuses is based on laboratory definition of concentration of slurry above which it loses properties of a Newtonian fluid. This concentration will answer concentration of operating fluid, which cannot be exceeded if it is required to secure with a constant of efficiency of a dust separation.

In the conditions of spent researches, for example, constant concentration of an aerosol on an entry in the apparatus, and at the assumption that water losses in the apparatus because of moistening of passing air and, accordingly, ablation in the form of drops, is compensated by volume of the trapped dust, the water discharge parameter is defined directly from the recommended time of duration of a cycle and differs for various types of a dust. Counted its maximum magnitude is in the interval 0.02–0.05 L/m^3, for example, is close to the magnitudes quoted in the literature.

For periodical regime dedusters this concentration defines directly a cycle of their work. In case of dedusters of continuous act with liquid circulation, the maximum extent of recirculation securing maintenance of a fixed level of efficiency of a dust separation, it is possible to count as the relation:

$$r = \frac{Q_{cir}}{Q_{ir}} \tag{3}$$

where Q $_{cir}$ – the charge of a recycling liquid, m³/h; Q $_{ir}$ – the charge of an inducted liquid on an irrigation, m³/h.

Assuming that all dust is almost completely trapped on a liquid surface, it is possible to write down a balance equation of weight of a dust as:

$$G(c_{on} - c_{in}) = L \cdot c_{on} \tag{4}$$

where—limiting concentration of a dust, g/m³.

Then in terms of for calculation of extent of recirculation it is possible to present (4) formula as:

$$r = \frac{G \cdot c_{o1}}{L \cdot c_r} \tag{5}$$

4.5 CONCLUSIONS

1. Excess of boundary concentration of slurry at which it loses properties of a Newtonian fluid, calls decrease in efficiency of a dust separation.
2. At known boundary concentration c_r it is possible to define boundary extent of recirculation of the irrigating liquid, securing stable efficiency of a dust separation.
3. Magnitude of boundary concentration depends on physical and chemical properties of system a liquid – a solid and changes over a wide range, from null to several tens percent. This magnitude can be defined now only laboratory methods.
4. Decrease in an overall performance of the apparatus at excess of boundary concentration is connected, first of all, with reduction of fractional efficiency of trapping of small corpuscles with sizes less than 5 microns.

5. On the basis of observations of work of an investigated deduster it is possible to assert that change of viscosity of an irrigated liquid influences conditions of a generating of an interphase surface and, especially, on intensity of formation of a drop layer.
6. Selection constructional and the operating conditions, securing high efficiency of a dust separation at small factor of water consumption, allow to recommend such dedusters for implementation in the industry.

KEYWORDS

- **boundary concentration**
- **cleaning**
- **fractional efficiency**
- **gas emissions**
- **interfacial area**
- **irrigation**
- **liquid**
- **properties**
- **recirculation**
- **research**
- **rheology**
- **the irrigating liquid**
- **the rotoklon**
- **viscosity**

REFERENCES

1. Valdberg, A. J. Wet-type collectors impact-sluggish, centrifugal and injector acts (in Rus.) Moscow: Petrochemistry Publishing House, 1981, 250 pp.
2. GOST 21235–75 (Standard of USSR) Talc and a talcum powder the ground. Specifications (in Rus.)

3. Egorov, N. N. Gas cooling in scrubbers (in Rus.). Moscow, Chemistry Publishing House, 1954, 245 pp.
4. Kutateladze, S. S., Styrikovich, M. A. Hydrodynamics of gas-liquid systems (in Rus.), Moscow, Energy Publishing House, 1976, 340 pp.
5. Kabsch-Korbutowicz, M., Majewska-Nowak, K., Removal of atrazine from water by coagulation and adsorption. Environ Protect Engineering Journal, 2003, 29(3), 15–24.
6. Kitano. T., Slinna T. "An empirical equation of the relative viscosity of polymer melts filled with various inorganic fillers". Rheological Acta Journal, 1981, 20(2), 7–14.
7. Margopin, E. V., Prihodko, V. P. Perfection of production engineering of wet clearing of gas emissions at aluminum factories (in Rus.), Moscow, Color. Metal information publishing house, 1977, 27–34 pp.
8. Jarzkbski, L., Giowiak. Moscow, Science (Nauka) publishing house, 1977, 350 pp.
9. Nukijama S., Tanasawa, Y. Trans. Soc. Mech. Eng. Japan. 1939, v. 5.
10. The patent 2317845 Russian Federations, The Rotoklon with adjustable sinusoidal guide vanes. R. R. Usmanova, V. S. Zhernakov, A. K. Panov; 27 February, 2008.
11. Ramm, V. M. Absorption of gases (in Rus.). Moscow, Chemistry Publishing House, 1976, 274 pp.
12. Rist, R. Aerosols: introduction in the theory (in Rus.), Moscow, World Publishing House, 1987, 357 pp.
13. Stepans, G. J., Zitser, I. M. The Inertia air cleaners (in Rus.), Moscow, Engineering Industry Publishing House, 1986, 274 pp.
14. Shwidkiy V. S. Purification of gases. Handbook (in Rus.) Ed. by V.S. Shwidkiy, Thermal Power Publishing House, Moscow, 2002, 375 pp.

CHAPTER 5

TRENDS IN POLYMER/ORGANOCLAY NANOCOMPOSITES

G. V. KOZLOV,[1] G. E. ZAIKOV,[2] and A. K. MIKITAEV[1]

[1]*Kh.M. Berbekov Kabardino-Balkarian State University, Chernyshevsky St., 173, 360004 Nalchik, Russian Federation, E-mail: i_dolbin@mail.ru*

[2]*N.M. Emanuel Institute of Biochemical Physics of Russian Academy of Sciences, Kosygin St., 4, 119991 Moscow, Russian Federation, E-mail: chembio@sky.chph.ras.ru*

CONTENTS

ABSTRACT

It has been shown that organoclay platelets aggregation in "packets" (tactoids) results in large-scale disorder intensification that reduces nanofiller anisotropy degree. In its turn, this factor decreases essentially nanocomposites reinforcement degree. The interfacial adhesion role in anisotropy level definition has been shown.

5.1 INTRODUCTION

Organoclay belongs to anisotropic nanofillers, for which their anisotropy degree, for example, ratio α of organoclay platelets (aggregates of platelets) length to thickness, has large significance [1]. The reinforcement degree E_n/E_m of nanocomposites polymer/organoclay can be estimated according to the equation [1]:

$$\frac{E_n}{E_m} = 1 + 2\alpha C_a \varphi_n \tag{1}$$

where E_n and E_m are elasticity moduli of nanocomposite and matrix polymer, respectively, C_a is an orientation factor, which is equal for organoclay to about 0.5 [1], φ_n is organoclay volume content.

In case of organoclay platelets aggregation, for example, their "packets" (tactoids) formation [2] such "packets" thickness increasing takes place in comparison with a separate platelet, that results in length/thickness ratio α reduction at platelet constant length and, as consequence, nanocomposites reinforcement degree decreasing is realized according to the Eq. (1). The present work purpose is the analytical study of organoclay aggregation, for example, large-scale disorder, influence on reinforcement degree on the example of nanocomposites plasticat of poly(vinyl chloride) – organomodified montmorillonite.

5.2 EXPERIMENTAL PART

The plasticat of poly(vinyl chloride) (PVC) of mark U30–13A, prescription 8/2 GOST 5960–72 was used as a matrix polymer. The modification product of montmorillonite (MMT) of deposit Gerpegezh (KBR, Russian Federation), modified by urea with content of 10 mass% with cation-changing capacity of 95 mg-eq/100 g of clay was applied as a nanofiller. Organoclay contents were varied within the limits of 1–10 mass%.

The nanocomposites PVC-MMT preparation was performed as follows. The components were mixed in a two-speed blender R 600/HC 2500 of firm Diosna, the design of which ensures intensive intermixing in turbulent regime with blends high homogenization and blowing by hot air. After components intensive intermixing the composition was cooled up to temperature 313 K and processed on a twin screw extruder Thermo Haake, model Reomex RTW 25/42, production of German Federal Republic, at temperature 418–438 K and screw rotation speed of 48 rpm.

Sheet nanocomposite was obtained by a hot rolling method at temperature (433±10) K during 5–15 min. The samples in the shape of a two-sided spade with sizes according to GOST 112–62–80 were cut out by punch. Uniaxial tension mechanical tests have been performed on the universal testing apparatus Gotech Testing Machine CT-TCS 2000, production of German Federal Republic, at temperature 293 K and strain rate of $\sim 2\times10^{-3}$ s^{-1}.

5.3 RESULTS AND DISCUSSION

The nanofiller initial particles aggregation is the main process, enhancing large-scale disorder level in polymer nanocomposites. For each nanofiller type this process has its specific character, but in case of anisotropic nanofillers (organoclay, carbon nanotubes) application this process always decreases their anisotropy degree, for example, aspect ratio α, that reduces nanocomposites reinforcement degree according to the Eq. (1). Let us consider the physical bases of the value α reduction at organoclay content increasing in the considered nanocomposites. As it is known [3], organoclay platelets number N_{pl} in "packet" (tactoid) can be determined as follows:

$$N_{pl} = 24 - 5.7b_{\alpha} \quad (2)$$

where b_{α} is the dimensionless parameter, characterizing the level of interfacial adhesion polymeric matrix-nanofiller, which is determined with the aid of the following percolation relationship [2]:

$$\frac{E_n}{E_m} = 1 + 11(c\varphi_n b_{\alpha})^{1.7} \quad (3)$$

where c is constant coefficient, which is equal to 1.955 for intercalated organoclay and 2.90 – for exfoliated one.

In its turn, the value φ_n can be determined according to the well-known formula [2]:

$$\varphi_n = \frac{W_n}{\rho_n} \tag{4}$$

where W_n is nanofiller mass content; ρ_n is its density, which for nanoparticles is determined as follows [2]:

$$\rho_n = 188(D_p)^{1/3},\ \text{kg/m}^3 \tag{5}$$

where D_p is the initial nanoparticle diameter, which is given in nm.

In case of organoclay parameter D_p is determined as mean arithmetical of its three basic sizes: length, width and thickness, which are equal to 100, 35 and 0.65 nm, respectively [2].

An alternative method of the value N_{pl} estimation gives the following equation [2]:

$$\chi = \frac{N_{pl}d_{pl}}{(N_{pl}-1)d_{001}+d_{pl}} \tag{6}$$

where χ is relative volume content of montmorillonite in tactoid ("effective particle" [4]); d_{pl} is thickness of organoclay separate platelet; d_{001} is interlayer spacing, for example, the distance between organoclay platelets in tactoid, which can be estimated according to the following formula [3]:

$$d_{001} = 1.27b_{\alpha} \tag{7}$$

In its turn, parameter χ is determined as follows [3]:

$$\chi = \frac{\varphi_n}{\varphi_n + \varphi_{if}} \tag{8}$$

where φ_{if} is a relative fraction of interfacial regions in nanocomposite, estimated with the aid of the following percolation relationship [2]:

$$\frac{E_n}{E_m} = 1 + 11(\varphi_n + \varphi_{if})^{1.7} \tag{9}$$

The comparison of N_{pl} value calculations according to the Eqs. (2) and (6) showed their close correspondence.

In Figure 5.1 the dependence $N_{pl}(\varphi_n)$ for nanocomposites PVC/MMT is adduced. As one can see, at quite enough small values $\varphi_n \leq 0.05$ fast growth of N_{pl} occurs, for example, strong aggregation of organoclay initial platelets, and at $\varphi_n > 0.05$ the value N_{pl} achieves the asymptotic branch: $N_{pl} \approx 22$. As it was noted above, the reduction of nanofiller anisotropy degree, characterized by parameter α, was defined by its aggregation, the level of which could be characterized by parameter N_{pl}. In Figure 5.2 the dependence $\alpha(N_{pl}^2)$ for the considered nanocomposites, where quadratic shape of dependence was chosen with the purpose of its linearization. As was to be expected, the organoclay anisotropy degree, characterized by the

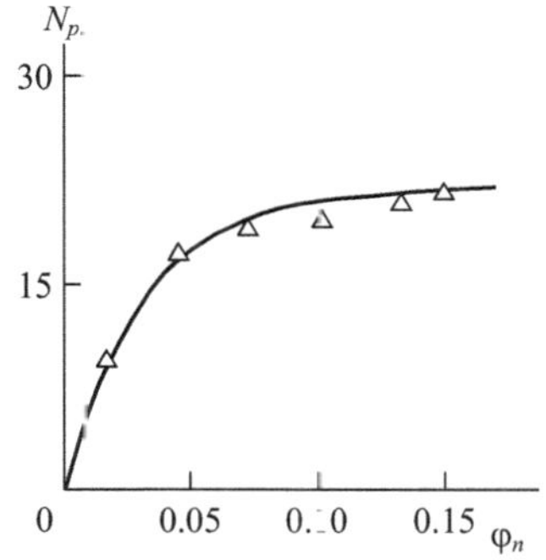

FIGURE 5.1 The dependence of organoclay platelets number per one tactoid N_{pl} on nanofiller volume content φ_n for nanocomposites PVC-MMT.

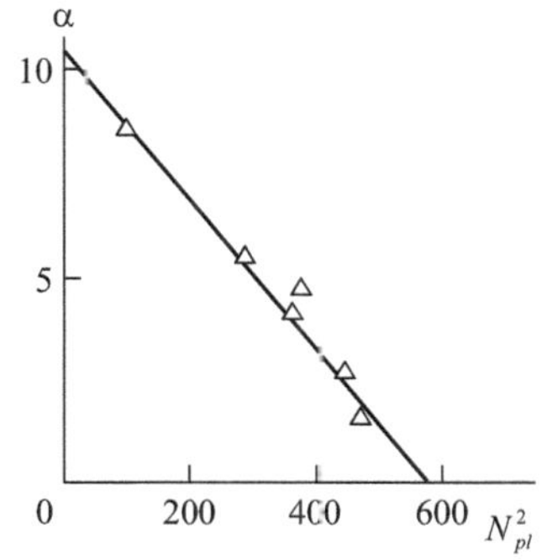

FIGURE 5.2 The dependence of organoclay anisotropy degree α on its platelets number per one tactoid Npl for nanocomposites PVC-MMT.

parameter α, reduction is observed at its platelets aggregation, characterized by the value N_{pl}, enhancement, which is expressed analytically by the following equation:

$$\alpha = 10.5 - 0.018 N_{pl}^2 \tag{10}$$

where the value α was estimated according to the Eq. (1).

Theoretical method of parameter α (α^T) estimation can be obtained as follows. Organoclay aggregates (tactoids) anisotropy degree can be determined according to the equation:

$$\alpha^T = \frac{L_{pl}}{t_{org}} \tag{11}$$

where L_{pl} is organoclay platelet length, which is equal to ~ 100 nm [2], t_{org} is its tactoid thickness.

In its turn, the value t_{org} is determined as follows:

$$t_{org} = d_{001} N_{pl} + 1 \tag{12}$$

Besides, it should be borne in mind, that experimental value α in the Eq. (1) is determined on the basis of reinforcement degree E_n/E_m, for example, on the basis of mechanical tests results. This means, that value α depends on conditions of stress transfer on interfacial boundary polymeric matrix-organoclay, for example, on the parameter b_α value. Then parameter α^T can be determined finally as follows:

$$\alpha^T = \frac{L_{pl} b_\alpha}{1.27 b_\alpha N_{pl} + 1} \tag{13}$$

In Figure 5.3, the comparison of experimental α and calculated according to the Eq. (13) α^T values of organoclay tactoids aspect ratio, characterizing large-scale disorder level for the considered nanocomposites, is adduced. As one can see, good enough correspondence of experiment and theory is obtained (average discrepancy of α and α^T makes up ~ 9%).

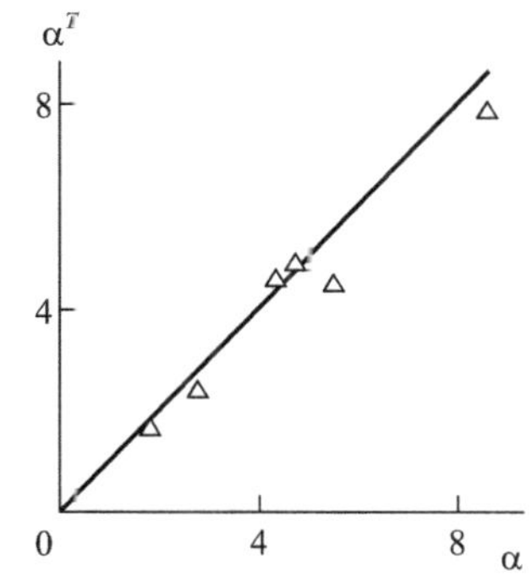

FIGURE 5.3 The comparison of experimental α and calculated according to the Eq. (13) α^T organoclay anisotropy degree for nanocomposites PVC-MMT.

The Eqs. (10) and (13) allow to predict reinforcement degree E_n/E_m on organoclay known structural characteristics. In Figure 5.4, the comparison of theoretical curves $E_n/E_m(\varphi_n)$, calculated according to the Eq. (1), where parameter α^T was determined according to the Eqs. (10) and (13), and corresponding the experimental data is adduced. As one can see, a good both qualitative (the theoretical curves are reflected experimental dependence maximum without existence of maximums for parameters N_{pl} and d_{001}) and quantitative correspondence of theory and experiment (their average discrepancy makes up less 2.5%).

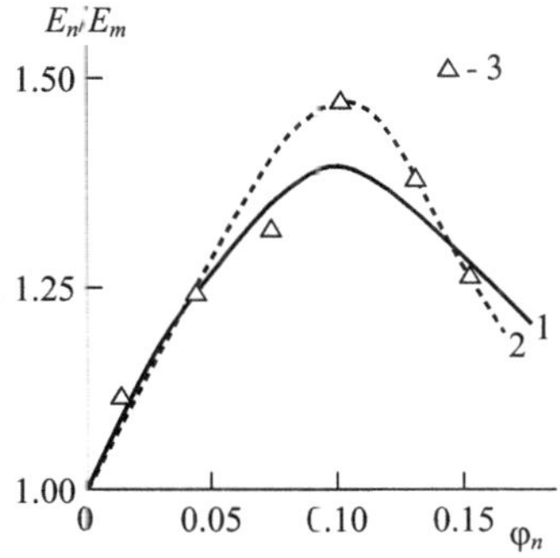

FIGURE 5.4 The comparison of calculated according to the Eq. (1) with usage of the Eqs. (10) (1) and (13) (2) for parameter α^T determination and experimental (3) dependences of reinforcement degree En/Em on nanofiller volume content φ_n for nanocomposites PVC-MMT.

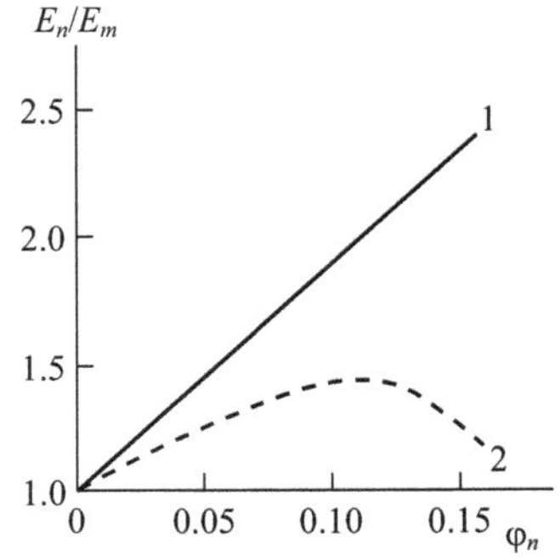

FIGURE 5.5 The comparison of calculated according to the Eq. (1) at the condition of organoclay minimum aggregation (1) and experimental (2) dependences of reinforcement degree E_n/E_m on nanofiller volume content φ_n for nanocomposites PVC-MMT.

Let consider in conclusion the influence of organoclay platelets aggregation or large-scale disorder on the considered nanocomposites reinforcement degree. In Figure 5.5, the experimental and calculated according to the Eq. (1) at organoclay aggregation minimum level (N_{pl}=9.70, b_α=2.51, α=8.6), corresponding to organoclay content W_n=1 mass %, dependences $E_n/E_m(\varphi_n)$ for nanocomposites PVC/MMT are adduced. As it follows from this comparison, both lower values of reinforcement degree and its decay at W_n>7 mass % are due precisely by organoclay platelets aggregation in “packets” (tactoids).

5.4 CONCLUSIONS

Thus, the present work results have demonstrated that organoclay platelets aggregation in “packets” (tactoids) results in large-scale disorder enhancement that reduces nanofiller anisotropy degree. In its turn, this factor decreases essentially reinforcement degree (or elasticity modulus) of nanocomposites polymer-organoclay, moreover at large enough organoclay contents (>7 mass %) reinforcement degree reduction at nanofiller content growth is observed. The important role of interfacial adhesion in anisotropy level determination has been shown.

KEYWORDS

- **aggregation**
- **anisotropy**
- **nanocomposite**
- **organoclay**
- **reinforcement degree**
- **tactoid**

REFERENCES

1. Schaefer, D. W., Justice, R. S. How nano are nanocomposites? *Macromolecules*, 2007, vol. 40, no. 24, 8501–8517.
2. Mikitaev, A. K., Kozlov, G. V., Zaikov, G. E. *Polymer Nanocomposites: Variety of Structural Forms and Applications,* New York: Nova Science Publishers, Inc., 2008.
3. Kozlov, G. V., Mikitaev, A. K. *Structure and Properties of Nanocomposites Polymer/Organoclay,* Saarbrücken: LAP LAMBERT Academic Publishing GmbH and Comp., 2013.
4. Sheng, N., Boyce, M. C., Parks, D. M., Rutledge, G. C., Abes, J. I., Cohen, R. E. Multiscale micromechanical modeling of polymer/clay nanocomposites and the effective clay particle, *Polymer*, 2004, vol. 45, no. 2, 487–506.

CHAPTER 6

SYNTHESIS OF THE HYBRID METAL-POLYMER NANOCOMPOSITE

S. ZH. OZKAN, G. P. KARPACHEVA, and I. S. EREMEEV

A.V. Topchiev Institute of Petrochemical Synthesis, Russian Academy of Sciences, 29, Leninsky prospect, 119991, Moscow, Russia

CONTENTS

ABSTRACT

During the IR heating of polyphenoxazine (PPhOA) in presence of cobalt (II) acetate $Co(CH_3CO_2)_2{\times}4H_2O$ in the inert atmosphere at the temperature T = 500–650°C the growth of the polymer chain via condensation of phenoxazine oligomers happens simultaneously with the dehydrogenation with the formation of conjugated C=N bonds. Hydrogen emitted during these processes contributes to the reduction of CO_2^+ to Co°. As a result the nanostructured composite material in which Co nanoparticles are dispersed into the polymer matrix is formed. According to TEM Co nanoparticles have size $4 < d < 14$ nm. The investigation of magnetic and thermal properties of Co/PPhOA nanocomposite has shown that the obtained nanomaterial is superparamagnetic and thermally stable.

6.1 INTRODUCTION

Metal-polymer nanocomposites combine useful properties of polymers and metal nanoparticles. Materials, based on polymers with a system of polyconjugation and magnetic nanoparticles attract a special attention due to their unique physico-chemical properties. Metal-polymer nanomaterials based on polymers with a system of conjugation are promising for application in organic electronics and electrorheology, creating microelectromechanical systems, supercapacitors, sensors, solar cells, displays, etc. Inclusion of magnetic nanoparticles into the nanocomposites makes them promising for use in systems of magnetic recording of information, for creating of electromagnetic screens, contrasting materials for magnetic resonancy tomography, etc.

In the current work a method of synthesis of the hybrid metal-polymer nanocomposite based on polyphenoxazine (PPOA) and Co nanoparticles was developed for the first time. PPOA is a half-ladder heterocyclic polymer, containing both nitrogen and oxygen atoms, which are included into the whole system of polyconjugation. The molecular mass of PPhOA is $M_w = 3.7 \times 10^4$. The selection of the polymer is described by its high thermal stability (up to 400°C in air and in the inert atmosphere at 1000°C the residue is 51%) [1].

The use of IR radiation for chemical and structural modifications instead of the generally used thermal impact is caused by the fact that due to the transition of the system into the vibrationally excited state it becomes possible to increase the rate of chemical reactions and thus to reduce the process time significantly [2–6].

During the IR heating of PPhOA in presence of cobalt (II) acetate $Co(CH_3CO_2)_2 \times 4H_2O$ in the inert atmosphere at the temperature T = 500–650°C the growth of the polymer chain via condensation of phenoxazine oligomers happens simultaneously with the dehydrogenation with the formation of conjugated C=N bonds.

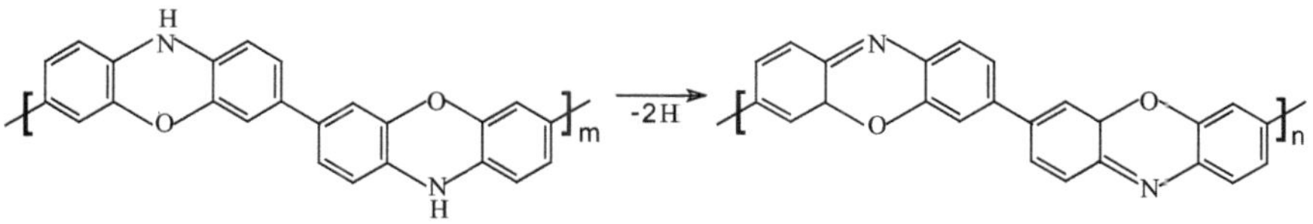

According to IR spectroscopy data the growth of the polymer chain is proved by the reduction of intensity of absorption band at 739 cm^{-1}, relating to nonplanar deformation vibrations of δ_{C-H} bonds of 1,2-substituted benzene ring of the end groups, for example, the number of polymers end groups significantly reduces. Absorption bands at 869 and 836 cm^{-1} are caused by nonplanar deformation vibrations of δ_{C-H} bonds of 1,2,4-substituted benzene ring (Figure 6.1). Presence of these bands indicates, that the growth of the polymer chain proceeds via C–C – connection type into para-positions of phenyl rings with respect to nitrogen [1].

The formation of C=N bonds is proved by the shift and broadening of the bands at 1587 and 1483 cm^{-1}, corresponding to stretching vibrations of ν_{C-C} bonds in aromatic rings. Absorption band at 3380 cm^{-1}, corresponding to the stretching vibrations of ν_{N-H} bonds in phenyleneamine structures practically disappears. The absorption band at 3420 cm^{-1}, associated with water appears (Figure 6.2).

Hydrogen emitted during these processes contributes to the reduction of CO_2^+ to Co^0. As a result the nanostructured composite material in which Co nanoparticles are dispersed into the polymer matrix is formed. The reflection bands of β-Co nanoparticles with a cubic face-centered lattice are identified in the diffraction pattern of the nanocomposite in the range of scattering angles 2θ = 68.35°, 80.55° (Figure 6.3). According to TEM Co nanoparticles have size 4 < d < 14 nm (Figure 6.4). According to AAS the content of cobalt in Co/PPOA nanocomposite, obtained at 550°C is 22.6 wt. %.

Metal-polymer Co/PPhOA nanocomposite is a black powder, insoluble in organic solvents.

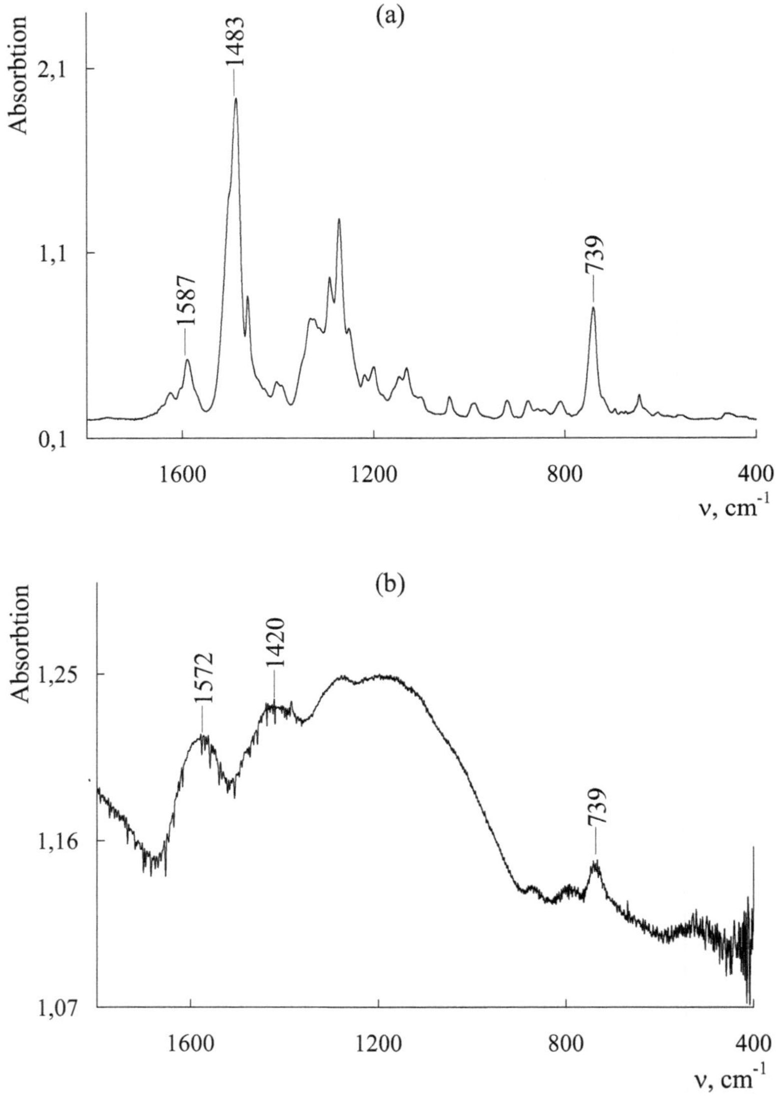

FIGURE 6.1 IR spectra of PPOA (a) and Co/PPOA nanocomposite, obtained at 500°C while heating for 10 minutes (b) in the absorption range of nonplanar deformation vibrations of $\delta_{C\text{-}H}$ bonds of aromatic rings.

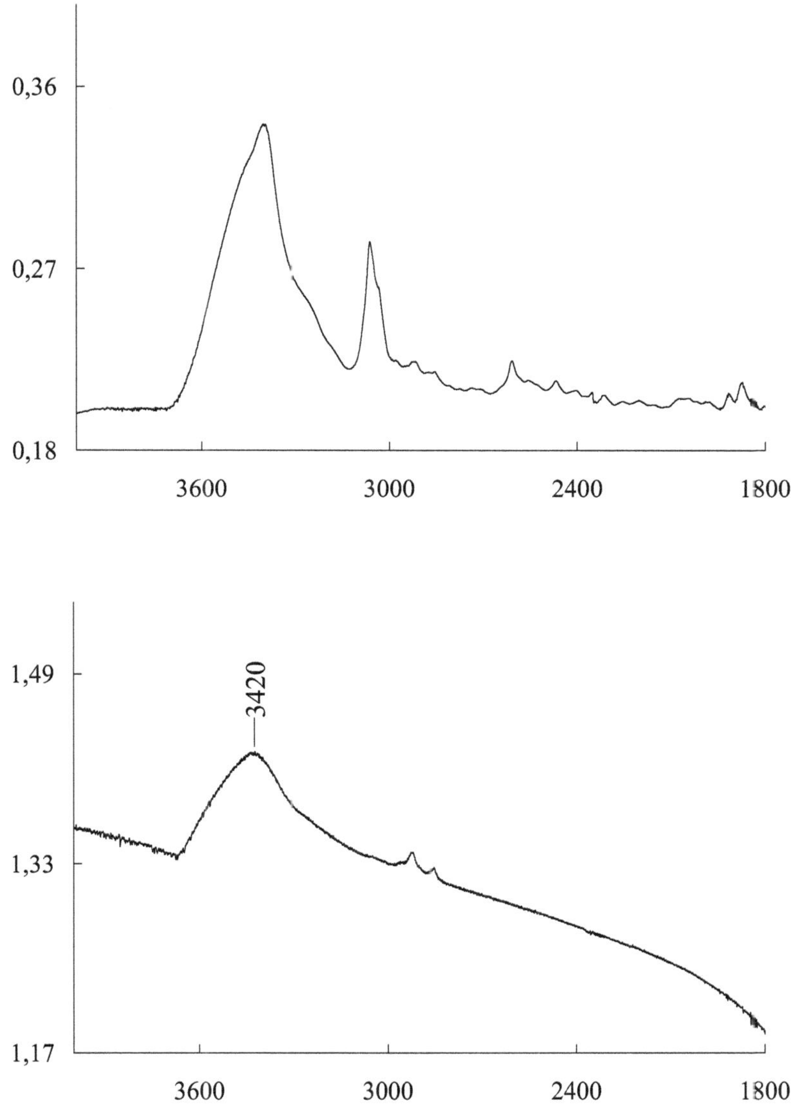

FIGURE 6.2 IR spectra of PPOA (a) and Co/PPOA nanocomposite, obtained at 500°C while heating for 10 minutes (b) in the absorption range of stretching vibrations of ν_{N-H} and ν_{C-H} bonds.

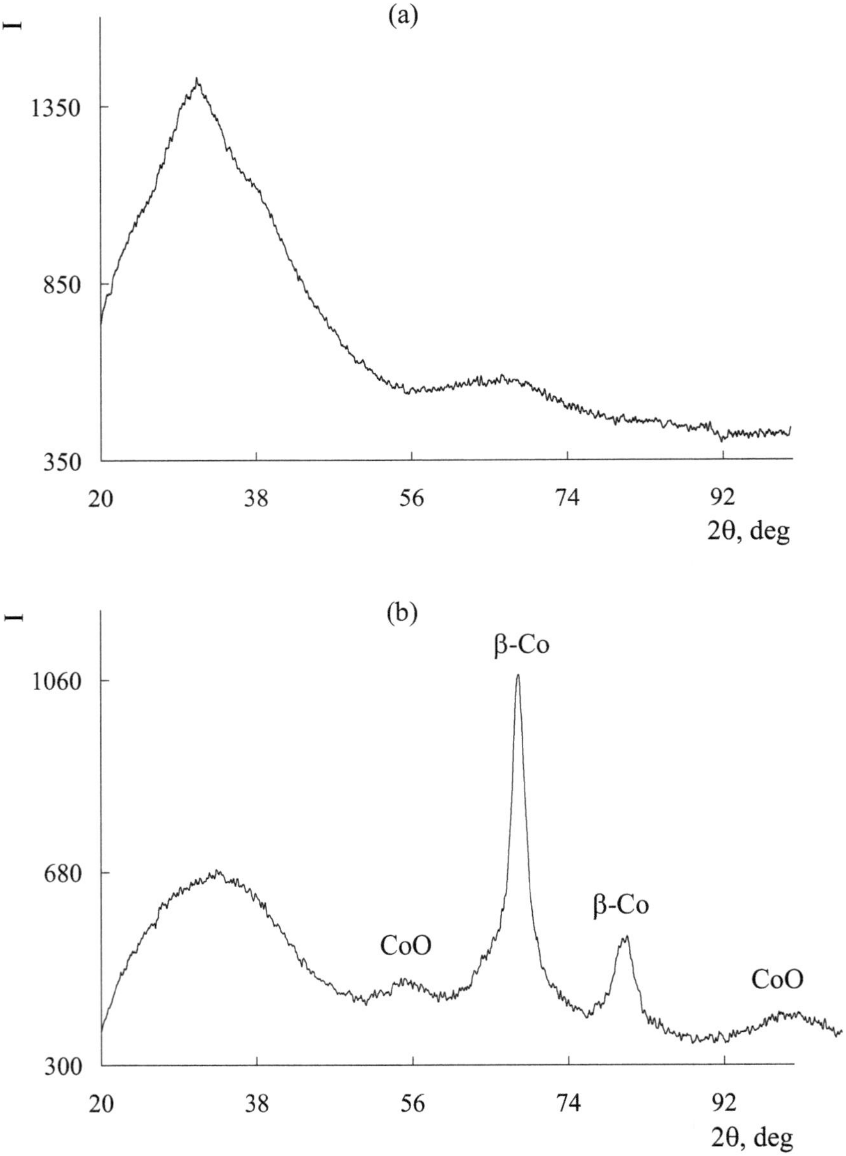

FIGURE 6.3 Continued

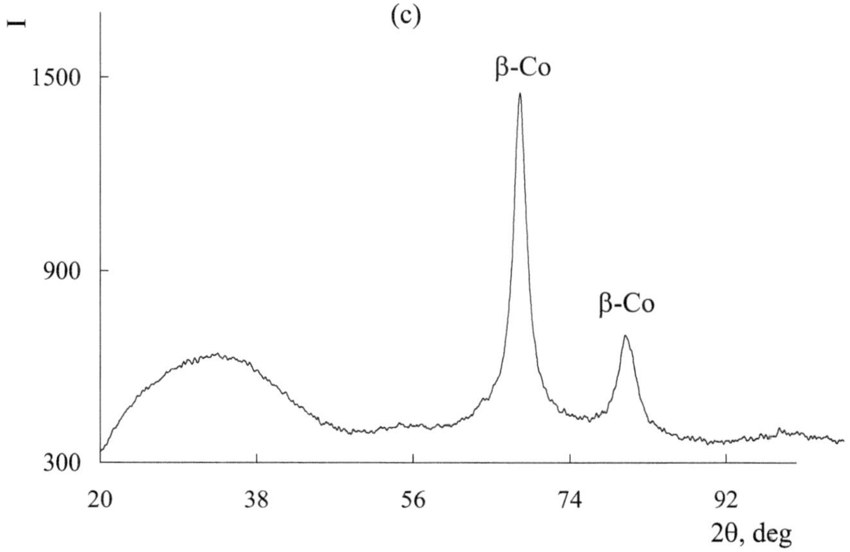

FIGURE 6.3 Diffraction patterns of PPOA (a) and Co/PPOA nanocomposite, obtained at 450 (b) and 500°C while heating for 10 minutes (c).

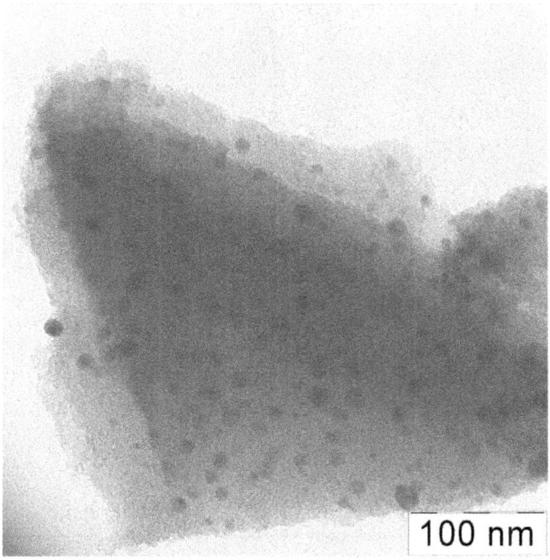

FIGURE 6.4 Microphotograph of Co/PPOA nanocomposite.

It was found that at temperatures below 500°C along with the Co nanoparticles, the nanoparticles of CoO are also present in the nanocomposite in the range of scattering angles 2θ = 55.61°, 65.32°, 99.05° (Figure 6.3b). The increase of heating time in the range of 5–30 minutes does not lead to the complete reduction of CoO to Co. At low concentrations of Co while loading, only nanoparticles of metallic Co are registered, but when [Co] = 30 wt. % the CoO nanoparticles prevail.

Investigation of magnetic properties at room temperature showed that the obtained Co/PPOA nanocomposites have hysteresis magnetization reversal (Figure 6.5). The values of the main magnetic characteristics of the nanocomposite, obtained at 500 and 650°C, are shown in Table 6.1.

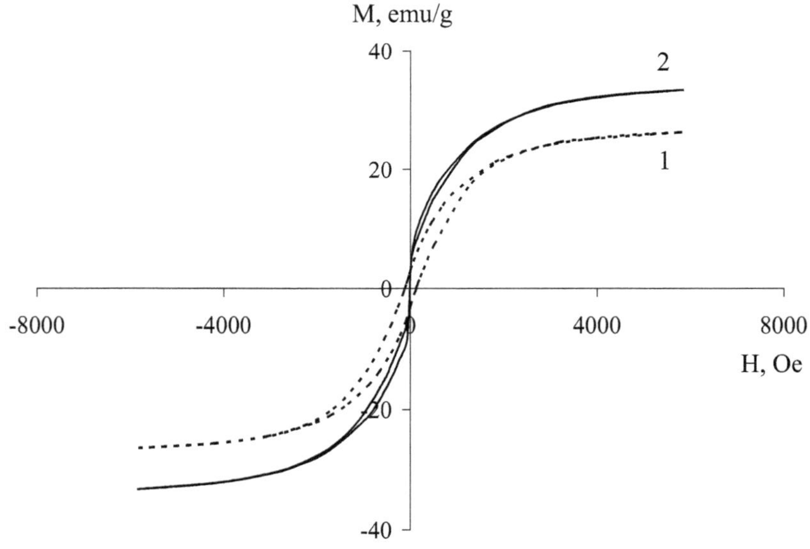

FIGURE 6.5 Magnetization of Co/PPOA nanocomposite, obtained at temperature 500 (1) and 650°C while heating for 10 minutes (2), as a function of the applied magnetic field at room temperature.

TABLE 6.1 Magnetic Characteristics of Co/PPOA Nanocomposite

Temperature of the sample, °C	H_C, Oe	M_S, emu/g	M_R, emu/g	M_R/M_S
500	134	26.33	3.05	0.116
650	0	33.33	0	0

It was found that after the increase of the temperature the coercive force H_C decreases together with the squareness coefficient of the hysteresis loop k_S, which is the ratio of the residual magnetization M_R to the saturation magnetization M_S. It happens due to the increase of the part of magnetic nanoparticles in the nanocomposite through the reduction of CoO to Co at high temperatures. Hybrid Co/PPOA nanocomposite is superparamagnetic. The obtained value $M_R/M_S = 0.116–0$ is characteristic of uniaxial, single-domain particles [5, 6].

Co/PPOA nanocomposite is characterized by the high thermal stability (Figure 6.6). 7% weight loss happens due to the presence of moisture in the nanocomposite, which is proved by DSC data (Figure 6.7). In the DSC thermogram of Co/PPOA nanocomposite there is an endothermic peak in this range of temperatures. After reheating this peak is absent.

After the removal of moisture the mass of the Co/PPOA nanocomposite does not change up to 300°C. In the inert atmosphere there is a gradual weight loss and at 1000°C the residue is 75%.

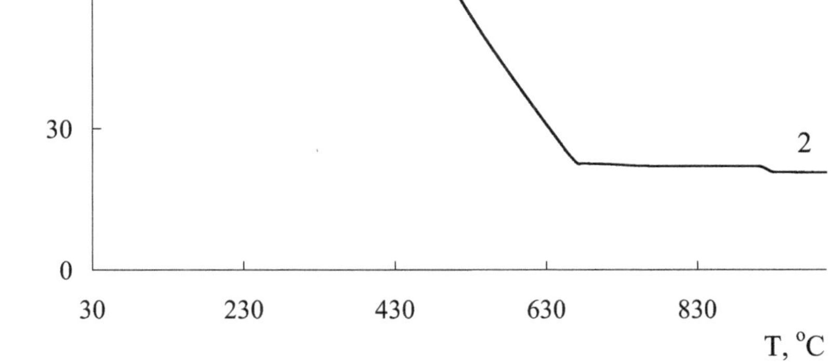

FIGURE 6.6 Weight decrease of Co/PPOA nanocomposite after heating to 1000°C at heating rate 10°C/min in nitrogen flow (1) and in air (2).

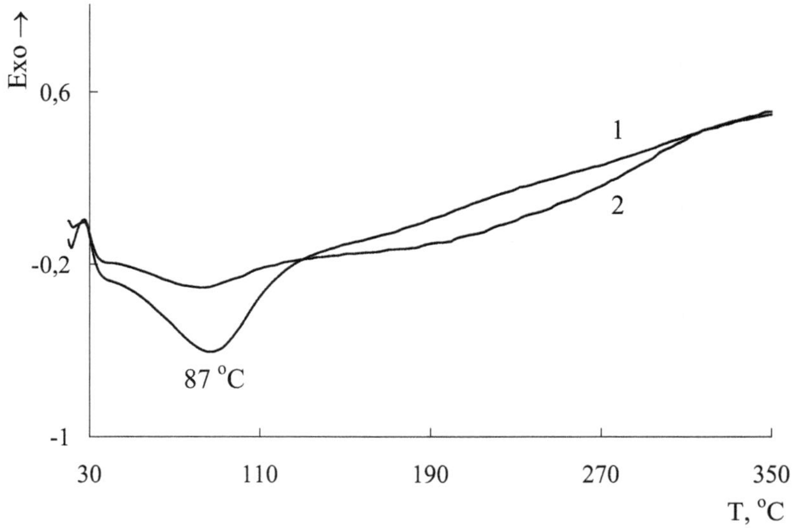

FIGURE 6.7 DSC thermograms of Co/PPOA nanocomposite while heating in nitrogen flow up to 350°C at heating rate 10°C/min (1 – first heating, 2 – second heating).

ACKNOWLEDGEMENTS

The work has been supported by the Russian Foundation for Basic Research, project № 14-03-31556 mol_a.

KEYWORDS

- **Co nanoparticles**
- **IR heating**
- **magnetic material**
- **metal-polymer nanocomposite**
- **polyphenoxazine**

REFERENCES

1. Ozkan, S. Zh., Karpacheva, G. P., Bondarenko, G. N. Polymers of phenoxazine: synthesis, structure. Proceedings of the Academy of Sciences. Chemical series. 2011, № 8, 1625–1630.
2. Ozkan, S. Zh., Kozlov, V. V., Karpacheva, G. P. Novel Nanocomposite based on Polydiphenylamine and Nanoparticles Cu and Cu_2O. J. Balkan Tribological Association. 2010, Book 3. V. 16. № 3, 393–398.
3. Ozkan, S. Zh., Karpacheva, G. P. Novel Composite Material Based on Polydiphenylamine and Fe_3O_4 Nanoparticles. Organic Chemistry, Biochemistry, Biotechnology and Renewable Resources. Research and Development. Volume 2 – Tomorrow and Perspectives. Editors: G. E. Zaikov, O. V. Stoyanov, E. L. Pekhtasheva. Nova Science Publishers, Inc: New York. 2013, Chapter 8, 93–96.
4. Ozkan, S. Zh., Dzidziguri, E. L., Krpacheva, G. P., Bondarenko, G. N. Metal-polymer nanocomposites based on polydiphenylamine and copper nanoparticles: synthesis, structure and properties. Russian Nanotechnologies. 2011, Vol. 6. № 11–12, 78–83.
5. Karpacheva, G., Ozkan, S. Polymer-metal hybrid structures based on polydiphenylamine and Co nanoparticles. Procedia Materials Science. 2013, V. 2, 52–59.
6. Ozkan, S. Zh., Dzidziguri, E. L., Chernavskii, P. A., Krpacheva, G. P., Efimov, M. N., Bondarenko, G. N. Metal-polymer nanocomposites based on polydiphenylamine and cobalt nanoparticles. Russian Nanotechnologies. 2013, Vol. 8. № 7–8, 34–40.

CHAPTER 7

CONTROL OF GAS EXHAUSTS OF FLARES IN SYNTHETIC RUBBER PRODUCTION

R. R. USMANOVA[1] and G. E. ZAIKOV[2]

[1]*Ufa State Technical University of Aviation, 12 Karl Marks str., Ufa 450100, Bashkortostan, Russia, E-mail: Usmanovarr@mail.ru*

[2]*N.M.Emanuel Institute of Biochemical Physics, Russian Academy of Sciences, 4 Kosygin str., Moscow 119334, Russia, E-mail: chembio@sky.chph.ras.ru*

CONTENTS

ABSTRACT

Actions for decrease in gas exhausts of flares in synthetic rubber production are developed. The device is developed for wet purification of the gas exhausts, confirmed high degree of purification both in laboratory, and in industrial conditions. The packaging scheme of gas purifying for synthetic rubber production is implanted. The complex of the made researches has formed the basis for designing of system of purification of air of industrial premises. Burning of gas exhausts on flares has allowed reducing pollution of air basin by toxic substances considerably.

7.1 INTRODUCTION

Hydrocarbons and their derivatives fall into the basic harmful exhausts of the petrochemical and oil refining enterprises. Actions for decrease in their pollutions are directed on elimination of losses of hydrocarbons at storage and transportation, and also on perfection of the control over hermetic sealing of the equipment and observance of a technological regime [1].

At many oil refining and petrochemical enterprises operate flares. They are intended for combustion formed at start-up of the equipment and in a process of manufacture of the gases, which further processing, is economically inexpedient or impossible. To flares make following demands:

- The completeness of burning excluding formation of aldehydes, acids and harmful products;
- Safe ignition, noiselessness and absence of a bright luminescence;
- Absence of a smoke and carbon black;
- Stability of a torch at change of quantity and composition of gas exhausts.

Burning of gas exhausts on flares allows reducing pollution of air basin by toxic substances considerably. However, salvaging of waste gases of the oil refining and petrochemical enterprises on flares is not a rational method of protection of environment. Therefore, it is necessary to provide decrease in exhausts of gases on a torch. Application of effective systems of purification of gas exhausts result in to reduction of number of torches at the petrochemical enterprises [2].

Actions for protection of air basin yes the oil refining and petrochemical enterprises should be directed on increase of culture of production; strict observance of a technological mode; improvement of technology for the purpose of gas-making decrease; the maximum use of formed gases; reduction of losses of hydrocarbons on objects of a manufacturing economy; working out and improvement of a quality monitoring and purification of pollutions.

7.2 ENGINEERING DESIGN AND EXPERIMENTAL RESEARCH OF NEW APPARATUSES FOR GAS CLEARING

Dynamic gas washer, according to Figures 7.1 and 7.2, contains the vertical cylindrical case with the bunker gathering slime, branch pipes of input and an output gas streams. Inside of the case it is installed conic vortex generator, containing.

Dynamic gas washer works as follows [3].

The gas stream containing mechanical or gaseous impurity, acts on a tangential branch pipe in the ring space formed by the case and rotor.

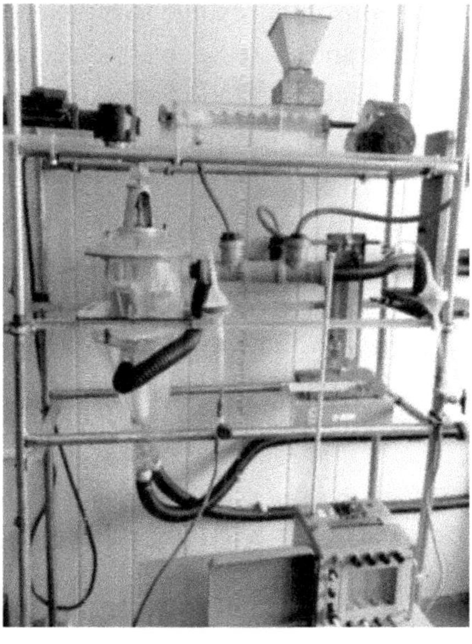

FIGURE 7.1 Experimental installation "dynamic gas washer."

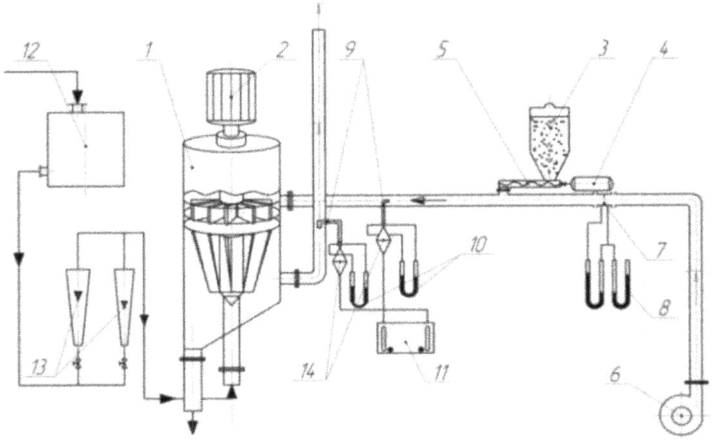

FIGURE 7.2 The circuit design of experimental installation: 1 – scrubber; 2 – the drive; 3 – the dust loading pocket; 4 – the electric motor; 5 – the batcher; 6 – the fan; 7 – a diaphragm; 8,10 – differential; 12 – the pressure tank; 13 – rotametr; 14 – sampling instruments.

The liquid acts in the device by means of an axial branch pipe. At dispersion liquids, the zone of contact of phases increases and, hence, the effective utilization of working volume of the device takes place more.

The invention is directed on increase of efficiency of clearing of gas from mechanical and gaseous impurity due to more effective utilization of action of centrifugal forces and increase in a surface of contact of phases. The centrifugal forces arising at rotation of a rotor provide crushing a liquid on fine drops that causes intensive contact of gases and caught particles to a liquid.

Owing to action of centrifugal forces, intensive hashing of gas and a liquid and presence of the big interphase surface of contact, there is an effective clearing of gas in a foamy layer. The aim was to determine the hydraulic resistance of irrigated unit when changing loads on the phases. The calculations take into account the angular velocity of rotation of the rotor blades and the direction of rotation of the swirl [4].

7.3 CLEARING OF GASES OF A DUST IN THE INDUSTRY

At the oil flares operate refining enterprise "Synthetic rubber," in Bashkortostan. They are intended for combustion formed at start-up of the equipment and in a process of manufacture of gases, (Figure 7.3).

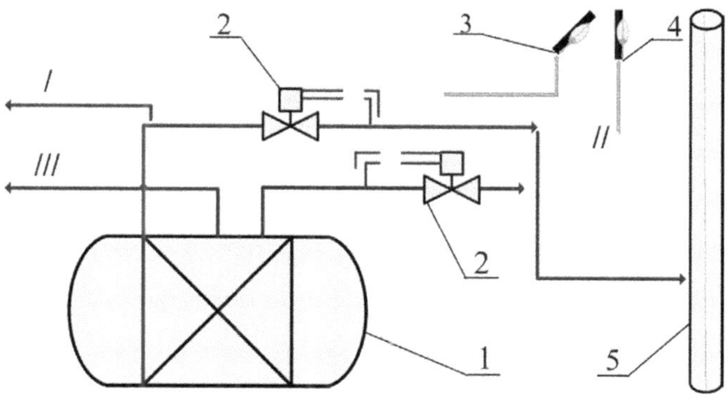

FIGURE 7.3 The Schema of a flare with scrubber dynamic: 1 – scrubber dynamic; 2 – governor valves; 3 – the ignite burner; 4 – a pilot-light burner; 5 – a torch pipe; *I* – waste gas; *II* – fuel gas; *III* – a condensate.

The dynamic scrubber is developed for decrease in gas exhausts of flares in synthetic rubber production.

Temperature of gases of baking ovens in main flue gas a copper-recovery 500–600°C, after exhaust-heat boiler 250°C. An average chemical compound of smoke gases (by volume): 17%$C0_2$; 16%N_2; 67% CO. Besides, in gas contains to 70 mg/m^3 SO_2; 30 mg/m^3 H_2S; 200 mg/m^3 F and 20 mg/m^3 CI. The gas dustiness on an exit from the converter reaches to 200/m^3 the Dust, as well as at a fume extraction with carbonic oxide after-burning, consists of the same components, but has the different maintenance of oxides of iron. In it than 1 micron, than in the dusty gas formed at after-burning of carbonic oxide contains less corpuscles a size less. It is possible to explain it to that at after-burning CO raises temperatures of gas and there is an additional excess in steam of oxides. Carbonic oxide before a gas heading on clearing burn in the special chamber. The dustiness of the cleared blast-furnace gas should be no more than 4 mg/m^3. The following circuit design (Figure 7.4) is applied to clearing of the blast-furnace gas of a dust.

Gas from a furnace mouth of a baking oven 1 on gas pipes 3 and 4 is taken away in the gas-cleaning plant. In raiser and down taking duct gas is chilled, and the largest corpuscles of a dust, which in the form of sludge are trapped in the inertia sludge remover, are inferred from it. In a centrifugal scrubber 5 blast-furnace gas is cleared of a coarse dust to final dust content 5–10/m^3 the dust drained from the deduster loading pocket periodically from a feeding system of water or steam for dust moistening. The final cleaning of the blast-furnace gas is carried out in a dynamic spray scrubber

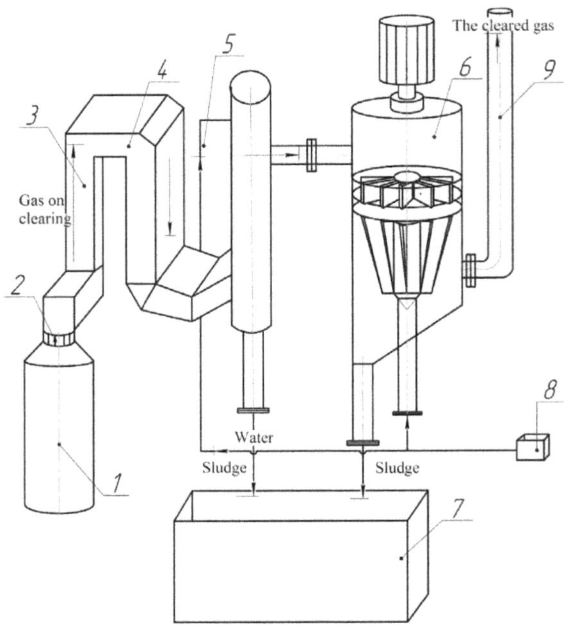

FIGURE 7.4 Process flow sheet of clearing of gas emissions: 1 – a Flare; 2 – water block; 3 – raiser; 4 – down-taking duct; 5 – Centrifugal scrubber; 6 – scrubber dynamic; 7 – forecastle of gathering of sludge; 8 – Hydraulic hitch; 9 – chimney.

where there is an integration of a finely divided dust. Most the coarse dust and drops of liquid are inferred from gas in the inertia mist eliminator. The cleared gas is taken away in a collecting channel of pure gas 9, whence is fed in an aerosphere. The clarified sludge from a gravitation filter is fed again on irrigation of apparatuses. The closed cycle of supply of an irrigation water to what in the capacity of irrigations the lime milk close on the physical and chemical properties to composition of dusty gas is applied is implemented. As a result of implementation of trial installation clearings of gas emissions the maximum dustiness of the gases, which are thrown out in an aerosphere, has decreased with 3950mg/m^3 to 840 mg/m^3, and total emissions of a dust from sources of limy manufacture were scaled down about 4800 to/a to 1300 to/a.

Such method gives the chance to make gas clearing in much smaller quantity, demands smaller capital and operational expenses, reduces an atmospheric pollution and allows to use water-recycling system [5].

TABLE 7.1 Results of Posttest Examination

Compound	Concentration at the inlet, g/m³	Concentration after clearing, g/m³
Dust	0.02	0.00355
NO_2	0.10	0.024
SO_2	0.03	0.0005
CO	0.01	0.0019

7.4 CONCLUSION

1. The solution of an actual problem on perfection of complex system of clearing of gas emissions and working out of measures on decrease in a dustiness of air medium of the industrial factories for the purpose of betterment of hygienic and sanitary conditions of work and decrease in negative affecting of dust emissions given.
2. Designs on modernization of system of an aspiration of smoke gases of a flare with use of the new scrubber which novelty is confirmed with the patent for the invention are devised. Efficiency of clearing of gas emissions is raised. Power inputs of spent processes of clearing of gas emissions at the expense of modernization of a flowchart of installation of clearing of gas emissions are lowered.
3. Ecological systems and the result of the implementation of the recommendations is to a high degree of purification of exhaust gases and improve the ecological situation in the area of production. The economic effect of the introduction of up to 3 million rubles/year.

KEYWORDS

- **dynamic scrubber**
- **flare**
- **purification of gas exhausts**
- **synthetic rubber**

REFERENCES

1. Belov, P. S., Golubeva, I. A., Nizova, S. A. *Production Ecology Chemicals from Petroleum Hydrocarbons and Gas,* Moscow: Chemistry, 1991, 256 p.
2. Tetelmin, V. V., Jazev, V. A. *Environment Protection in the Oil and Gas Complex,* Dolgoprudnyj: Oil and Gas Engineering, 2009, 352 p.
3. Usmanova, R. R. Dynamic gas washer. Patent for the invention of the Russian Federation № 2339435. 20 November 2008. The bulletin. № 33.
4. Shvydky, V. S., Ladygichev, M. G. *Clearing of Gases. The Directory,* Moscow: Heat power engineering, 2002, 640 p.
5. Straus, V. *Industrial Clearing of Gases.* Moscow: Chemistry, 1981, 616 p.

CHAPTER 8

OXIDATION OF POLYPROPYLENE-GRAPHITE NANOCOMPOSITES

T. V. MONAKHOVAA,[1] P. M. NEDOREZOVA,[2]
S. V. POL'SHCHIKOV,[2] A. A. POPOVA,[3] A. L. MARGOLIN,[1] and
A.YA. GORENBERG[2]

[1]*Emanuel Institute of Biochemical Physics, Russian Academy of Sciences, Moscow, Russia; E-mail: monakhova@sky.chph.ras.ru*

[2]*Semenov Institute of Chemical Physics, Russian Academy of Sciences, Moscow, Russia*

[3]*Plekhanov Russian University of Economics, Moscow, Russia*

CONTENTS

ABSTRACT

The reactivity of isotactic polypropylene (PP)–fine-grained graphite nanocomposites in the reactions of thermooxidation and chemiluminescence is studied. It is demonstrated that, even at low (less than 1%) concentrations, graphite retards the oxidation of polypropylene and speeds up the termination of PP peroxy macroradicals. It has been concluded that the mechanism of the protective action of graphite in the oxidation of PP involves inhibition due to the interaction of graphite nanoparticles with peroxy PP macroradicals. The average size of the reactive graphite particles was estimated from kinetic data on the reaction of graphite with radicals.

8.1 INTRODUCTION

Polymer–graphite nanocomposites are attracting growing interest due to a significant improvement in their mechanical and electrical properties [1]. For example, additives of graphite increase the conductivity of polymers by 12 orders of magnitude, impart to them new properties, for example, piezoelectric. The modification of polyolefins by introducing carbon fillers offers prospects for creating multifunctional polymeric materials with high heat resistance, electrically and thermally conductive, and with improved tribological and adhesion properties. A substantial modification of polymers is achieved by introducing even small amounts of nanosized fillers [2, 3].

One of the known methods for preparing polymer composites with fillers of various types is polymerization filling (*in situ* polymerization). The technique enables to obtain composites with high and uniform distribution of the filler throughout the volume. This method of introducing fillers into the polyolefine matrix was proposed in [4, 5] and then developed for producing composites of polypropylene (PP) with graphite [6, 7] and graphite oxide [8], being currently applied to a wide variety of polymers [1].

It is known that carbon fillers retard the oxidation of polymers [9–11], which is accounted for by the termination of kinetic chains on their surface [12, 13]. Obvious factors that determine the effectiveness of inhibition are the high specific surface area of the filler, uniformity of distribution over the volume, and its strong interaction with the polymer [14]. This leaves open the question of the mechanism of inhibition. It can be assumed, for example, that when free radicals in the polymer encounter the carbon surface they are temporarily adsorbed thereon [15]. This mechanism reduces the efficiency of radicals in the propagation of the oxidation reaction chain, but does not affect their loss. The second mechanism involves the direct destruction of radicals by reaction with double bonds or impurities on the surface of the carbon filler. As demonstrated recently, this mechanism accounts for the slowdown of PP oxidation in the presence of fullerenes, which attach peroxy radicals via double bonds [16]. In graphite, such a mechanism is hardly realizable, since due to the planar structure of graphene layers in graphite, their double bonds form a conjugated aromatic system with electron delocalization, a factor that hinders the addition of radicals.

In this paper, we attempt to estimate the mechanism of the protective action of graphite in the thermal oxidation of PP by comparing the kinetics of PP oxidation and the loss of peroxy radicals monitored through the decay of PP chemiluminescence.

8.2 EXPERIMENTAL PART

Fine-grained graphite (FGG) was prepared by mechanical grinding of MPG-6 artificial graphite ($S_{sp} = 8\ m^2/g$) in an inert atmosphere until a specific surface of $S_{sp} \approx 480\ m^2/g$ was reached. Isotactic PP-FGG composites were prepared by in situ polymerization in a liquid propylene medium in the presence of a highly effective homogeneous catalytic systems based on rac-$Me_2Si(2\text{-}Me\text{-}4PhInd)_2ZrCl_2$ activated with polymethylalumoxane (PPP-1/FGG samples) and a $TiCl_4$–Et_2AlCl catalytic system (PPP-2/FGG sample) as described in [6] and [7], respectively. The metallocene catalyst used produces PP with high molecular weight and degree of isotacticity [17]. The titanium-chloride-based catalyst fixed on a graphite surface also produces isotactic PP with a high molecular weight (MW) [18]. Propylene was polymerized at a temperature of 60°C in a 200 cm^3 steel reactor equipped with a high-speed stirrer (3000 rpm). The nanocomposites were synthesized as follows: FGG powder, preliminary evacuated at 200°C, was introduced into the reactor and flushed with liquid propylene, after which the required amount of the respective cocatalyst and catalyst was added. After polymerization completion, the composite material powder was discharged from the reactor and washed from remnants of catalyst components with a mixture of ethanol and HCl (10% solution) and then repeatedly washed with ethanol and dried to constant weight in vacuo at 60°C.

The IR spectra of PP in the composites thereof were recorded on a Vertex-70 FTTIR spectrometer (Bruker). The stereoregularity of isotactic PP in the composite samples was determined from the ratio of the optical densities of the absorption bands at 973 and 998 cm^{-1} (D_{998}/D_{973}). This ratio characterizes the presence of isotactic sequences of length exceeding 11–13 monomeric units in the polymer chain [19].

The thermal properties (melting point and enthalpy) of the nanocomposites were determined by differential scanning calorimetry (DSC) on a DSC-7 instrument (Perkin-Elmer) at a heating/cooling rate of 10 K/min. Table 8.1 shows some characteristics of the composites obtained using the different catalyst systems. The distribution of filler particles throughout the composite material was analyzed by scanning electron microscopy (SEM) on a JSMM-5300LV (Jeol) instrument.

The thermal oxidation of the composites was studied in the kinetic regime at 130°C and an oxygen pressure of 300 Torr [15]. The kinetics of oxygen uptake was investigated on a high sensitivity manometric set. The absorber of volatile products was solid KOH.

To monitor the chemiluminescence (CL) from the composites, thin films of the samples were irradiated in the air at 23°C with light from a mercury low-pressure lamp, 80% of the radiation of which is 254 nm monochromatic light. The irradiated samples were placed in the CL

TABLE 8.1 Characteristics of PP/FGG Composites

Sample	FGG content wt %	MM_w, 10-5 g/mol	MWD*	D_{998}/D_{973}	T_{m1}	ΔH_{m1}	X**, %	T_{m2}	ΔH_{m2}
PP1/FGG	0.8	6–7	2	0.88	162.2	90.5	55	162	88.4
PP2/FGG	0.6	8–9	3	0.85	163.4	82.3	50	161	74.6

* Molecular weight distribution.

** X is the degree of crystallinity calculated by the formula $X = \Delta H_m/165.100\%$.

chamber of the SNK-7 unit [20], and the kinetics of CL intensity decay at 23°C was measured (starting 1 min after irradiation).

8.3 RESULTS AND DISCUSSION

8.3.1 DISTRIBUTION OF GRAPHITE IN NANOCOMPOSITES PP/FGG

A typical micrograph of FGG powder is displayed in Figure 8.1. As can be seen, the particles have a wide size distribution in the nanometer range. The SEM micrograph of a low temperature cleavage of a PP-1/FGG sample shows that the distribution of the graphite particles throughout the resultant composite is a nearly homogeneous.

The calculated particle size distribution for the FGG powder is shown in Figure 8.2. As can be seen from this figure, the experimental distribution can be quantitatively described by the sum of two normal distributions ρ(L). Based on this distribution, the average FGG particle size can be calculated by the equation

$$\langle L \rangle = \int L\rho(L)dL \tag{1}$$

FIGURE 8.1 SEM microphotograph of FGG powder.

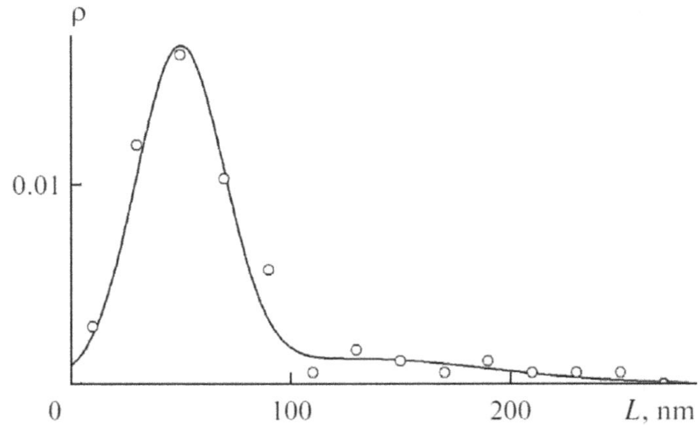

FIGURE 8.2 Size L distribution of FGG powder particles (symbols) and its approximation by the sum of two normal distributions (curve).

Substituting the distribution $\rho(L)$ into Eq. (1) and integrating over all values of L yields an average particle size of $L = 65$ nm. On the other hand, the average particle size of graphite can be evaluated from its specific surface area S_{sp}. Assuming that the representative particle is spherical with diameter L, we can calculate S_{sp} as the ratio of the surface area of the sphere to its mass:

$$S_{sp} = 6/Ld \tag{2}$$

where d is the graphite density (for crystalline graphite, $d = 2.26$ g/cm^3).

According to Eq. (2), the specific surface area of FGG of $S_{sp} = 480$ m^2/g corresponds to an average particle size of $L = 5$ nm, which is 11 times less than that determined by SEM. This difference may be accounted for by the aggregation of individual powder graphite particles into larger particles. It is known that graphite particles are easily attracted to each other, because they have a high surface energy and high surface tension [21]. Because of the random coalescence of particles, bonds between them are weak and, therefore, disintegrate during mixing the graphite powder with the polymer. Since part of the surface of the particles in the agglomerates is not available because of the particles contacting with each other, the total surface area of the individual particles is higher than $S_{sp} = 480$ m^2/g. Therefore, the average particle size may be less than 5 nm.

8.3.2 OXIDATION OF PP-GRAPHITE NANOCOMPOSITES

Oxidation is the main cause of the rapid destruction of PP during its service and processing. The oxidation of polyolefins proceeds by the radical chain mechanism with degenerate branching on hydroperoxide [15]. It is known that, at the initial stage of uninhibited oxidation of polyolefins, the oxygen uptake kinetics is described by the parabolic law:

$$N_{0_2} = b(t-\tau)^2 \tag{3}$$

where is the amount of absorbed oxygen; t and τ are the current time and the induction period of oxidation, respectively; b is the kinetic parameter equal given by

$$b = \alpha\sigma k_2^2 k_4 [RH]^3 / 8k_6 \tag{4}$$

Here, k_2 and k_6 are, respectively, the rate constants of propagation and termination of oxidation chains; k_4 is the hydroperoxide decomposition rate constant; α is the hydroperoxide yield per mole of oxygen absorbed; σ is the probability of degenerate branching of oxidation chains; and *[RH]* is the concentration of reactive bonds.

To analyze the kinetics of oxidation, the kinetics data were calculated according to Eq. (3):

$$(N_{O_2})^{1/2} = \sqrt{b}(t-\tau) \tag{5}$$

In the absence of linear termination, $\tau = 0$, and anamorphosis of the kinetic curve is a straight line passing through the coordinate origin. In the case of linear termination, the kinetic curve anamorphosis is shifted along the t axis by the value of τ [15]. The value of was determined as the slope of the linear anamorphosis in the coordinates of Eq. (5); τ was calculated from intercept of the straight line at the t axis.

The induction period is directly related to the linear termination rate constant k [15]:

$$\tau = k / \alpha\sigma k_2 k_4 [RH]^2 \tag{6}$$

The anamorphosis of the kinetic curves of oxidation in the coordinates of Eq. (5) are displayed in Figure 8.3.

As can be seen, the kinetics of the oxidation of the PP/FGG nanocomposites has features characteristic of the inhibited oxidation of PP: the parabolic law of oxidation holds, with the parameter *b* being approximately the same for all the samples. A significant increase in the induction period is observed at the content of FGG 1.6–3.6 wt %: τ=140 min for pure PP and τ = 152, 310, 390 min at 0.8, 1.6, 3.6 wt% of FGG content respectively. This means that FGG is a very effective inhibitor of the oxidation of PP/FGG nanocomposites. However, at low FGG content (0.8%), the induction period is almost the same as that for pure PP. It makes sense that, under these conditions, carbon particles manifest their ability to initiate the oxidation of PP, as was observed previously in the oxidation of graphite–PP composites [12] and in the oxidation fullerene–PP nanocomposites [16].

Thus, these results confirm the known data on the stabilizing and initiating action of graphite on PP oxidation. The observed kinetic pattern is characteristic mainly of PP oxidation inhibitors, being observed in the oxidation of graphite–PP and fullerene–PP composites.

Additional information on the mechanism of the action of graphite was obtained by studying chemiluminescence of the samples.

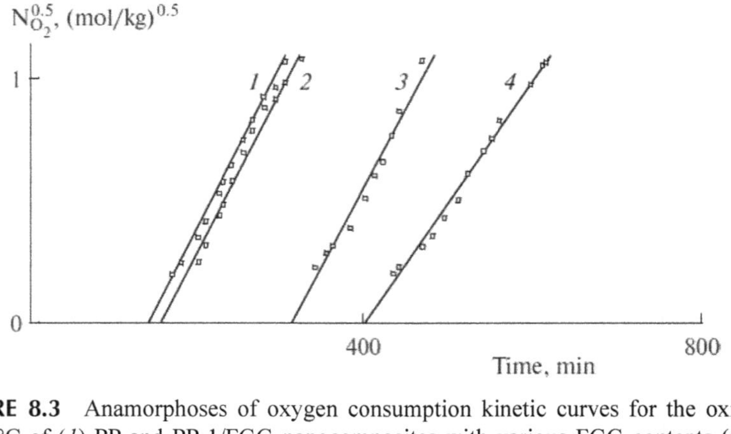

FIGURE 8.3 Anamorphoses of oxygen consumption kinetic curves for the oxidation at 130°C of (*1*) PP and PP-1/FGG nanocomposites with various FGG contents (wt %): (*2*) 0.8, (*3*) 1.6, and (*4*) 3.6.

8.3.3 CHEMILUMINESCENCE OF GRAPHITE–PP NANOCOMPOSITES

Light induced PP chemiluminescence arises due to the termination of PP peroxy macroradicals formed under the action of light and oxygen [22]. It is known that, at low concentrations of radicals, this reaction is first order in the concentration of radicals (linear termination of oxidation chains), resulting in the chemiluminescence intensity I from PP at room temperature being proportional to the product of the rate constant of peroxy macroradicals decay and their concentration [22]:

$$I = \varphi k[RO_2] \tag{7}$$

where φ is the chemiluminescence quantum yield (number of photons per radical loss event), k is the rate constant for the loss of free radicals, and $[RO_2]$ is the concentration of peroxy radicals.

In the linear termination reaction, the radical concentration varies as

$$[RO2] = [RO2]_o.exp(-kt) \tag{8}$$

where $[RO2]_o$ is the initial concentration of radicals.

Combining Eqs. (7) and (8) readily yields the equation for the time evolution of the CL intensity:

$$ln(I/I_o) = -kt \tag{9}$$

where $I_o = \varphi k[RO_2]_o$ is the initial CL intensity.

Figure 8.5 shows typical kinetic curves in the coordinates of Eq. (9) for chemiluminescence decay upon a short irradiation of PP. The behavior of the initial portions of the curves is determined by the conditions of photoinitiation, being dependent on the initial nonequilibrium distribution of the radicals over their reactivity and on the characteristics of the relevant relaxation processes [23]. The progress of the relaxation processes leads to the establishment of the exponential decay of the chemiluminescence intensity with a fixed rate constant k, which can be determined from the slope of the asymptote of the kinetic curve in the $\ln(I/I_o)$–t coordinates.

The results of measurements of the rate constants *k* of the peroxy radicals decay for PP and its compositions with graphite are summarized in Table 8.2.

These data show that, even at a low graphite content (0.8%), the rate constants for the loss of graphite peroxy radicals is approximately 2-fold higher than that for pure PP. This effect is observed for powders and

TABLE 8.2 Values of the Rate Constants *k* for the Decay of PP Peroxy Radicals at 23°C in the Air

№	Sample	FGG content, wt %.	k.104, s–1
1	PP-1, powder	0	3.4
	PP-1, film	0	3.4
2	PP-1/FGG, powder	0.8	6.3
3	PP-1/FGG, film	0.8	8.4
4	PP-2/FGG, film	0.6	7.2

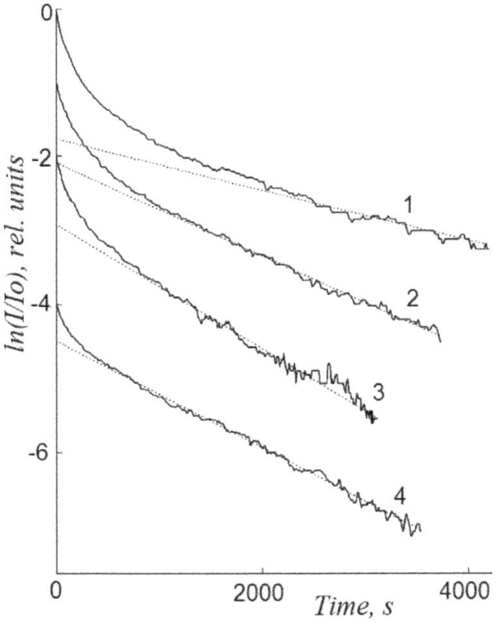

FIGURE 8.4 Kinetics of chemiluminescence decay in the air at 23°C: (*1*) PP, (*2, 3*) PP-1/FGG, and (*4*) PP-2/FGG with FGG content of (*2, 3*) 0.8 and (*4*) 0.6 wt % in the form of (*1, 2*) a powder and (*3, 4*) a film. For better perception curves *1–4* are moved apart along the ordinate.

composite films prepared on the different polymerization catalysts. Thus, these findings confirm the inhibitory effect of graphite and indicate that the mechanism of its protective action is due to its ability to reduce the concentration of PP peroxy radicals in the course of oxidation. In other words, the effect of graphite is similar to that of fullerene [16].

This conclusion requires further discussion. As noted in the introduction, the delocalization of the electrons in the flat graphene layers of graphite should impede their reaction with radicals, just as it does in other aromatic systems, such as benzene. Nevertheless, it was recently discovered that the double C=C bond of graphite is capable of attaching methyl radicals [24]. It can be assumed that the reaction involves bonds in defects, where the overlapping of p orbitals is violated, for example, in convex or concave areas of the graphene plane. Defects in the structure of graphite facilitate the splitting of layers during grinding. Therefore, the surface of milled graphite particles must be enriched in defects capable of reacting with radicals.

To produce an inhibitory effect, filler particles should meet another condition: the distance between them and radicals must be small enough to ensure that, during its lifetime, the radical would have time to diffuse to the particle's surface (a necessary condition for the reaction with graphite). The distance between particles is determined by their molar concentration C_m, which is calculated for spherical particles of diameter L by the equation

$$C_m = \frac{10C_w}{\frac{4p}{3}\left(\frac{L}{2}\right)^3 d(1-X)} \quad (10)$$

where C_w is the weight concentration of graphite in the composite, d is the specific density of graphite, X is degree of crystallinity of the polymer (graphite particles are located in the amorphous part of the polymer).

On the other hand, the concentration C_m required to terminate free radicals can be estimated from data on the kinetics of the free radical decay. In the presence of graphite, the rate of radical decay is the sum of the rate of the first order reaction of free radicals with each other with rate constant k_o and the rate of consumption of free radicals on the graphite surface with rate constant k_g:

$$k = k_o + k_g C_m \quad (11)$$

The value of k_g can be calculated based on the segmental diffusion model [25]:

$$k_g = 4\pi\left(\frac{L}{2} + a\right)\frac{a^2 k_2[RH]}{6} \tag{12}$$

where a is the segmental motion amplitude.

The system of Eqs. (10)–(12) can be easily solved numerically. To evaluate the concentration and size of the particles, we used the commonly accepted values of a = 3 nm and $k_2[RH]$ = 0.01 s^{-1} for PP at room temperature, preset values of d = 2.26 g/cm^3 and X = 55% (Table 8.1), and the experimental values of k_o=3.4×10^{-4} and k = 6.3×10^{-4} s^{-1} at C_w = 0.8% (Table 8.3). At these values, Eqs. (9)–(12) give C_m = 5.1×10^{-4} Mole/kg and an average graphite particle size of L = 4 nm. The latter value is consistent with the above estimate L < 5 nm based on the value of S_{sp}.

8.4 CONCLUSIONS

Thus, it was experimentally shown that graphite nanoparticles react with PP peroxy radicals. The reaction between graphite and radicals occurs at defects on the particle surface. Estimates show that the concentration and size of the graphite particles are high enough to trap PP radicals.

KEYWORDS

- **decay of radicals**
- **fullerene**
- **graphene**
- **graphite**
- **nanocomposite**
- **oxidation**
- **polypropylene**

REFERENCES

1. Sengupta, R., Bhattacharya, M., Bhowmick, A. K., Bandyopadhyay, S. Progress Polym. Sci. 36, 638 (2011).
2. Kim, H., Abdala, A. A., Macosco, C. Macromolecules 43, 6515 (2010).
3. Badamshina, E. R., Gafurova, M. P. Polymer Sci., Ser. B 50, 215 (2008).
4. Kostandov, L. A., Enikolopov, N. S. Dyachkovsky, F. S. et al., US Patent No. 4241112 (1980).
5. Dyachkovsky, F. S., Novokshonova, L. A. Russ. Chem. Rev. 53, 117 (1984).
6. Galashina, N. M., Nedorezova, P. M. Tsvetkova, V. I., D'yachkovskii, F. S., Enikolopov, N. S. Dokl. Akad. Nauk SSSR 278, 620 (1984).
7. Nedorezova, P. M., Shevchenko, V. G., Shchegolikhin, A. N., Tsvetkova, V. I., Korolev, Yu. M. Polymer Sci., Ser. A 46, 242 (2004).
8. Huang, Y., Qin, Y., Zhou, Y., Niu, H., Yu, Z. Z., Dong, J. Y. Chem. Mater. 22, 4096 (2010).
9. Fang, Z., Song, P., Tong, L., Guo, Z. Thermochim. Acta 473, 106 (2008).
10. Zhao, L., Song, P., Cao, Z., Fang, Z., Guo, Z., J. Nanomater. 2012, 1 (2012).
11. Gopakumar, T. G., Page, D. J. Y. S. Polym. Eng. Sci. 44, 1162 (2004).
12. Monakhova, T. V., Nedorezova, P. M., Bogaevskaya, T. A., Tsvetkova, V. I., Shlyapnikov, Yu. A. Vysokomol. Soedin. Ser. A 30, 2415 (1988).
13. Margolin, A. L., Velichko, V. A., Sorokina, A. V. et al., Vysokomol. Soedin., Ser. A 27, 1313 (1985).
14. Verdejo, R., Bernal, M. M., Romasanta L. J., Lopez_Manchado, M. A. J. Mater. Chem. 21, 3301 (2011).
15. Shlyapnikov, Yu. A., Kiryushkin, S. G., Mar'in, A. P. *Antioxidative Polymer Stabilization* (Khimiya, Moscow, 1988) [in Russian].
16. Polschikov, S. V., Nedorezova, P. M., Monakhova, T. V. et al., Polym. Sci., Ser. B 55, 286 (2013).
17. Spaleck, W., Kuber, F., Winter, A. et al., Organometallics 13, 954 (1994).
18. Nedorezova, P. M., Tsvetkova, V. I., Kolbanev, I. V., D'yachkovskii, F. S., Vysokomol. Soedin. A 31, 2657 (1989).
19. Kissin, Yu. V. *Isospecific Polymerization of Olefins* (Springer, New York, Berlin, Heidelberg, Tokyo, 1985, p. 439.
20. Shlyapintokh, V. Ya., Karpukhin, O. N., Postnikov, L. M. *Chemiluminescence Methods for Studying Slow Chemical Processes* (Nauka, Moscow, 1966), p. 300 [in Russian].
21. Quan, H., Zhang, B., Zhao, Q., Yuen, R. K. K., Li, R. K. Y., Composites, Pt. A 40, 1506 (2009).
22. Margolin, A. L., Shlyapintokh, V. Ya. Polym. Deg. Stabil. 66, 279 (1999).
23. Margolin, A. L. Kinet. Catal. 49, 161 (2008).
24. Mandeltort, L., Choudhury, P., Johnson, J. K., Yates, J. T. Jr., J. Phys. Chem. C 116, 18347 (2012).
25. Emanuel, N. M., Buchachenko, A. L. *The Chemical Physics of Molecular Destruction and Stabilization of Polymers* (V.N.U. Sci., Utrecht, 1987; Nauka, Moscow, 1988).

CHAPTER 9

NANOMATERIALS: AN ENGINEERING INSIGHT

A. AFZALI

University of Guilan, Rasht, Iran

CONTENTS

ABSTRACT

One of the most interesting and perspective directions in material engineering of the last years is development of technology of nano-materials consisting from two or more phases with precise interphase border and nanostructured materials based on interpenetrated polymer network. Nowadays many countries are leaders in nanotechnology; not only in fundamental academic researches but mainly in industrial introduction of scientific and technological achievements. Some important results in the nano-materials engineering are summarized in this chapter.

9.1 INTRODUCTION

It worth mentioning that the essence of nanotechnology is the ability to work at the molecular level to create large structures with fundamentally new molecular organization. Materials with features on the scale of nanometers often have properties different from their macroscale counterparts. The prospect of a new materials technology that can function as a low-cost alternative to high-performance composites has, thus, become irresistible around the world. By this means nanotechnology presents a new approach to material science and engineering as wall as for design of new devices and processes.

Nowadays, the economic, security, military, and environmental implications of molecular manufacturing are extreme. Unfortunately, conflicting definitions of nanotechnology and blurry distinctions between significantly different fields have complicated the effort to understand those differences and to develop sensible, effective policy for each.

The risks of today's nanoscale technologies cannot be treated the same as the risks of longer-term molecular manufacturing. It is a mistake to put them together in one basket for policy consideration – each is important to address, but they offer different problems and will require far different solutions. As used today, the term nanotechnology usually refers to a broad collection of mostly disconnected fields. Essentially, anything sufficiently small and interesting can be called nanotechnology. Much of it is harmless. For the rest, much of the harm is of familiar and limited quality. Molecular manufacturing, by contrast, will bring unfamiliar risks and new classes of problems.

9.2 BACKGROUND

As an example, desktop nanofactories will use vast arrays of tiny machines to fasten single molecules together quickly and precisely, allowing engineers, designers, and potentially anyone else to make powerful products at the touch of a button. Although such a contraption has been envisioned in some detail for almost two decades, and although the basic concept goes back to 1959, when the physicist R. Feynman first articulated it, it's only in recent years that technology has advanced to the point where we can begin to see the practical steps that might bring it into reality.

As is known, composite materials are two- or multi-phase with well-defined interphase border. Such materials contain the reinforcing elements immersed into a polymeric, ceramic or metal matrix. Mechanical properties of composites depend on structure and properties of the interphase border. Phases of usual composite materials have micron and submicron sizes.

Important among these nanoscale materials are ofcourse the nanocomposites, in which the constituents are mixed on a nanometer-length scale. They often have properties that are superior to conventional microscale composites and can be synthesized using surprisingly simple and inexpensive techniques.

The tendency to reduction of the phase's sizes of a filler (a strengthening element) is attributable to decrease in its microscopic deficiency (the size of one of nanocomposite phases does not exceed 100 nm). Due to the nanometer size of the particles, which is smaller than the wavelength of visible light, the reinforced polymer remains transparent. Other characteristics of the composites include high barrier performance and improved thermal stability, which make these compounds suitable for many applications. Because of this technology of nanocomposites is one of the most perspective directions in a material engineering. Specifically nanocomposite technology is applicable to a vide range of polymers. Cutting across the material classes of thermoplastic, thermosets and elastomers.

There are many academic and industrial sectors specialized in providing applied and theoretical research and development to the scientific and technological fields of material, chemical and environmental engineering, with a focus on the development, marketing, commercialization and manufacturing of advanced nanocomposites. Research and Developments currently underway in the following:

- nanostructured composites based on interpenetrated polymer network;
- epoxy-rubber composites with nanoheterogenic structure;
- nanocomposites based on hybrid organo-silicate matrix;
- polymer nanocomposites with very low permeability and high resistance to aggressive environments;
- metal matrix nanocomposites produced by super-deep penetration (SDP) method.
- polymer matrix nanocomposites and nanomembranes produced by SDP method;
- biodegradable materials based on nanocellulose.

9.2.1 NANOSTRUCTURED COMPOSITES BASED ON INTERPENETRATED POLYMER NETWORK

This project is oriented to prepare nanocomposites based on interpenetrated polymer network (IPN), such as polyurethanes, epoxies and acrylate by way of creating nanoparticles of $SiO_{2,}$ TiO_2 and other metal oxides during a technological stage from a liquid phase. Using as inter-penetrating polymer networks principle in production of composite materials provides a unique possibility to regulate their both micro- and nano-structured properties. Formulation of a new class of nanocomposite materials is characterized by the absence of contaminants for a network polymers technology. As a main component of such technology scientists are using branched (dendro)-aminosilanes that at the first stage are curing agents for many oligomers).

The proposed dendro-aminosilane hardeners give the possibility to introduce the siloxane fragments into aromatic structure of diphenylol-propane based epoxy-amine network polymers. Additional hydrolysis of aminosilane oligomer creates the secondary nano-structured network polymer that improves the service properties of the compound. Branched (dendro) polyamine hardeners are novel direction in epoxy and cyclocar-bonate and acryl resins chemistry.

The new hardeners give rise to formation of IPN of a polymerized resin with a polysiloxane network by the hydrolytic polycondensation of silane groups. IPN network may be formed on the base of epoxy-cyclocarbonate oligomers. It was found that at least 0.1 equivalent weight of silane per epoxy resin equivalent weight may result in IPN formation. It has been known that epoxy resin has low resistance to acetone and methanol attack. IPN film provides increasing the resistance.

9.2.2 POLYMER NANOCOMPOSITES WITH VERY LOW PERMEABILITY AND HIGH RESISTANCE TO AGGRESSIVE ENVIRONMENTS

Novel chemically resistant polymer materials were elaborated with adding nano-size inorganic active fillers that react with aggressive medium into which they are introduced, forming a new phase of high-strength hydrate complexes. This enhanced bonding occurs upon the penetration of aggressive media into active nano-fillers containing polymer material. The chemical resistant properties of the forming polymer materials are activated by harsh environmental conditions where polymer systems without additives remain defenseless to chemical corrosion.

9.2.3 NOVEL METALLIC AND POLYMER MATRIX NANOREINFORCED MATERIALS PRODUCED BY METHOD OF SUPER-DEEP PENETRATION

These materials can be used for replacement base steels in metal-cutting and stamp tools. The application of the new SDP technology allows to increase the service life of the tools up to 1.5–5.0 times compared to the common used tools. The technology can be applied for the volume strengthening practically any type of instrumental steels.

Use of SDP effect allows receiving composite materials on the basis of aluminum with the set anisotropy of physical and chemical properties. Electrical conduction of aluminum can be changed in several times. The new technology of volumetric reorganization of aluminum and creation zones of nano-structures will find wide application by manufacture of electric installations and electronic devices. The economic effect from use of the new aluminum material in electric installations and electronic control systems will be hundred millions dollars USA.

Use of new physical effect SDP allows as well to obtain the polymeric nanomembranes by using as a penetrated agent water solution of soluble salts.

9.3 MEMBRANE FILTRATION

This can be a very efficient and economical way of separating components that are suspended or dissolved in a liquid. The membrane is a physical barrier that allows certain compounds to pass through, depending on their physical and/or chemical properties. Polymeric membrane materials are intrinsically limited by a tradeoff between their permeability and their selectivity. One approach to increase the selectivity is to include dispersions of inorganic nanoparticles, such as zeolites, carbon molecular sieves, or carbon nanotubes, into the polymeric membranes – these membranes are classified as mixed-matrix membranes.

9.3.1 MANUFACTURING FIBROUS FILTERING MATERIALS

Current environmental problems are caused by human activities in the last 150 years. They are having serious negative impacts on us and this is likely to continue for a very long time. Effective solutions are urgently needed to protect our environment. Due to their high specific surface area, electrospun nanofibers are expected to be used to collect pollutants via physical blocking or chemical adsorption.

The first nonwovens using melted organic polymers were manufactured in the 1950s, using a method similar to air-blowing of the polymer melt. Application of this latter method enabled super-thin fibers to be obtained with a diameter smaller than 5 nm. The melt-blown technique of manufacturing nonwovens from super-fine fibers was developed by Wente at the Naval Research Laboratory in USA [1]. Buntin, a worker at Exxon Research and Engineering introduced the melt-blown technique for processing PP into the industry [2]. Recently, the Nonwoven Technologies Inc., USA has announced the possibility of manufacturing melt-blown PP nonwovens composed from nano-size fibers of a diameter equal to 300nm. To enhance the filtration efficiency, the melt -blown nonwovens are subjected to the process of activation, mainly using the corona discharge method [3].

An overview of flash-spinning technologies is presented by Wehman. Flashspun nonwovens made from fibers with very low linear density, which can be obtained using splittable fibers as a raw material for production of conventional webs [4].

Subsequently, webs can be subjected to the classical needle punching or spunlace process during which sacrificial polymer is removed and fibers of low linear density are obtained. The flash-spinning process can be also accomplished using such bicomponent melt-blown technology, which is based on spinning two incompatible polymers together and forming a web, which is then subjected to the splitting process.

Induction of electric charges is another mechanism used in filtering material technology. Induction consists of electric charge generation in a conductor placed in an electric field. Therefore, fine-fibers made from conductive solutions or melts, charged during electrostatic extrusion, belong to this group. Formation of nanofibers by the electrospinning method results from the reaction of a polymer solution drop subjected to an external electric field. This method enables manufacturing fibers with transversal dimensions of nanometers. Gilbert in 1600 made the first observation concerning the behavior of an electroconductive fluid under the action of an electrical field. He pointed out that a spherical drop of water on a dry surface is drowning up, taking the shape of a cone when a piece of rubbed amber is held above it. One of the first investigations into the phenomena of interaction of an electric field with a fluid drop was carried out by Zeleny [5]. He used the apparatus presented in Figure 9.1 in his experiment.

The apparatus include an open-end capillary tube of metal or glass. The conductive fluid is delivered to this tube using the reservoir C. A plate B is mounted opposite to the capillary tube in a distance of h. The capillary tube and the plate are maintained at a given potential difference V using a high voltage generator. Formhals [6] used this kind of technology for spinning thin polymer filaments.

The electric charges, which diffuse in the liquid, forced by the electric field, cause a strong deformation of the liquid surface in order to minimize the system's total energy. The electric forces exceed the forces of surface tension in the regions of the maximum field strength and charge density, and the liquid forms a cone at the nozzle outlet. A thin stream of liquid particles is torn off from the end of the cone. Taylor [7] proved that for a given type of fluid, a critical value of the applied voltage exists, at which the drop of fluid, flowing from the capillary tube, is transformed into a cone under

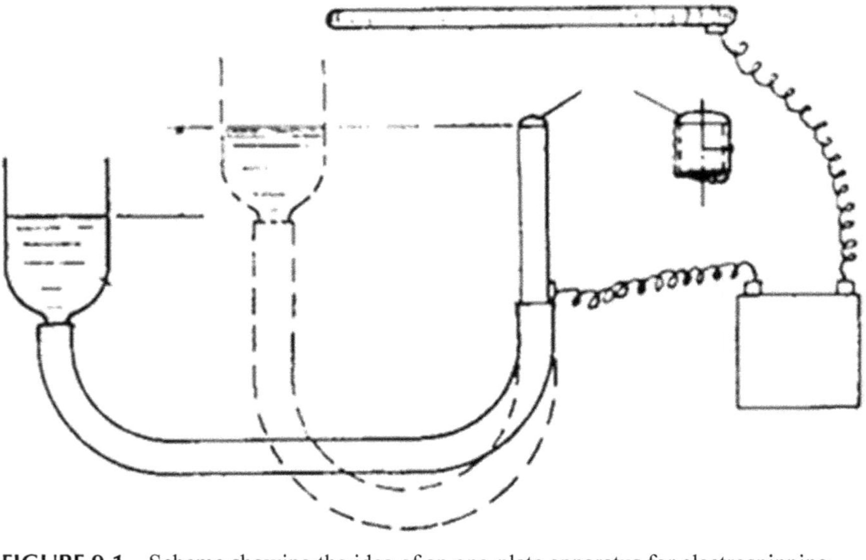

FIGURE 9.1 Scheme showing the idea of an one-plate apparatus for electrospinning.

the influence of the electric field, and loss its stability. The critical value of this potential depends on the surface tension of the fluid and of the initial radius of drop. Zeleny's and Taylor's investigations have been an inspiration for many researchers who carried out observations of the behavior of different kinds of polymers in the electric field. These observations were the basis for the development of manufacturing technologies for a new generation of fibers with very small transversal dimensions. Schmidt demonstrated the possibility of application of electrospun polycarbonate fibers to enhance the dust filtration efficiency [8].

In the 1980's, the Carl Freudenberg company used the electrospinning technology first commercially. Trouilhet and Weghmann presented a wide range of applications of electrospun webs especially in the filtration area. In that time the electrospinning method for manufacturing filtering materials did not find common application. The revival of this technology has been observed for the last five-four years. In 2000, Donaldson Inc., USA realized dust filters with a thin layer of nanofibers [9, 10].

A basic set for electrospinning consists form three major components, such as: a high voltage generator, a metal or glass capillary tube, and a collecting plate electrode, similar to the set designed by Zeleny. Such type of set is characterized by low productivity, usually less than 1 mLh^{-1}. To solve this problem, the array of multiply capillary tubes should be developed. Experiments carried out indicate that due to the interference between the electrical fields developed around such system an uniform electrical field strength cannot be ensured at the tip of each tube. For such a system, high probability of the tube clogging appears. To avoid such problems during the electrospinning process, some authors proposed to spin the fibers directly form the polymer solution surface. A new method with high productivity was developed by Jirsak at the Technical University of Liberec [11]. The proposed invention was commercialized by Elmarco company. The idea is very simple. The set is composed from two electrodes. The bottom electrode formed in the shape of a roll is immersed in the solution of a polymer, as shown in Figure 9.2.

A thin layer of polymer solution covers the rotating electrode, and multiple jets are formed due to the action of the electrical field. The Elmarco company offers a wide assortment of spun-bonded nonwovens covered by nanofiber membranes made of polyamide, polyurethane and

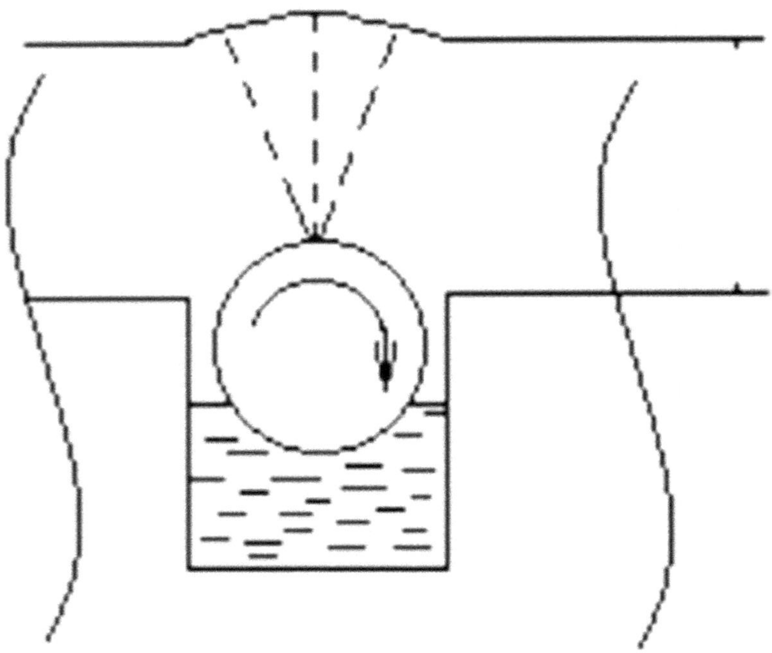

FIGURE 9.2 The idea of the electrospinning method developed by Jirsak.

polyvinyl alcohol. A further approach related to spinning directly from the solution surface was invented by Yarin and Zussman [12]. The proposed system is composed from two layers: a bottom layer in the form of ferromagnetic suspension and an upper layer in the form of polymer solution. The two- layer system is subjected to the magnetic field provided by a permanent magnet. The scheme of this apparatus is presented in Figure 9.3.

Vertical spikes of magnetic suspension appear as the result of action of the magnetic field, what causes the perturbation of the free surface of the polymer solution. Under the action of the electrical field, perturbations of the free surface become the sites of jetting directed upward.

9.3.1.1 Basic Research on Electrospun Nanofibers in Filtration

Usually, the particle filtration occurs via multiple collection mechanism such as sieving, direct interception, inertial impaction, diffusion, and electrostatic collection. In practice, sieving is not an important mechanism

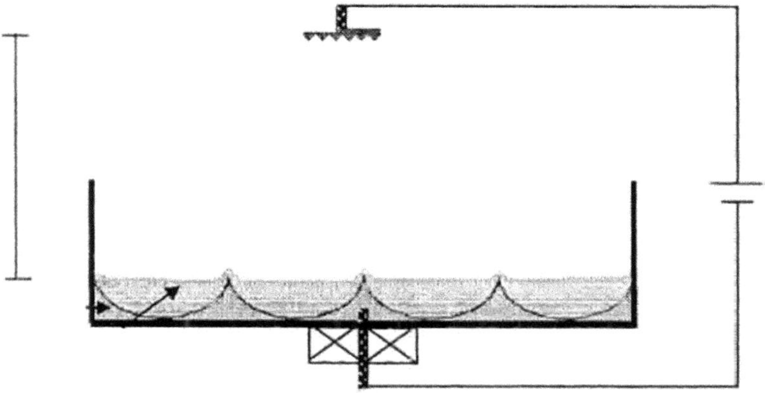

FIGURE 9.3 The idea of manufacturing electrospun nonwovens directly from the surface of a polymer solution: (a) ferromagnetic suspension, (b) polymer solution, (c) upper electrode, (d) lower electrode, (e) high voltage generator, (f) permanent magnet.

in most air filtration application. Additionally, commercial nanofibers are electrically neutral. So, the remaining important mechanisms in mechanical filtration are direct interception, inertial impaction, and diffusion. The reasonable approximations of filtering media performance have been made using single-fiber filtration theory.

9.3.1.2 Membranes Filtration

Membrane filtration is a mechanical filtration technique, which uses an absolute barrier to the passage of particulate material as any technology currently available in water treatment. The term "membrane" covers a wide range of processes, including those used for gas/gas, gas/liquid, liquid/liquid, gas/solid, and liquid/solid separations. Membrane production is a large-scale operation. There are two basic types of filters: depth filters and membrane filters.

Depth filters have a significant physical depth and the particles to be maintained are captured throughout the depth of the filter. Depth filters often have a flexuous three-dimensional structure, with multiple channels and heavy branching so that there is a large pathway through which the liquid must flow and by which the filter can retain particles. Depth filters have the advantages of low cost, high throughput, large particle retention capacity, and the ability to retain a variety of particle sizes. However, they can endure from entrainment of the filter medium, uncertainty regarding effective pore size, some ambiguity regarding the overall integrity of the filter, and the risk of particles being mobilized when the pressure differential across the filter is large.

The second type of filter is the membrane filter, in which depth is not considered momentous. The membrane filter uses a relatively thin material with a well-defined maximum pore size and the particle retaining effect takes place almost entirely at the surface. Membranes offer the advantage of having well-defined effective pore sizes, can be integrity tested more easily than depth filters, and can achieve more filtration of much smaller particles. They tend to be more expensive than depth filters and usually cannot achieve the throughput of a depth filter. Filtration technology has developed a well-defined terminology that has been well addressed by commercial suppliers.

The term membrane has been defined in a number of ways. The most appealing definitions to us are the following:

"A selective separation barrier for one or several components in solution or suspension."

A thin layer of material that is capable of separating materials as a function of their physical and chemical properties when a driving force is applied across the membrane.

Membranes are important materials, which form part of our daily lives. Their long history and use in biological systems has been extensively studied throughout the scientific field. Membranes have proven themselves as promising separation candidates due to advantages offered by their high stability, efficiency, low energy requirement and ease of operation. Membranes with good thermal and mechanical stability combined with good solvent resistance are important for industrial processes [13].

The concept of membrane processes is relatively simple but nevertheless often unknown. Membranes might be described as conventional filters but with much finer mesh or much smaller pores to enable the separation of tiny particles, even molecules. In general, one can divide membranes into two groups: porous and nonporous. The former group is similar to classical filtration with pressure as the driving force; the separation of a mixture is achieved by the rejection of at least one component by the membrane and passing of the other components through the membrane (Figure 9.4). However, it is important to note that nonporous membranes do not operate on a size exclusion mechanism.

Membrane separation processes can be used for a wide range of applications and can often offer significant advantages over conventional separation such as distillation and adsorption since the separation is based on a physical mechanism. Compared to conventional processes, therefore, no chemical, biological, or thermal change of the component is involved for most membrane processes. Hence membrane separation is particularly attractive to the processing of food, beverage, and bioproducts where the processed products can be sensitive to temperature (vs. distillation) and solvents (vs. extraction).

Synthetic membranes show a large variety in their structural forms. The material used in their production determines their function and their driving forces. Typically the driving force is pressure across the membrane barrier (see Table 9.1) [14–16]. Formation of a pressure gradient across the membrane allows separation in a bolter-like manner. Some other forms of separation that exist include charge effects and solution diffusion. In this separation, the smaller particles are allowed to pass through as permeates whereas the larger molecules (macromolecules) are retained. The retention or permeation of these species is ordained by the pore architecture as well as pore sizes of the membrane employed. Therefore, based on the

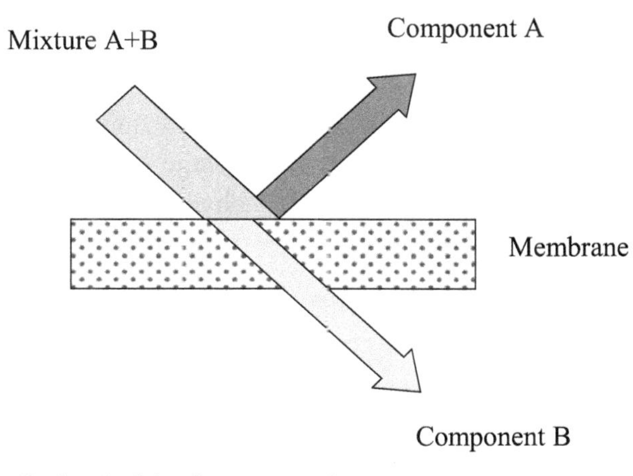

FIGURE 9.4 Basic principle of porous membrane processes.

TABLE 9.1 Driving Forces and Their Membrane Processes

Driving force	Membrane process
Pressure difference	Microfiltration, Ultrafiltration, Nanofiltration, Reverse osmosis
Chemical potential difference	Pervaporation, Pertraction, Dialysis, Gas separation, Vapour permeation, Liquid Membranes
Electrical potential difference	Electrodialysis, Membrane electrophoresis, Membrane electrolysis
Temperature difference	Membrane distillation

pore sizes, these pressure driven membranes can be divided into reverse osmosis (RO), nanofiltration (NF), ultrafiltration (UF), and microfiltration (MF), are already applied on an industrial scale to food and bioproducts processing [17–19].

A. Microfiltration (MF) Membranes

MF membranes have the largest pore sizes and thus use less pressure. They involve removing chemical and biological species with diameters ranging between 100 to 10000 nm and components smaller than this, pass through as permeates. MF is primarily used to separate particles and bacteria from other smaller solutes [16].

B. Ultrafiltration (UF) Membranes

UF membranes operate within the parameters of the micro- and nanofiltration membranes. Therefore, UF membranes have smaller pores as compared to MF membranes. They involve retaining macromolecules and colloids from solution which range between 2–100 nm and operating pressures between 1 and 10 bar., for example, large organic molecules and proteins. UF is used to separate colloids such as proteins from small molecules such as sugars and salts [16].

C. Nanofiltration (NF) Membranes

NF membranes are distinguished by their pore sizes of between 0.5–2 nm and operating pressures between 5 and 40 bar. They are mainly used for the removal of small organic molecules and di- and multivalent ions. Additionally, NF membranes have surface charges that make them suitable for retaining ionic pollutants from solution. NF is used to achieve separation between sugars, other organic molecules, and multivalent salts on the one hand from monovalent salts and water on the other. Nanofiltration, however, does not remove dissolved compounds [16].

D. Reverse Osmosis (RO) Membranes

RO membranes are dense semi-permeable membranes mainly used for desalination of sea water [38]. Contrary to MF and UF membranes, RO membranes have no distinct pores. As a result, high pressures are applied to increase the permeability of the membranes [16]. The properties of the various types of membranes are summarized in Table 9.2.

The NF membrane is a type of pressure-driven membrane with properties in between RO and UF membranes. NF offers several advantages such as low operation pressure, high flux, high retention of multivalent anion salts and an organic molecular above 300, relatively low investment and low operation and maintenance costs. Because of these advantages, the applications of NF worldwide have increased [20]. In recent times, research in the application of nanofiltration techniques has been extended from separation of aqueous solutions to separation of organic solvents to homogeneous catalysis, separation of ionic liquids, food processing, etc. [21].

Figure 9.5 presents a classification on the applicability of different membrane separation processes based on particle or molecular sizes. RO process is often used for desalination and pure water production, but it is the UF and MF that are widely used in food and bioprocessing.

While MF membranes target on the microorganism removal, and hence are given the absolute rating, namely, the diameter of the largest

TABLE 9.2 Summary of Properties of Pressure Driven Membranes

	MF	UF	NF	RO
Permeability($L/h.m^2.bar$)	1000	10–1000	1.5–30	0.05–1.5
Pressure (bar)	0.1–2	0.1–5	3–20	5–1120
Pore size (nm)	100–10000	2–100	0.5–2	<0.5
Separation Mechanism	Sieving	Sieving	Sieving, charge effects	Solution diffusion
Applications	Removal of bacteria	Removal of bacteria, fungi, viruses	Removal of multivalent ions	Desalination

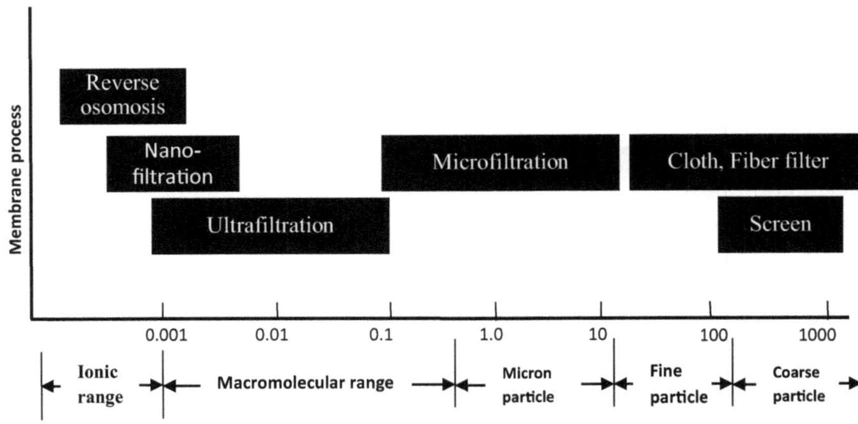

FIGURE 9.5 The applicability ranges of different separation processes based on sizes.

pore on the membrane surface, UF/NF membranes are characterized by the nominal rating due to their early applications of purifying biological solutions. The nominal rating is defined as the molecular weight cut-off (MWCO) that is the smallest molecular weight of species, of which the membrane has more than 90% rejection (see later for definitions). The separation mechanism in MF/UF/NF is mainly the size exclusion, which is indicated in the nominal ratings of the membranes. The other separation mechanism includes the electrostatic interactions between solutes and membranes, which depends on the surface and physiochemical properties of solutes and membranes [17]. Also, the principal types of membrane are shown schematically in Figure 9.6 and and membrane process characteristics are shown in Figure 9.7.

9.3.2 THE RELATIONSHIP BETWEEN NANOTECHNOLOGY AND FILTRATION

Nowadays, nanomaterials have become the most interested topic of materials research and development due to their unique structural properties (unique chemical, biological, and physical properties as compared to larger particles of the same material) that cover their efficient uses in various fields, such as ion exchange and separation, catalysis, bimolecular isolation and purification as well as in chemical sensing [22]. However, the

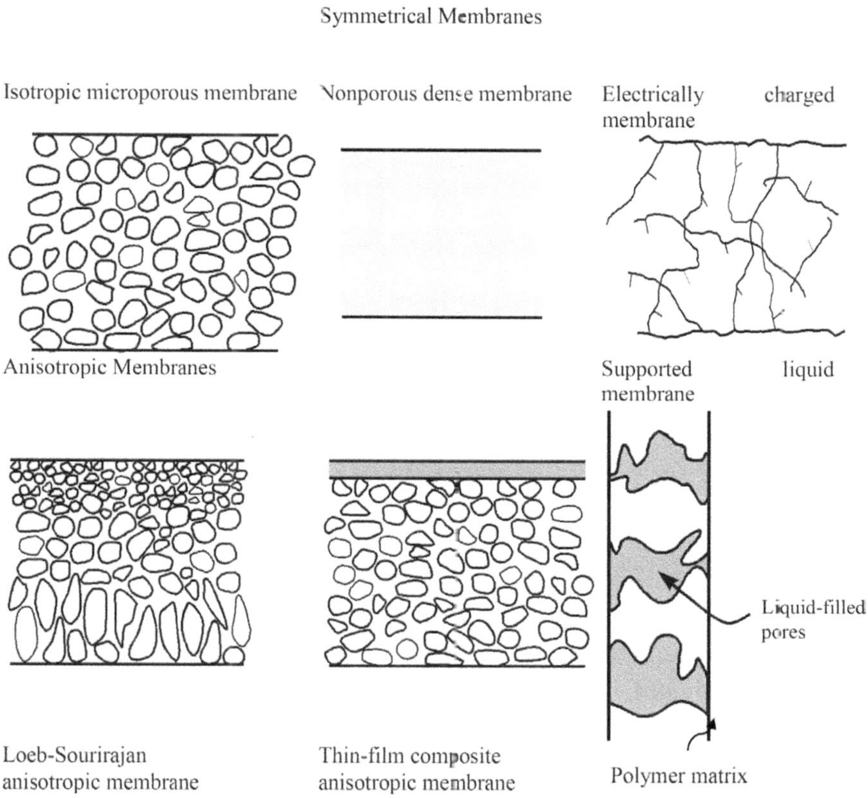

FIGURE 9.6 Schematic diagrams of the principal types of membranes.

understanding of the potential risks (health and environmental effects) posed by nanomaterials hasn't increased as rapidly as research has regarding possible applications.

One of the ways to enhance their functional properties is to increase their specific surface area by the creation of a large number of nanostructured elements or by the synthesis of a highly porous material.

Classically, porous matter is seen as material containing three-dimensional voids, representing translational repetition, while no regularity is necessary for a material to be termed "porous." In general, the pores can be classified into two types: open pores that connect to the surface of the material, and closed pores that are isolated from the outside. If the material exhibits mainly open pores, which can be easily transpired, then one can consider its use in functional applications such as adsorption, catalysis and

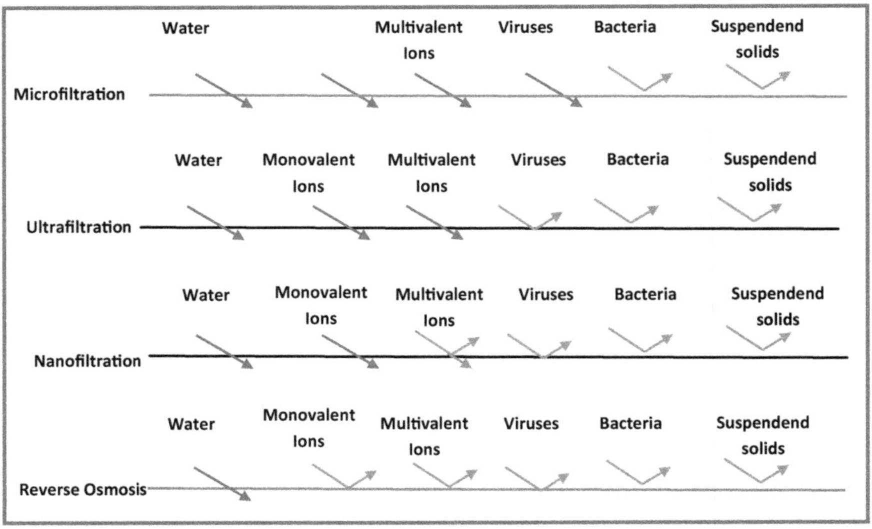

FIGURE 9.7 Membrane process characteristics.

sensing. In turn, the closed pores can be used in sonic and thermal insulation, or lightweight structural applications. The use of porous materials offers also new opportunities in such areas as coverage chemistry, guest–host synthesis and molecular manipulations and reactions for manufacture of nanoparticles, nanowires and other quantum nanostructures. The International Union of Pure and Applied Chemistry (IUPAC) (Figure 9.8) defines porosity scales as follows:

- Microporous materials 0–2-nm pores
- Mesoporous materials 2–50-nm pores
- Macroporous materials >50-nm pores

This definition, it should be noted, is somewhat in conflict with the definition of nanoscale objects, which typically have large relative porosities (> 0.4), and pore diameters between 1 and 100 nm. In order to classify porous materials according to the size of their pores the sorption analysis is one of the tools often used. This tool is based on the fact that pores of different sizes lead to totally different characteristics in sorption isotherms. The correlation between the vapor pressure and the pore size can be written as the Kelvin equation:

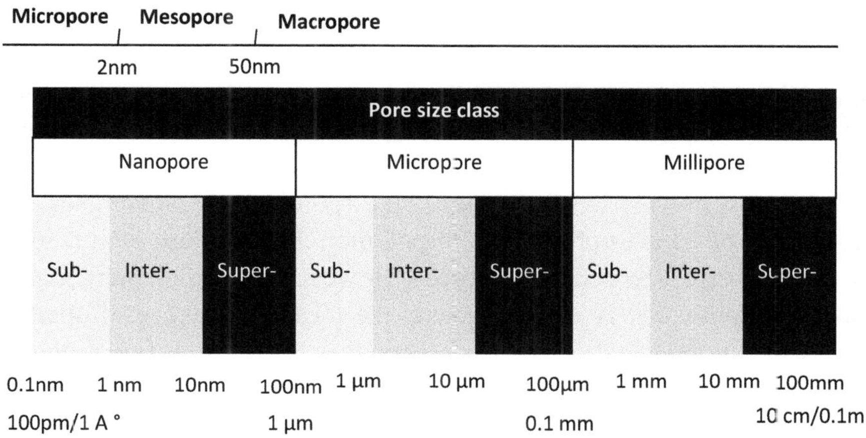

FIGURE 9.8 New pore size classification as compared with the current IUPAC nomenclature.

$$r_p\left(\frac{p}{p_0}\right)=\frac{2\gamma V_L}{RT\ln(3.\,{}^{p}\!/_{p_0}}+t\left(3.\frac{p}{p_0}\right) \tag{1}$$

Therefore, the isotherms of microporous materials show a steep increase at very low pressures (relative pressures near zero) and reach a plateau quickly. Mesoporous materials are characterized by a so-called capillary doping step and a hysteresis (a discrepancy between adsorption and desorption). Macroporous materials show a single or multiple adsorption steps near the pressure of the standard bulk condensed state (relative pressure approaches one) [22].

Nanoporous materials exuberate in nature, both in biological systems and in natural minerals. Some nanoporous materials have been used industrially for a long time. Recent progress in characterization and manipulation on the nanoscale has led to noticeable progression in understanding and making a variety of nanoporous materials: from the merely opportunistic to directed design. This is most strikingly the case in the creation of a wide variety of membranes where control over pore size is increasing dramatically, often to atomic levels of perfection, as is the ability to modify physical and chemical characteristics of the materials that make up the pores [23].

The available range of membrane materials includes polymeric, carbon, silica, zeolite and other ceramics, as well as composites. Each type of membrane can have a different porous structure, as illustrated in Figure 9.9. Membranes can be thought of as having a fixed (immovable) network of pores in which the molecule travels, with the exception of most polymeric membranes [24, 25]. Polymeric membranes are composed of an amorphous mix of polymer chains whose interactions involve mostly Van der Waals forces. However, some polymers manifest a behavior that is consistent with the idea of existence of opened pores within their matrix. This is especially true for high free volume, high permeability polymers, as has been proved by computer modeling, low activation energy of diffusion, negative activation energy of permeation, solubility controlled permeation [26, 27]. Although polymeric membranes have often been viewed as nonporous, in the modeling framework discussed here it is convenient to consider them nonetheless as porous. Glassy polymers have pores that can be considered as 'frozen' over short times scales, while rubbery polymers have dynamic fluctuating pores (or more correctly free volume elements) that move, shrink, expand and disappear [28].

Three nanotechnologies that are often used in the filtering processes and show great potential for applications in remediation are:

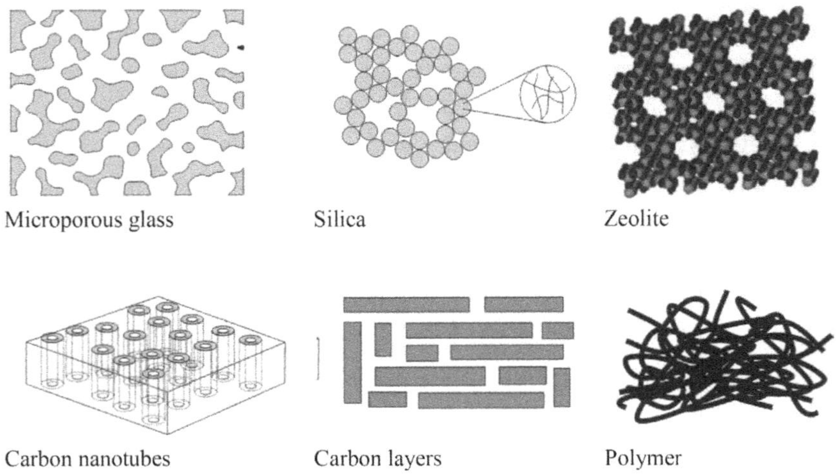

FIGURE 9.9 Porous structure within various types of membranes.

1. Nanofiltration (and its sibling technologies: reverse osmosis, ultra-filtration, and microfiltration), is a fully-developed, commercially-available membrane technology with a large number of vendors. Nanofiltration relies on the ability of membranes to discriminate between the physical size of particles or species in a mixture or solution and is primarily used for water pre-treatment, treatment, and purification). There are almost 600 companies in worldwide which offering membrane systems.
2. Electrospinning is a process utilized by the nanofiltration process, in which fibers are stretched and elongated down to a diameter of about 10 nm. The modified nanofibers that are produced are particularly useful in the filtration process as an ultra-concentrated filter with a very large surface area. Studies have found that electrospun nanofibers can capture metallic ions and are continually effective through re-filtration.
3. Surface modified membrane is a term used for membranes with altered makeup and configuration, though the basic properties of their underlying materials remain intact.

9.3.3 TYPES OF MEMBRANES

As it mentioned, membranes have achieved a momentous place in chemical technology and are used in a broad range of applications. The key property that is exploited is the ability of a membrane to control the permeation rate of a chemical species through the membrane. In essence, a membrane is nothing more than a discrete, thin interface that moderates the permeation of chemical species in contact with it. This interface may be molecularly homogeneous, that is completely uniform in composition and structure or it may be chemically or physically heterogeneous for example, containing holes or pores of finite dimensions or consisting of some form of layered structure. A normal filter meets this definition of a membrane, but, generally, the term filter is usually limited to structures that separate particulate suspensions larger than 1–10 μm [29].

The preparation of synthetic membranes is however a more recent invention which has received a great audience due to its applications [30].

Membrane technology like most other methods has undergone a developmental stage, which has validated the technique as a cost-effective treatment option for water. The level of performance of the membrane technologies is still developing and it is stimulated by the use of additives to improve the mechanical and thermal properties, as well as the permeability, selectivity, rejection and fouling of the membranes [31]. Membranes can be fabricated to possess different morphologies. However, most membranes that have found practical use are mainly of asymmetric structure. Separation in membrane processes takes place as a result of differences in the transport rates of different species through the membrane structure, which is usually polymeric or ceramic [32].

The versatility of membrane filtration has allowed their use in many processes where their properties are suitable in the feed stream. Although membrane separation does not provide the ultimate solution to water treatment, it can be economically connected to conventional treatment technologies by modifying and improving certain properties [33].

The performance of any polymeric membrane in a given process is highly dependent on both the chemical structure of the matrix and the physical arrangement of the membrane [34]. Moreover, the structural impeccability of a membrane is very important since it determines its permeation and selectivity efficiency. As such, polymer membranes should be seen as much more than just sieving filters, but as intrinsic complex structures which can either be homogenous (isotropic) or heterogeneous (anisotropic), porous or dense, liquid or solid, organic or inorganic [34, 35].

9.3.3.1 Isotropic Membranes

Isotropic membranes are typically homogeneous/uniform in composition and structure. They are divided into three subgroups, namely: microporous, dense and electrically charged membranes. Isotropic microporous membranes have evenly distributed pores (Figure 9.10a). Their pore diameters range between 0.01–10 μm and operate by the sieving mechanism. The microporous membranes are mainly prepared by the phase inversion method albeit other methods can be used. Conversely, isotropic dense membranes do not have pores and as a result they tend to be thicker than

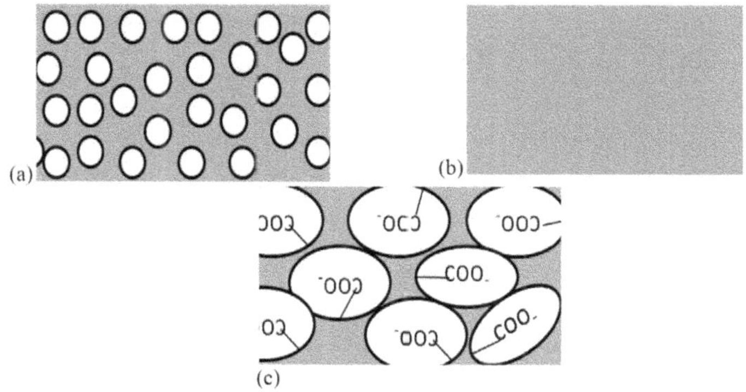

FIGURE 9.10 Schematic diagrams of isotropic membranes: (a) microporous; (b) dense; and (c) electrically charged membranes.

the microporous membranes (Figure 9.10b). Solutes are carried through the membrane by diffusion under a pressure, concentration or electrical potential gradient. Electrically charged membranes can either be porous or non-porous. However, in most cases they are finely microporous with pore walls containing charged ions (Figure 9.10c) [20, 28].

9.3.3.2 Anisotropic Membranes

Anisotropic membranes are often referred to as Loeb-Sourirajan, based on the scientists who first synthesized them [36, 37]. They are the most widely used membranes in industries. The transport rate of a species through a membrane is inversely proportional to the membrane thickness. The membrane should be as thin as possible due to high transport rates are eligible in membrane separation processes for economic reasons. Contractual film fabrication technology limits manufacture of mechanically strong, defect-free films to thicknesses of about 20 μm. The development of novel membrane fabrication techniques to produce anisotropic membrane structures is one of the major breakthroughs of membrane technology. Anisotropic membranes consist of an extremely thin surface layer supported on a much thicker, porous substructure. The surface layer and its substructure may be formed in a single operation or separately [29]. They are represented by non-uniform structures, which consist of a thin active skin layer and a highly porous support layer. The active layer enjoins the efficiency of the

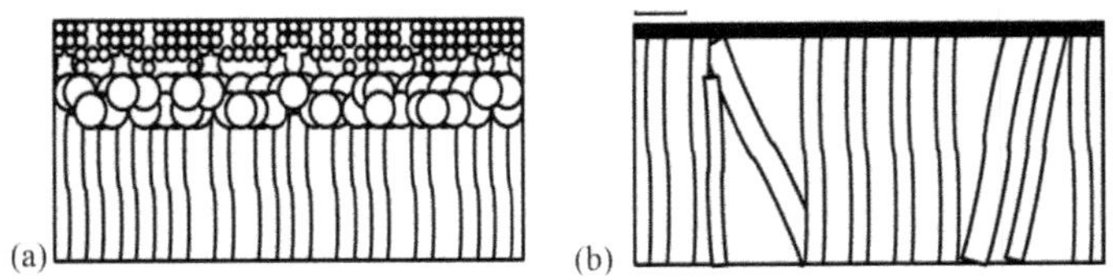

FIGURE 9.11 Schematic diagrams of anisotropic membranes: (a) Loeb-Sourirajan and (b) thin film composite membranes.

membrane, whereas the porous support layer influences the mechanical stability of the membrane. Anisotropic membranes can be classified into two groups, namely: (i) integrally skinned membranes where the active layer is formed from the same substance as the supporting layer, (ii) composite membranes where the polymer of the active layer differs from that of the supporting sub-layer [37]. In composite membranes, the layers are usually made from different polymers. The separation properties and permeation rates of the membrane are determined particularly by the surface layer and the substructure functions as a mechanical support. The advantages of the higher fluxes provided by anisotropic membranes are so great that almost all commercial processes use such membranes [29].

9.3.3.3 Porous Membrane

In Knudsen diffusion (Figure 9.12a), the pore size forces the penetrant molecules to collide more frequently with the pore wall than with other incisive species [38]. Except for some special applications as membrane reactors, Knudsen-selective membranes are not commercially attractive because of their low selectivity [39]. In surface diffusion mechanism (Figure 9.12b), the pervasive molecules adsorb on the surface of the pores so move from one site to another of lower concentration. Capillary condensation (Figure 9.12c) impresses the rate of diffusion across the membrane. It occurs when the pore size and the interactions of the penetrant with the pore walls induce penetrant condensation in the pore [40]. Molecular-sieve membranes in Figure 9.12d have gotten more attention because of their higher productivities and selectivity than solution-diffusion membranes. Molecular sieving membranes are means to polymeric membranes. They have ultra microporous (<7Å) with sufficiently small pores to barricade

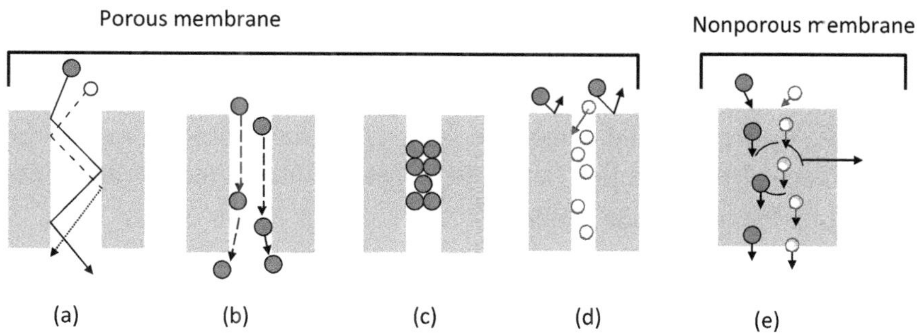

FIGURE 9.12 Schematic representation of membrane-based gas separations. (a) Knudsen-flow separation, (b) surface-diffusion, (c) capillary condensation, (d) molecular-sieving separation, and (e) solution-diffusion mechanism.

some molecules, while allowing others to pass through. Although they have several advantages such as permeation performance, chemical and thermal stability, they are still difficult to process because of some properties like fragile. Also they are expensive to fabricate.

9.3.3.4 Nonporous (Dense) Membrane

Nonporous, dense membranes consist of a dense film through which permeants are transported by diffusion under the driving force of a pressure, concentration, or electrical potential gradient. The separation of various components of a mixture is related directly to their relative transport rate within the membrane, which is determined by their diffusivity and solubility in the membrane material. Thus, nonporous, dense membranes can separate permeants of similar size if the permeant concentrations in the membrane material differ substantially. Reverse osmosis membranes use dense membranes to perform the separation. Usually these membranes have an anisotropic structure to improve the flux [29].

The mechanism of separation by non-porous membranes is different from that by porous membranes. The transport through nonporous polymeric membranes is usually described by a solution–diffusion mechanism (Figure 9.12a). The most current commercial polymeric membranes operate according to the solution–diffusion mechanism. The solution–diffusion mechanism has three steps: (1) the absorption or adsorption at the upstream boundary, (2) activated diffusion through the membrane, and (3) desorption or evaporation on the other side. This solution–diffusion mechanism is driven by a difference in the thermodynamic activities existing at the upstream and downstream faces of the membrane as well as the intermolecular forces acting between the permeating molecules and those making up the membrane material.

The concentration gradient causes the diffusion in the direction of decreasing activity. Differences in the permeability in dense membranes are caused not only by diffusivity differences of the various species but also by differences in the physicochemical interactions of the species within the polymer. The solution–diffusion model assumes that the pressure within a membrane is uniform and that the chemical potential gradient across the membrane is expressed only as a concentration gradient. This mechanism controls permeation in polymeric membranes for separations.

9.3.4 INTRODUCTION TO MORPHOLOGY AND POROSITY

Union of Pure and Applied Chemists (IUPAC) defines morphology as the "shape, optical appearance, or form of phase domains in substances, such as high polymers, polymer blends, composites, and crystals." Since this is a very broad and diffuse definition, two classes of morphology are set apart in this work. Shape and bulk morphology are distinguished, because both are very useful in the description of the porous networks. The former

TABLE 9.3 Examples of Size-Dependent Shape Morphology

Size range	Shape morphology
Nanometer	Polymer brush
	Micelle
	Microgel
	Pores
Micrometer	Powders
	polyHIPE(high internal phase emulsion) pores
Macroscopic	Beads
	Discs
	Membranes

TABLE 9.4 Examples of Cross-Link-Dependent Bulk Morphology

Content of cross-links	Bulk morphology
None	Soluble polymer
	Supported polymer brush
Low	Swellable polymer gels
High	Polymer networks
Extra-high	Hypercross-linked
	Polymers

concerns the particle size, shape and pore structure, the latter classifies the polymers by the molecular architecture of the pore walls. Polymers have the advantage that they can be prepared in almost any micro- or macroscopic size and shape. This allows extensive tuning of the shape morphology to the desired application, while keeping the bulk morphological parameters unchanged and vice versa.

Classically, porous matter is seen as a material that has voids through and through. The voids show a translational repetition in 3-D space, while no regularity is necessary for a material to be termed " porous." A common and relatively simple porous system is one type of dispersion classically described in colloid science, namely foam or, better, solid foam. In correlation with this, the most typical way to think about a porous material is as a material with gas-solid interfaces as the most dominant characteristic. This already indicates that classical colloid and interface science as the creation of interfaces due to nucleation phenomena (in this case nucleation of wholes), decreasing interface energy, and stabilization of interfaces is of elemental importance in the formation process of nanoporous materials [23, 41–43].

These factors are often omitted because the final products are stable (they are metastable). This metastability is due to the rigid character of the void-surrounding network, which is covalently cross-linked in most cases. However, it must be noticed that most of the porous materials are not stable by thermodynamic means. As soon as kinetic energy boundaries are overcome, materials start to breakdown [23].

Porous materials have been extensively exploited for use in a broad range of applications: for example, as membranes for separation and purification [44], as high surface-area adsorbents, as solid supports for sensors [45] and catalysts, as materials with low dielectric constants in the fabrication of microelectronic devices [46], and as scaffolds to guide the growth of tissues in bioengineering [47].

Porous materials occur widely and have many important applications. They can, for example, offer a convenient method of imposing fine structure on adsorbed materials.

They can be used as substrates to support catalysts and can act as highly selective sieves or cages that only allow access to molecules up to a certain size.

Food is often finely structured. Many biologically active materials are porous, as are many construction and engineering materials. Porous geological materials are of great interest; high porosity rock may contain water, oil or gas; low porosity rock may act as a cap to porous rock, and is of importance for active waste sealing.

9.3.5 POROSITY

Porosity φ is the fraction of the total soil volume that is taken up by the pore space. Thus, it is a single-value quantification of the amount of space available to fluid within a specific body of soil. Being simply a fraction of total volume, φ can range between 0 and 1, typically falling between 0.3 and 0.7 for soils. With the assumption that soil is a continuum, adopted here as in much of soil science literature, porosity can be considered a function of position.

9.3.6 POROSITY IN NATURAL SOILS

The porosity of a soil depends on several factors, including (1) packing density, (2) the breadth of the particle size distribution (polydisperse vs. monodisperse), (3) the shape of particles, and (4) cementing. Mathematically considering an idealized soil of packed uniform spheres, φ must fall between 0.26 and 0.48, depending on the packing. Spheres randomly thrown together will have φ near the middle of this range, typically 0.30 to 0.35. Sand with grains nearly uniform in size (monodisperse) packs to about the same porosity as spheres. In polydisperse sand, the fitting of small grains within the pores between large ones can reduce φ, conceivably below the 0.26 uniform-sphere minimum. Figure 9.13 illustrates this concept. The particular sort of arrangement required to reduce φ to 0.26 or less is highly improbable, however, so φ also typically falls within the 0.30–0.35 for polydisperse sands. Particles more irregular in shape tend to have larger gaps between their nontouching surfaces, thus forming media of greater porosity. In porous rock such as sand-stone, cementation or welding of particles not only creates pores that are different in shape from those of particulate media, but also reduces the porosity as solid material takes up space that would otherwise be pore space. Porosity in such a case can easily be less than 0.3, even approaching 0. Cementing material can also have the opposite effect. In many soils, clay and organic substances cement particles together into aggregates. An individual aggregate might have 0.35 porosity within it but the medium as a whole has additional pore space in the form of gaps between aggregates, so that φ can be 0.5 or greater. Observed porosities can be as great as 0.8 to 0.9 in a peat (extremely high organic matter) soil.

Porosity is often conceptually partitioned into two components, most commonly called textural and structural porosity. The textural component is the value the porosity would have if the arrangement of the particles were random, as described above for granular material without cementing. That is, the textural porosity might be about 0.3 in a granular medium.

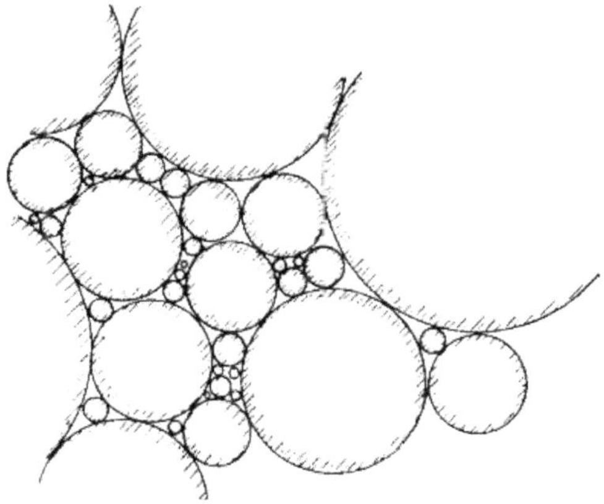

FIGURE 9.13 Dense packing of polydisperse spheres.

The structural component represents nonrandom structural influences, including macropores and is arithmetically defined as the difference between the textural porosity and the total porosity.

The texture of the medium relates in a general way to the pore-size distribution, as large particles give rise to large pores between them, and therefore is a major influence on the soil water retention curve. Additionally, the structure of the medium, especially the pervasive-ness of aggregation, shrinkage cracks, worm-holes, etc. substantially influences water retention.

9.3.7 MEASUREMENT OF POROSITY

The technology of thin sections or of tomographic imaging can produce a visualization of pore space and solid material in a cross-sectional plane. The summed area of pore space divided by total area gives the areal porosity over that plane. An analogous procedure can be followed along a line through the sample, to yield a linear porosity. If the medium is isotropic, either of these would numerically equal the volumetric porosity as defined above.

The volume of water contained in a saturated sample of known volume can indicate porosity. The mass of saturated material less the oven-dry mass of the solids, divided by the density of water, gives the volume of water. This divided by the original sample volume gives porosity.

An analogous method is to determine the volume of gas in the pore space of a completely dry sample. Sampling and drying of the soil must be conducted so as not to com-press the soil or otherwise alter its porosity. A pycnometer can measure the air volume in the pore space. A gas-tight chamber encloses the sample so that the internal gas-occupied volume can be perturbed by a known amount while the gas pressure is measured. This is typically done with a small piston attached by a tube connection. Boyle's law indicates the total gas volume from the change in pressure resulting from the volume change. This total gas volume minus the volume within the piston, connectors, gaps at the chamber walls, and any other space not occupied by soil, yields the total pore volume to be divided by the sample volume.

To avoid having to saturate with water or air, one can calculate porosity from measurements of particle density and bulk density. From the definitions of bulk density as the solid mass per total volume of soil and particle density as the solid mass per solid volume, their ratio ρ_b/ρ_p is the complement of φ, so that:

$$\varnothing = 1 - \rho_b / \rho_p \tag{2}$$

Often the critical source of error is in the determination of total soil volume, which is harder to measure than the mass. This measurement can be based on the dimensions of a minimally disturbed sample in a regular geometric shape, usually a cylinder. Significant error can result from irregularities in the actual shape and from unavoidable compaction. Alternatively, the measured volume can be that of the excavation from which the soil sample originated. This can be done using measurements of a regular geometric shape, with the same problems as with measurements on an extracted sample. Additional methods, such as the balloon or sand-fill methods, have other sources of error.

9.3.8 PORES AND PORE-SIZE DISTRIBUTION: THE NATURE OF A PORE

Because soil does not contain discrete objects with obvious boundaries that could be called individual pores, the precise delineation of a pore unavoidably requires artificial, subjectively established distinctions. This contrasts with soil particles, which are easily defined, being discrete material objects with obvious boundaries. The arbitrary criterion required to partition pore space into individual pores is often not explicitly stated when pores or their sizes are discussed. Because of this inherent arbitrariness, some scientists argue that the concepts of pore and pore size should be avoided. Much valuable theory of the behavior of the soil-water-air system, however, has been built on these concepts, defined using widely, if not universally, accepted criteria.

9.3.9 POROUS MATERIALS

Porous materials are solid forms of matter permeated by interconnected or non-interconnected pores (voids) of different kinds: channels, cavities or interstices; that can be divided into several classes.

According to the nomenclature suggested by the International Union of Pure and Applied Chemists (IUPAC), porous materials are usually classified into three different categories depending on the lateral dimensions of their pores: microporous (<2 nm), mesoporous (between 2 and 50 nm) and macroporous (>50 nm) [48].

Liquid and gaseous molecules have been known to exhibit characteristic transport behaviors in each type of porous material. For example, mass transport can be obtained via viscous flow and molecular diffusion in a macroporous material, through surface diffusion and capillary flow in a mesoporous material and by activated diffusion in a microporous material.

Pores from the nanoscopic to the macroscopic scale are generated depending on the method. A summary of selected methods that can be applied to styrene-codivinylbenzene polymers is given in Table 9.5 [49].

The internal structural architecture of the void space potentially controls the physical and chemical properties, such as reactivity, thermal and electric conductivity, as well as the kinetics of numerous transport processes. The characterization of porous materials, therefore, has been of great practical interest in numerous areas including catalysis, adsorption, purification, separation, etc., where the essential aspects for such applications are pore accessibility, narrow pore size distribution (PSD), relatively high specific surface area and easily tunable pore sizes.

Ordered porous materials are judged to be much more interesting because of the control over pore sizes and pore shapes. Their disordered counterparts exhibit high polydispersity in pore sizes, and the shapes of the pores are irregular. Ordered porous materials seem to be much more

TABLE 9.5 Overview of Methods of Generating Porosity During Polymer Synthesis

Method	Porogene	Porosity	
		Accessibility	Typical size
Foaming	Gas, solvent, supercritical solvent	Open/closed	100 nm–1 mm
Phase separation	Solvent	Open	1 μm–1 mm
High internal phase emulsion polym.	Emulsion	Open	10 μm–100 μm
Soft templating by	Molecules (solvent)	Micelles	Bicontinuous microemulsion
Hard templating by	Colloidal crystals	Porous solids	Open

homogeneous. In many cases a material possesses more than one porosity. This could be for microporous materials: an additional meso- or macroporosity caused by random grain packing

For mesoporous materials: an additional macroporosity caused by random grain packing, or an additional microporosity in the continuous network. For macroporous materials: an additional meso- and microporosity, these factors should be taken into consideration when materials are classified according to their homogeneity. A material possessing just one type of pore, even when the pores are disordered, might be more homogenous than one having just a fraction of nicely ordered pores.

Ordered porous solids contain a regularly arranged pore system and it is desired to design materials with different cylindrical, window-like, spherical or slit-like pore shapes and sizes [50].

9.3.10 PROPERTIES OF POROUS MATERIALS

There are a number of important properties of porous materials such as:

- porosity
- specific surface area
- permeability
- breakthrough capillary pressure
- diffusion properties of liquids in pores
- pore size distribution
- radial density function

9.3.11 MACROPOROUS MATERIALS AND THEIR USES

Macroporous materials are formed from the packing of monodisperse spheres (polystyrene or silica) into a three-dimensional ordered arrangement, to form face-centered cubic (FCC) or hexagonal close-packed (HCP) structures. The spaces between the packed spheres create a macroporous structure.

Glass or rubbery polymer that includes a large number of macropores (50 nm–1 μm in diameter) that persist when the polymer is immersed in solvents or in the dry state.

Macroporous polymers are often network polymers produced in bead form. However, linear polymers can also be prepared in the form of macroporous polymer beads. They swell only slightly in solvents.

Macroporous polymers can be used, for instance, as precursors for ion-exchange polymers, as adsorbents, as supports for catalysts or reagents, and as stationary phases in size-exclusion chromatography columns.

Macroporous materials have many applications in the field of engineering due to their large effective surface area, and can be used for purposes such as filters, catalysts, supports, heat exchangers, and fuel cells. Although microporous and mesoporous materials also possess large surface areas, their small pore diameters (less than 10nm) do not allow larger molecules to pass through them. Hence, macroporous materials with larger pore diameters are preferred and are of more practical use. Through colloidal crystal templating techniques, three-dimensionally ordered macroporous materials with uniform pore size can be successfully synthesized, thus improving the efficiency of transport of materials through the pores. Furthermore, photonic crystals possessing optical band-gaps can be synthesized from these macroporous structures by infiltrating the macroporous material with a precursor fluid, followed by removal of the original spheres through calcination.

Photonic crystals are materials in which the dielectric constants vary periodically in space. Due to the alternating dielectric properties, photonic crystals are hence able to control the propagation of photons, by creating a frequency (band-gap) in which light is not able to propagate. Photonic crystals themselves have great potential use in the engineering field. Due to their ability to localize photons, photonic crystals can be used as wave-guides in optical fibres, which would prove very valuable in optical communications for the transfer of information. With the advent of information technology and the need for faster, quicker and more efficient data transmission, the importance and potential of photonic crystals are ever more apparent.

9.3.12 MESOPOROUS MATERIALS

Mesoporous solids consists of inorganic or inorganic/organic hybrid units of long-range order with amorphous walls, tunable textural and structural properties with highly controllable pore geometry and narrow pore size distribution in the 2–50 nm range [51].

The pores can have different shapes such as spherical or cylindrical and be arranged in varying structures, see Figure 9.14. Some structures have pores that are larger than 50 nm in one dimension, see, for example, the two first structures in Figure 9.14, but there the width of the pore is in the meso range and the material is still considered to be mesoporous.

Mesoporous materials can have a wide range of compositions but mainly consists of oxides such as SiO_2, TiO_2, ZnO_2, Fe_2O or combinations of metal oxides, but also mesoporous carbon can be synthesized [52–55]. Most commonly is to use a micellar solution and grow oxide walls around the micelles. Both organic metal precursors such as alkoxides [56, 57] as well as inorganic salts such as metal chloride salts [53] can be used. Alternatively a mesoporous template can be used to grow another type of mesoporous material inside it. This is often used for synthesizing, for example, mesoporous carbon [55, 58, 59].

9.3.13 MICROPOROUS MATERIALS

Porous materials are networks of solid material, which contain void spaces. These materials can be further classified depending on the size of the pores present in the material. Microporous solids are materials that contain permanent cavities with diameters of less than 2 nm. Mesoporous materials contain pores ranging from 2 nm to 50 nm and macroporous materials contain pores of greater than 50 nm [60]. The field of microporous materials contains several classes, which are well known [61], including naturally occurring zeolites, activated carbons and silica. Synthetic microporous solids have recently emerged as a potentially important class of materials.

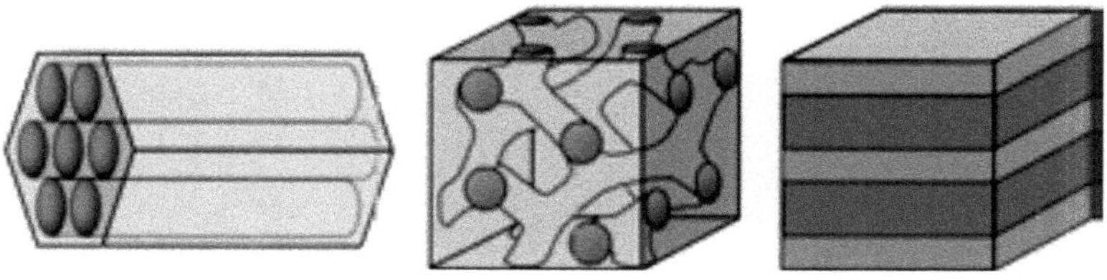

FIGURE 9.14 Different pore structures of mesoporous materials.

These include Metal Organic Frameworks (MOFs), Microporous Organic Polymers (MOPs) including Covalent Organic Frameworks (COFs) and Polymers of Intrinsic Microporosity (PIMs) [62]. It is the very large surface areas and very small pore sizes of these materials, which make them of specific interest. These two factors permit microporous materials to be useful in applications such as heterogeneous catalysis, separation chemistry, and potential uses in hydrogen or other gas storage [63, 64]. Most synthetic strategies to prepare microporous materials consist of linking together smaller units with di-topic or poly-topic functionalities in order to form extended networks much like is displayed in the general diagram of Figure 9.15.

Whether a network formed is crystalline or amorphous is generally governed by whether the covalent bonds being formed involve reversible chemistry or not. Crystalline networks are typically formed under reversible reaction conditions that allow error corrections during network formation and produce thermodynamically stable networks. These types of reactions are commonly condensation reactions [65, 66]. On the other hand, when irreversible reactions are employed such as cross couplings, the networks tend to form in a disordered manner [67, 68], resulting in amorphous materials. This is because a carbon-carbon bond is formed irreversibly under conditions such as a Sonagashira or Yamamoto couplings

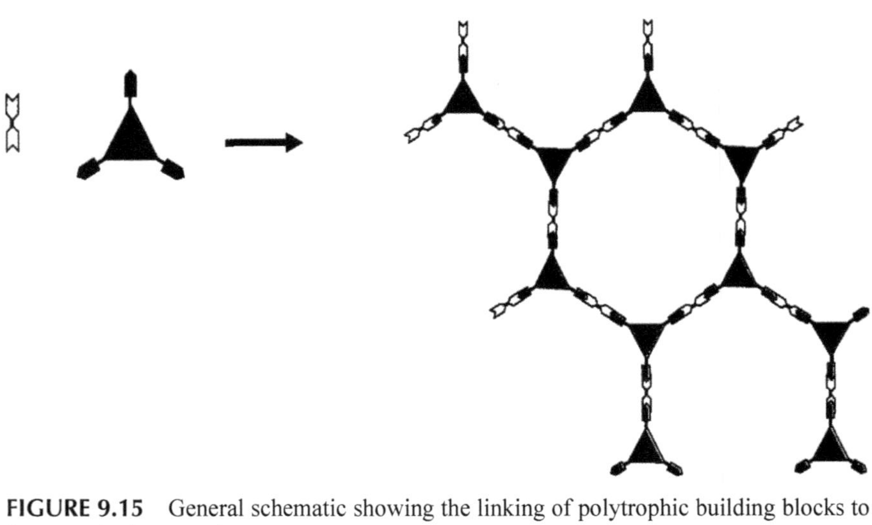

FIGURE 9.15 General schematic showing the linking of polytrophic building blocks to form synthetic networks.

resulting in amorphous networks. This is of course unless some other templating measure is taking into account while considering the reaction conditions [69].

Certain factors such as temperature, solvent and solvent-to-head-space ratio play an important role in the formation of a crystalline framework [66]. Certain solvents can be employed in order to form ordered networks by means of their ability to dissolve the monomer building blocks. If the concentration of a monomer in solution is controlled by a solvent in which it is slightly soluble, then a network is more likely to form under thermodynamic, instead of kinetic control [68]. Solvents could also be used based on their molecular size to act as templates for pores to form around [69]. While this idea of MOP/COF templating is generally understood in qualitative terms (which solvents produce crystalline networks) there has been no research into the quantitative effects (what solvent ratio is required to produce a well-structured network).

Finally, if a material is to exhibit microporosity, it must be composed of somewhat rigid building blocks in order to impart rigidity within the network and provide directionality for the formation of the network. This rigidity prevents collapse of the network upon itself and results in free volume, which becomes the pores within a framework.

9.3.14 NANOPOROUS POLYMERS

In the past few decades, nanomaterials have received substantial attention and efforts from academic and industrial world, due to the distinct properties at the nanoscale. Nanoporous materials as a subset of nanomaterials possess a set of unique properties: large specific surface volume ratio, high interior surface area, exclusive size sieving and shape selectivity, nanoscale space confinement, and specific gas/fluid permeability. Moreover, pore-filled nanoporous materials can offer synergistic properties that can never be reached by pure compounds. As a result, nanoporous materials are of scientific and technological importance and also considerable interest in a broad range of applications that include templating, sorting, sensing, isolating and releasing.

Nanoporous materials can be classified by pore geometry (size, shape, and order) or distinguished by type of bulk materials. Nanoporous materials are considered uniform if the pore size distribution is relatively narrow and the pore shape is relatively homogenous. The pores can be cylindrical, conical, slit-like, or irregular in shape. They can be well ordered with an alignment as opposed to a random network of tortuous pores. Nanoporous materials cover a wide variety of materials, which can be generally divided into inorganic, organic and composite materials. The majority of investigated nanoporous materials have been inorganic, including oxides, carbon, silicon, silicate, and metal. On the other side, polymers have been identified as materials that offer low cost, less toxicity, easy fabrication process, diverse chemical functionality, and extensive mechanical properties [70, 71].

Naturally, the success of inorganic materials to form nanoporous materials has promoted the development of analogous polymers. More importantly, advances in polymer synthesis and novel processing techniques have led to various nanoporous polymers.

9.3.15 NANOPOROUS MATERIALS CONNECTION

A whole variety of nanoporous materials in nature can be found in many different functions. The most common task for nanoporous materials in nature is to make inorganic material much lighter while preserving or improving the high structural stability of these compounds. Often, by filling the voids between inorganic matters the desired properties of the hybrid materials exceed the performances of the pure compounds by several orders of magnitude. Nanoporous materials in nature are organic-inorganic hybrids. Naturally occurring materials exhibit synergistic properties. Neither the organic material filling the void nor the inorganic network materials are able to achieve comparable performances by themselves [72–74].

It is seen that complex mechanisms are involved in the formation of these hierarchical materials. Similar to the structure motives on different length-scales cells, vesicles, supra-molecular structures, and biomolecules are involved in the structuring process of inorganic matter occurring in nature. This process is commonly known as biomineralization [73].

It is often not seen in this relation, but it will be shown later that ordered porous materials, and therefore artificial materials, are constructed according to very similar principles. A completely different area where nanoporous materials are highly important is in the lungs, where foam with a high surface area permits sufficient transfer of oxygen to the blood. Even the most recent developments in nanoporous materials, such as their application as photonic materials are already present in nature, the color of butterfly wings, for instance, originates from photonic effects [75, 76].

It can be concluded that nature applies the concept of nanoporous materials (either filled or unfilled) as a powerful tool for constructing all kinds of materials with advanced properties. So it is not surprising how much research has recently been devoted to porous materials in different areas such as chemistry, physics, and engineering. The current interests in nanoporous materials are now far behind their size-sieving properties.

9.3.16 CLASSIFICATION BY NETWORK MATERIAL

One of the most important goals in the field of nanoporous materials is to achieve any possible chemical composition in the network materials "hosting" the pores. It makes sense to divide the materials into two categories: (a) inorganic materials and (b) organic materials. Among the inorganic materials, the larger group, it could be found:

(i) Inorganic oxide-type materials. This is the field of the most commonly known porous silica, porous titania, and porous zirconia materials.
(ii) A category of its own is given for nanoporous carbon materials. In this category are the highly important active carbons and some examples for ordered mesoporous carbon materials.
(iii) Other binary compounds such as sulfides, nitrides. Into this category also fall the famous AlPO materials.
(iv) There are already some examples in addition to carbon where just one element (for instance, a metal) could be prepared in a nanoporous state. The most appropriate member of this class of materials is likely to be nanoporous silicon, with its luminescent properties [77, 78]. There are far fewer examples of nanoporous organic materials, such as polymers [79].

9.3.17 SUMMARY OF CLASSIFICATIONS

Three main criteria could be defined as:

a) Size of pores
b) Type of network material
c) State of order: ordered or disordered materials

9.3.18 ORIGIN AND CLASSIFICATION OF PORES IN SOLID MATERIALS

Solid materials have a cohesive structure, which depends on the interaction between the primary particles. The cohesive structure leads indispensably to a void space, which is not occupied by the composite particles such as atoms, ions, and line particles. Consequently, the state and population of such voids strongly depends on the inter-particle forces. The inter-particle forces are different from one system to another; chemical bonding, van der Waals force, electrostatic force, magnetic force, surface tension of adsorbed films on the primary particles, and so on. Even the single crystalline solid, which is composed of atoms or ions has intrinsic voids and defects. Therefore, pores in solids are classified into intra-particle pores and inter-particle pores (Table 9.6) [80].

9.3.19 TRAPARTICLE PORES

9.3.19.1 Intrinsic in Traparticle Pore

Zeolites are the most representative porous solids whose pores arise from the intrinsic crystalline structure. Zeolites have a general composition of Al, Si, and O, where Al-O and Si-O tetrahedral units cannot occupy the space perfectly and therefore produce cavities. Zeolites have intrinsic pores of different connectivities according to their crystal structures [81]. These pores may be named intrinsic crystalize pores. The carbon nanotube has also an intrinsic crystalline pore [82].

Although all crystalline solids other than zeolites have more or less intrinsic crystalline pores, these are not so available for adsorption or diffusion due to their isolated state and extremely small size.

TABLE 9.6 Classification of Pores From Origin, Pore Width *w*, and Accessibility

Origin and structure		
Intraparticle pore	Intrinsic intraparticle pore	Structurally intrinsic type injected intrinsic type
	Extrinsic intraparticle pore	Pure type
		Pillared type
Interparticle pore	Rigid interparticle *pore* (Agglomerated)	
	Flexible interparticle pore (Aggregated)	
Pore width		
Macropore	$w < 50$ nm	
Mesopore	2nm< w <50nm	
Micropore	W< 2 nm	
	Supermicropare, 0.7 < W 2 nm	
	Ultramicropore, W < 0.7 nm	
	(Ultrapore, w < 0.35 nm in this review)	
Accesssibility to surroundings		
Open pore	Communicating with external surface	
Closed pore	No communicating with surroundings	
Latent pore	Ultrapore and closed pore	

There are other types of pores in a single solid particle. α-FeOOH is a precursor material for magnetic tapes, a main component of surface deposits and atmospheric corrosion products of iron-based alloys, and a mineral. The α-FeOOH microcrystal is of thin elongated plate [83].

These new created intrinsic intra-particle pores should have their own name different from the intrinsic intra-particle pore. The latter is called a structurally intrinsic intra-particle pore, while the former is called an injected intrinsic intra-particle pore.

9.3.19.2 Extrinsic Intra-Particle Pore

When a foreign substance is impregnated in the parent material in advance this is removed by a modification procedure [84]. This type of pores should be called extrinsic intra-particle pores. Strictly speaking, as the material does not contain other components, extrinsic pure intra-particle pore is recommended. However, the extrinsic intra-pore can be regarded as the inter-particle pore in some cases.

It can be introduced a pore-forming agent into the structure of solids to produce voids or fissures which work as pores. This concept has been applied to layered compounds in which the interlayer bonding is very weak; some inserting substance swells the interlayer space. Graphite is a representative layered compound; the graphitic layers are weakly bound to each other by the van der Waals force [85]. If it heated in the presence of intercalants such as K atoms, the intercalants are inserted between the interlayer spaces to form a long periodic structure. K-intercalated graphite can adsorb a great amount of H_2 gas, while the original graphite cannot [86]. The interlayer space opened by intercalation is generally too narrow to be accessed by larger molecules. Intercalation produces not only pores, but also changes the electronic properties. Montmorillonite is a representative layered clay compound, which swells in solution to intricate hydrated ions or even surfactant molecules [87, 88]. Then some pillar materials such as metal hydroxides are intricate in the swollen interlayer space under wet conditions. As the pillar compound is not removed upon drying, the swollen structure can be preserved even under dry conditions.

The size of pillars can be more than several nm, being different from the above intercalants. As the graphite intercalation compounds and pillar ones need the help of foreign substances, they should be distinguished from the intrinsic intra-particle pore system. Their pores belong to extrinsic intra-particle pores. As the intercalation can be included in the pillar formation, it is better to say that both the pillared and intercalated compounds have pillared in traparticle pores [89].

9.3.20 INTER-PARTICLE PORES

Primary particles stick together to form a secondary particle, depending on their chemical composition, shape and size. In colloid chemistry, there are two gathering types of primary particles. One is aggregation and the other is agglomeration.

The aggregated particles are loosely bound to each other and the assemblage can be readily broken down. Heating or compressing the assemblage of primary particles brings about the more tightly bound agglomerate [90].

There are various interaction forces among primary particles, such as chemical bonding, van der Waals force, magnetic force, electrostatic force, and surface tension of the thick adsorbed layer on the particle surface. Sintering at the neck part of primary particles produces stable agglomerates having pores. The aggregate bound by the surface tension of adsorbed water film has flexible pores. Thus, inter-particle pores have wide varieties in stability, capacity, shape, and size, which depend on the packing of primary particles. They play an important role in nature and technology regardless of insufficient understanding. The inter-particle pores can be divided into rigid and flexible pores. The stability depends on the surroundings. Almost all inter-particle pores in agglomerates are rigid, whereas those in aggregates are flexible. Almost all sintered porous solids have rigid pores due to strong chemical bonding among the particles. The rigid inter-particle porous solids have been widely used and have been investigated as adsorbents or catalysts. Silica gel is a representative of inter-particle porous solids. Ultrafine spherical silica particles form the secondary particles, leading to porous solids [91, 92].

9.3.21 STRUCTURE OF PORES

The pore state and structure mainly depend on the origin. The pores communicating with the external surface are named open pores, which are accessible for molecules or ions in the surroundings. When the porous solids are insufficiently heated, some parts of pores near the outer shell are collapsed inducing closed pores without communication to the surroundings. Closed pores also remain by insufficient evolution of gaseous substance. The closed pore is not associated with adsorption and permeability of molecules, but it influences the mechanical properties of solid materials, the new concept of latent pores is necessary for the best description of the pore system. This is because the communication to the surroundings often depends on the probe size, in particular, in the case of molecular resolution porosimetry. The open pore with a pore width smaller than the probe molecular size must be regarded as a closed pore. Such effectively closed pores and chemically closed pores should be designated the latent pores [93]. The combined analysis of molecular resolution porosimetry and small angle X-ray scattering (SAXS) offers an effective method for separate determination of open and closed (or latent) pores, which will be described later. The porosity is defined as the ratio of the pore volume to the total solid volume [94].

The geometrical structure of pores is of great concern, but the three-dimensional description of pores is not established in less-crystalline porous solids. Only intrinsic crystalline intra-particle pores offer a good description of the structure. The hysteresis analysis of molecular adsorption isotherms and electron microscopic observation estimate the pore geometry such as cylinder (cylinder closed at one end or cylinder open at both ends), cone shape, slit shape, interstice between closed-packing spheres and inkbottle. However, these models concern with only the unit structures. The higher order structure of these unit pores such as the network structure should be taken into account. The simplest classification of the higher order structures is one-, two- and three-dimensional pores. Some zeolites and aluminophosphates have one-dimensional pores and activated carbons have basically two-dimensional slit-shaped pores with complicated network structures [95].

The IUPAC has tried to establish a classification of pores according to the pore width (the shortest pore diameter), because the geometry determination of pores is still very difficult and molecular adsorption can lead to the reliable parameter of the pore width. The pores are divided into three categories: macropores, mesopores, and micropores, as mentioned above. The fact that nanopores are often used instead of micropores should be noted.

These sizes can be determined from the aspect of N, adsorption at 77 K, and hence N_2 molecules are adsorbed by different mechanisms – multilayer adsorption, capillary condensation, and micropore filling for macropores, mesopores, and micropores, respectively. The critical widths of 50 and 2 nm are chosen from empirical and physical reasons. The pore width of 50 nm corresponds to the relative pressure of 0.96 for the N_2 adsorption isotherm. Adsorption experiments above that are considerably difficult and applicability of the capillary condensation theory is not sufficiently examined. The smaller critical width of 2 nm corresponds to the relative pressure of 0.39 through the Kelvin equation, where an unstable

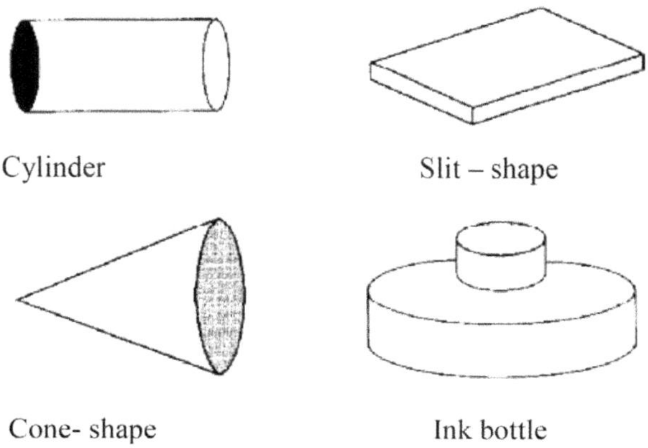

FIGURE 9.16 Pore shapes.

behavior of the N, adsorbed layer (tensile strength effect) is observed. The capillary condensation theory cannot be applied to pores having a smaller width than 2 nm. The micropores have two subgroups, namely ultra-micropores (0.7 nm) and super-micropores (0.7 nm $< w <$ 2 nm). The statistical thickness of the adsorbed N2 layer on solid surfaces is 0.354 nm. The maximum size of ultra-micropores corresponds to the bilayer thickness of nitrogen molecules, and the adsorbed N_2 molecules near the entrance of the pores often block further adsorption. The ultra-micropore assessment by N_2 adsorption has an inevitable and serious problem. The micropores are divided into two groups.

Recently the molecular statistical theory has been used to examine the limitation of the Kelvin equation and predicts that the critical width between the micropore and the mesopore is 1.3–1.7 nm (corresponding to 4–5 layers of adsorbed N_2), which is smaller than 2 nm [96].

9.3.22 POROSITY AND PORE SIZE MEASUREMENT TECHNIQUES ON POROUS MEDIA

Crushing measure the volume of the porous material, crush it to remove the void space, and remeasure the volume.

- Optically this may involve filling the pores with a material such as black wax or Wood's metal, sectioning and inspecting with a microscope or scanning electron microscope.

- Imbibition weighing before and after filling the pores with a liquid.
- Gas Adsorption measure the change in pressure as a gas is adsorbed by the sample.
- Mercury intrusion measure the volume of mercury forced into the sample as a function of pressure.
- Thermoporosimetry fill the pores with a liquid, freeze it, then measure the heat evolved as the sample is warmed, until all the liquid is melted.
- NMR cryoporometry fill the pores with a liquid, freeze it, then measure the amplitude of the NMR signal from the liquid component as the sample is warmed, until all the liquid is melted.
- Small Angle Neutron Scattering (SANS) scatter neutrons from the pores, then the smaller the dimensions of the variations in density distribution, the larger the angle through which the neutrons will be scattered.

Many of these methods give results that quite frequently differ from one another. This is often because they are in fact measuring different things some measurements are directly on the pores themselves. Others (such as mercury intrusion) are in effect measuring the necks that give access to the pores [97].

9.3.23 CARBON NANOTUBES-POLYMER MEMBRANE

Iijima discovered carbon nanotubes (CNTs) in 1991 and it was really a revolution in nanoscience because of their distinguished properties. CNTs have the unique electrical properties and extremely high thermal conductivity [98, 99] and high elastic modulus (>1 TPa), large elastic strain – up to 5%, and large breaking strain – up to 20%. Their excellent mechanical properties could lead to many applications [100]. For example, with their amazing strength and stiffness, plus the advantage of lightness, perspective future applications of CNTs are in aerospace engineering and virtual biodevices [101].

CNTs have been studied worldwide by scientists and engineers since their discovery, but a robust, theoretically precise and efficient prediction of the mechanical properties of CNTs has not yet been found. The problem is, when the size of an object is small to nanoscale, their many physical properties cannot be modeled and analyzed by using constitutive laws from traditional continuum theories, since the complex atomistic processes affect the results of their macroscopic behavior. Atomistic simulations can give more precise modeled results of the underlying physical properties. Due to atomistic simulations of a whole CNT are computationally infeasible at present, a new atomistic and continuum mixing modeling method is needed to solve the problem, which requires crossing the length and time scales. The research here is to develop a proper technique of spanning multi-scales from atomic to macroscopic space, in which the constitutive laws are derived from empirical atomistic potentials which deal with individual interactions between single atoms at the micro-level, whereas Cosserat continuum theories are adopted for a shell model through the application of the Cauchy-Born rule to give the properties which represent the averaged behavior of large volumes of atoms at the macro-level [102, 103]. Since experiments of CNTs are relatively expensive at present, and often unexpected manual errors could be involved, it will be very helpful to have a mature theoretical method for the study of mechanical properties of CNTs. Thus, if this research is successful, it could also be a reference for the research of all sorts of research at the nanoscale, and the results can be of interest to aerospace, biomedical engineering [104].

Subsequent investigations have shown that CNTs integrate amazing rigid and tough properties, such as exceptionally high elastic properties, large elastic strain, and fracture strain sustaining capability, which seem inconsistent and impossible in the previous materials. CNTs are the strongest fibers known. The Young's Modulus of SWNT is around 1TPa, which is 5 times greater than steel (200 GPa) while the density is only 1.2~1.4 g/cm^3. This means that materials made of nanotubes are lighter and more durable.

Beside their well-known extra-high mechanical properties, single-walled carbon nanotubes (SWNTs) offer either metallic or semiconductor characteristics based on the chiral structure of fullerene. They possess superior thermal and electrical properties so SWNTs are regarded as the most promising reinforcement material for the next generation of high performance structural and multifunctional composites, and evoke great interest in polymer based composites research. The SWNTs/polymer composites are theoretically predicted to have both exceptional mechanical and functional properties, which carbon fibers cannot offer [105].

9.3.23.1 Carbon Nanotubes

Nanotubular materials are important "building blocks" of nanotechnology, in particular, the synthesis and applications of CNTs [82, 106, 107]. One application area has been the use of carbon nanotubes for molecular separations, owing to some of their unique properties. One such important property, extremely fast mass transport of molecules within carbon nanotubes associated with their low friction inner nanotube surfaces, has been demonstrated via computational and experimental studies [108, 109]. Furthermore, the behavior of adsorbate molecules in nano-confinement is fundamentally different than in the bulk phase, which could lead to the design of new sorbents [110].

Finally, their one-dimensional geometry could allow for alignment in desirable orientations for given separation devices to optimize the mass transport. Despite possessing such attractive properties, several intrinsic limitations of carbon nanotubes inhibit their application in large scale separation processes: the high cost of CNT synthesis and membrane formation (by microfabrication processes), as well as their lack of surface functionality, which significantly limits their molecular selectivity [111]. Although outer-surface modification of carbon nanotubes has been developed for nearly two decades, interior modification via covalent chemistry is still challenging due to the low reactivity of the inner-surface. Specifically, forming covalent bonds at inner walls of carbon nanotubes requires a transformation from sp^2 to sp^3 hybridization. The formation of sp^3 carbon is energetically unfavorable for concave surfaces [112].

Membrane is a potentially effective way to apply nanotubular materials in industrial-scale molecular transport and separation processes. Polymeric membranes are already prominent for separations applications due to their low fabrication and operation costs. However, the main challenge for utilizing polymer membranes for future high-performance separations is to overcome the tradeoff between permeability and selectivity. A combination of the potentially high throughput and selectivity of nanotube materials with the process ability and mechanical strength of polymers may allow for the fabrication of scalable, high-performance membranes [113, 114].

9.3.23.2 Structure of Carbon Nanotubes

Two types of nanotubes exist in nature: multi-walled carbon nanotube (MWNTs), which were discovered by Iijima in 1991 [82] and SWNTs, which were discovered by Bethune et al. in 1993 [115, 116].

Single-wall nanotube has only one single layer with diameters in the range of 0.6–1nm and densities of 1.33–1.40 g/cm^3 [117] MWNTs are simply composed of concentric SWNTs with an inner diameter is from 1.5 to 15 nm and the outer diameter is from 2.5 nm to 30 nm [118]. SWNTs have better defined shapes of cylinder than MWNT, thus MWNTs have more possibilities of structure defects and their nanostructure is less stable. Their specific mechanical and electronic properties make them useful for future high strength/modulus materials and nanodevices. They exhibit low density, large elastic limit without breaking (of up to 20–30% strain before failure), exceptional elastic stiffness, greater than 1000GPa and their extreme strength which is more than twenty times higher than a high-strength steel alloy. Besides, they also posses superior thermal and elastic properties: thermal stability up to 2800°C in vacuum and up to 750°C in air, thermal conductivity about twice as high as diamond, electric current carrying capacity 1000 times higher than copper wire [119]. The properties of CNTs strongly depend on the size and the chirality and dramatically change when SWCNTs or MWCNTs are considered [120].

CNTs are formed from pure carbon bonds. Pure carbons only have two covalent bonds: sp^2 and sp^3. The former constitutes graphite and the latter constitutes diamond. The sp^2 hybridization, composed of one s orbital and two p orbitals, is a strong bond within a plane but weak between planes. When more bonds come together, they form six-fold structures, like honeycomb pattern, which is a plane structure, the same structure as graphite [121].

Graphite is stacked layer by layer so it is only stable for one single sheet. Wrapping these layers into cylinders and joining the edges, a tube of graphite is formed, called nanotube [122].

Atomic structure of nanotubes can be described in terms of tube chirality, or helicity, which is defined by the chiral vector, and the chiral angle, θ. Figure 9.17 shows visualized cutting a graphite sheet along the dotted lines and rolling the tube so that the tip of the chiral vector touches its tail.

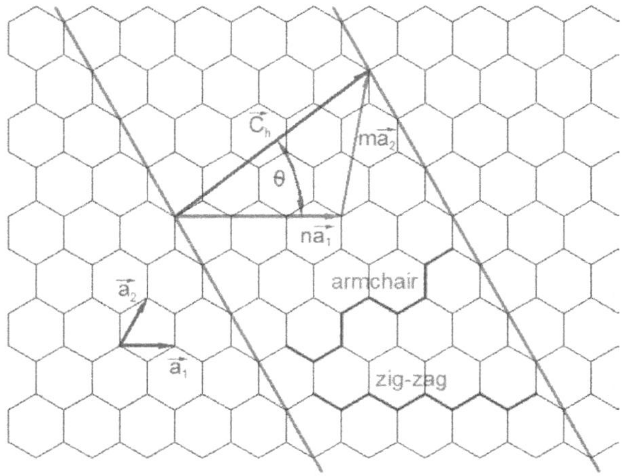

FIGURE 9.17 Schematic diagram showing how graphite sheet is 'rolled' to form CNT.

The chiral vector, often known as the roll-up vector, can be described by the following equation [123]:

$$C_h = na_1 + ma_2 \tag{3}$$

As shown in Figure 9.17, the integers (n, m) are the number of steps along the carbon bonds of the hexagonal lattice. Chiral angle determines the amount of "twist" in the tube. Two limiting cases exist where the chiral angle is at 0° and 30°. These limiting cases are referred to as zig-zag (0°) and armchair (30°), based on the geometry of the carbon bonds around the circumference of the nanotube. The difference in armchair and zig-zag nanotube structures is shown in Figure 9.18. In terms of the roll-up vector, the zig-zag nanotube is (n, 0) and the armchair nanotube is (n, n). The roll-up vector of the nanotube also defines the nanotube diameter since the inter-atomic spacing of the carbon atoms is known [105].

Chiral vector C_h is a vector that maps an atom of one end of the tube to the other. C_h can be an integer multiple a_1 of a_2, which are two basis vectors of the graphite cell. Then we have $C_h = a_1 + a_2$, with integer n and m, and the constructed CNT is called a (n, m) CNT, as shown in Figure 9.19. It can be proved that for armchair CNTs n = m, and for zig-zag CNTs m = 0. In Figure 9.19, the structure is designed to be a (4,0) zigzag SWCNT.

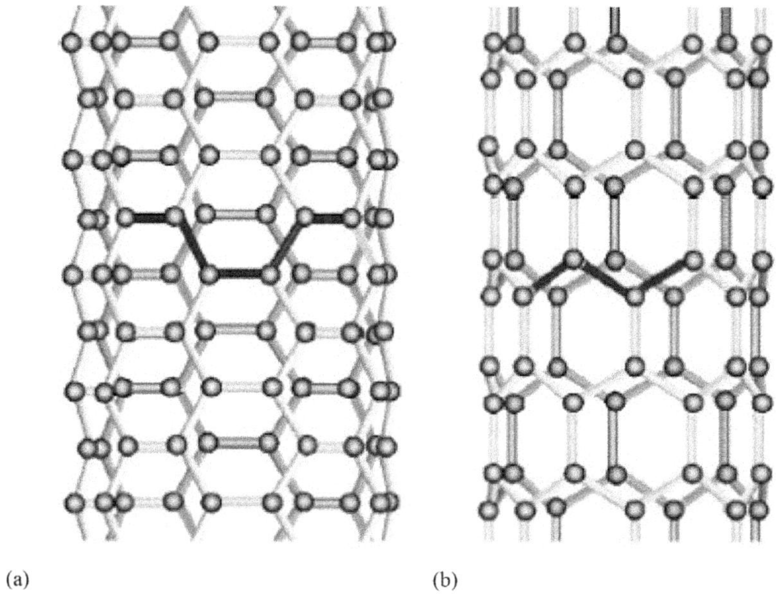

FIGURE 9.18 Illustrations of the atomic structure (a) an armchair and (b) a zig-zag nanotube.

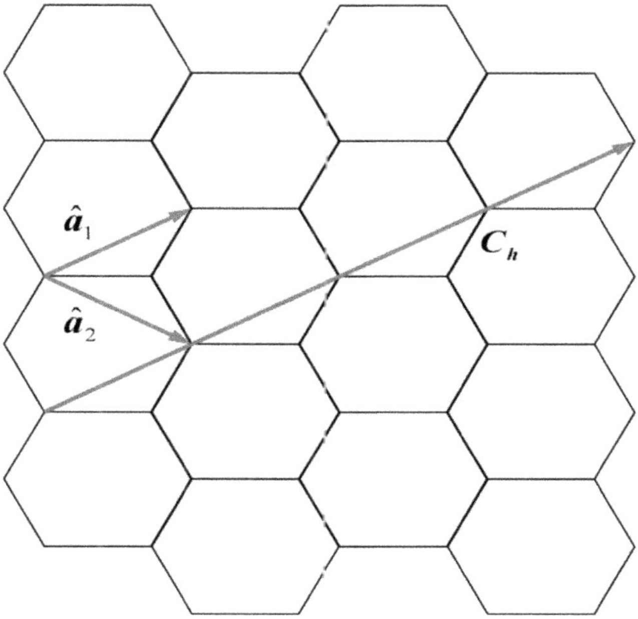

FIGURE 9.19 Basis vectors and chiral vector.

MWCNT can be considered as the structure of a bundle of concentric SWCNTs with different diameters. The length and diameter of MWCNTs are different from those of SWCNTs, which means, their properties differ significantly. MWCNTs can be modeled as a collection of SWCNTs, provided the interlayer interactions are modeled by Van der Waals forces in the simulation. A SWCNT can be modeled as a hollow cylinder by rolling a graphite sheet as presented in Figure 9.20.

If a planar graphite sheet is considered to be an undeformed configuration, and the SWCNT is defined as the current configuration, then the relationship between the SWCNT and the graphite sheet can be shown to be:

$$e_1 = G_1, e_2 = \mathrm{R}\sin\frac{\mathrm{G}_2}{\mathrm{R}}, e_3 = \mathrm{R}\cos\frac{\mathrm{G}_2}{\mathrm{R}} - \mathrm{R} \quad (4)$$

The relationship between the integer's n, m and the radius of SWCNT is given by:

$$R = a\sqrt{m^2 + mn + n^2} / 2\pi \quad (5)$$

where $a = \sqrt{3}a_0$, and a_0 is the length of a non-stretched C-C bond which is 0.142 nm [124].

As a graphite sheet can be 'rolled' into a SWCNT, we can 'unroll' the SWCNT to a plane graphite sheet. Since a SWCNT can be considered as a rectangular strip of hexagonal graphite monolayer rolling up to a cylindrical tube, the general idea is that it can be modeled as a cylindrical shell, a cylinder surface, or it can pull-back to be modeled as a plane sheet deforming into curved surface in three-dimensional space. A MWCNT can be modeled as a combination of a series of concentric SWCNTs with inter-layer intra-atomic reactions. Provided the continuum shell theory captures the deformation at the macro-level, the inner micro-structure

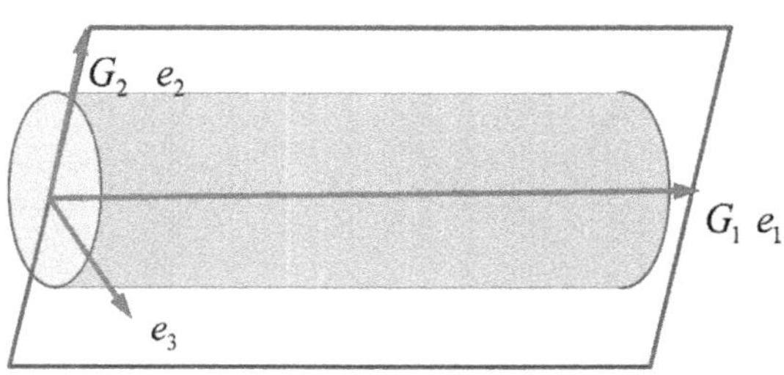

FIGURE 9.20 Illustration of a graphite sheet rolling to SWCNT.

can be described by finding the appropriate form of the potential function which is related to the position of the atoms at the atomistic level. Therefore, the SWCNT can be considered as a generalized continuum with microstructure [104].

9.3.23.3 CNT Composites

CNT composite materials cause significant development in nanoscience and nanotechnology. Their remarkable properties offer the potential for fabricating composites with substantially enhanced physical properties including conductivity, strength, elasticity, and toughness. Effective utilization of CNT in composite applications is dependent on the homogeneous distribution of CNTs throughout the matrix. Polymer-based nanocomposites are being developed for electronics applications such as thin-film capacitors in integrated circuits and solid polymer electrolytes for batteries. Research is being conducted throughout the world targeting the application of carbon nanotubes as materials for use in transistors, fuel cells, big TV screens, ultra-sensitive sensors, high-resolution Atomic Force Microscopy (AFM) probes, super-capacitor, transparent conducting film, drug carrier, catalysts, and composite material. Nowadays, there are more reports on the fluid transport through porous CNTs/polymer membrane.

9.3.23.4 Structural Development in Polymer/CNT Fibers

The inherent properties of CNT assume that the structure is well preserved (large-aspect-ratio and without defects). The first step toward effective reinforcement of polymers using nano-fillers is to achieve a uniform dispersion of the fillers within the hosting matrix, and this is also related to the as-synthesized nano-carbon structure. Secondly, effective interfacial interaction and stress transfer between CNT and polymer is essential for improved mechanical properties of the fiber composite. Finally, similar to polymer molecules, the excellent intrinsic mechanical properties of CNT can be fully exploited only if an ideal uniaxial orientation is achieved. Therefore, during the fabrication of polymer/CNT fibers, four key areas need to be addressed and understood in order to successfully control the micro-structural development in these composites. These are: (i) CNT pristine structure, (ii) CNT dispersion, (iii) polymer/CNT interfacial interaction and (iv) orientation of the filler and matrix molecules (Figure 9.21). Figure 9.21 Four major factors affecting the micro-structural development in polymer/CNT composite fiber during processing [125].

Achieving homogenous dispersion of CNTs in the polymer matrix through strong interfacial interactions is crucial to the successful development of CNT/polymer nanocomposite [126]. As a result, various chemical or physical modifications can be applied to CNTs to improve its dispersion and compatibility with polymer matrix. Among these approaches acid treatment is considered most convenient, in which hydroxyl and carboxyl groups generated would concentrate on the ends of the CNT and at defect sites, making them more reactive and thus better dispersed [127, 128].

The incorporation of functionalized CNTs into composite membranes are mostly carried out on flat sheet membranes [129, 130]. For considering the potential influences of CNTs on the physicochemical properties of dope solution [131] and change of membrane formation route originated from various additives [132], it is necessary to study the effects of CNTs on the morphology and performance.

9.3.23.5 General Fabrication Procedures for Polymer/CNT Fibers

In general, when discussing polymer/CNT composites, two major classes come to mind. First, the CNT nano-fillers are dispersed within a polymer

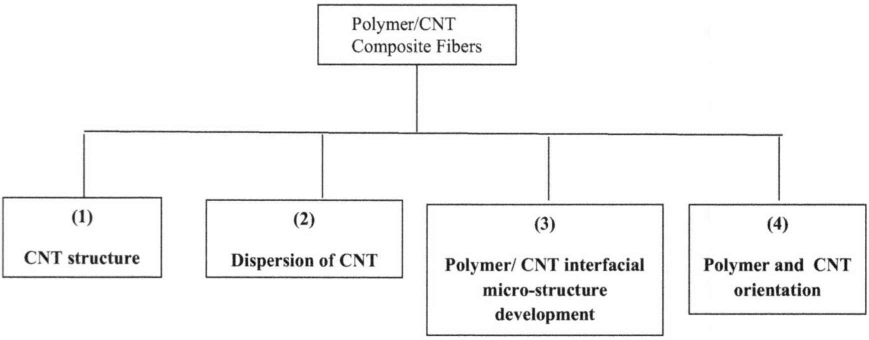

FIGURE 9.21 Four major factors affecting the micro-structural development in polymer/CNT composite fiber during processing.

at a specified concentration, and the entire mixture is fabricated into a composite. Secondly, as grown CNT are processed into fibers or films, and this macroscopic CNT material is then embedded into a polymer matrix [133]. The four major fiber-spinning methods (Figure 9.22) used for polymer/CNT composites from both the solution and melt include dry-spinning [134], wet-spinning [135], dry-jet wet spinning (gel-spinning), and electrospinning [136]. An ancient solid-state spinning approach has been used for fabricating 100% CNT fibers from both forests and aero gels. Irrespective of the processing technique, in order to develop high-quality fibers many parameters need to be well controlled.

All spinning procedures generally involve:

(i) fiber formation;
(ii) coagulation/gelation/solidification;
(iii) drawing/alignment.

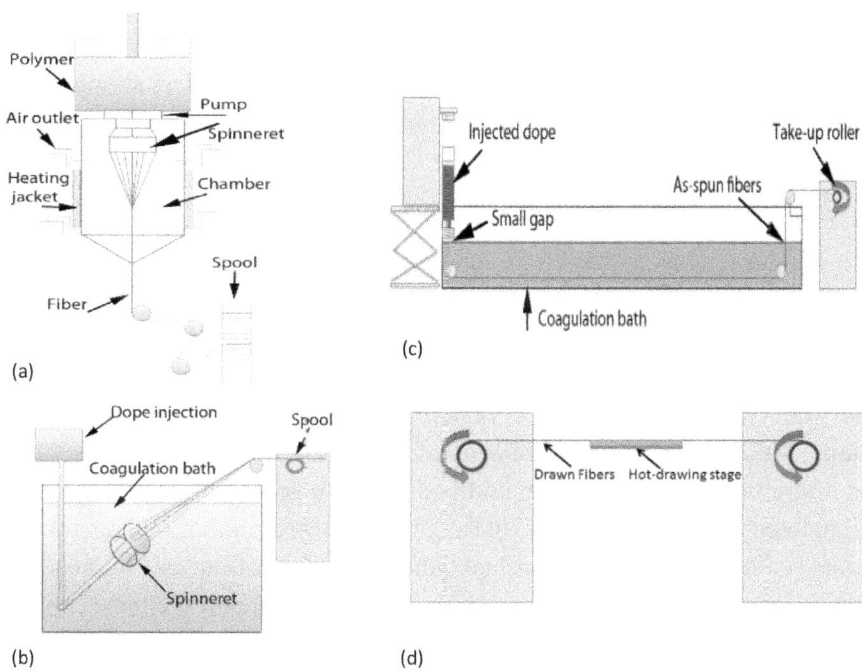

FIGURE 9.22 Schematics representing the various fiber processing methods (a) dry-spinning; (b) wet-spinning; (c) dry-jet wet or gel-spinning; and (d) post-processing by hot-stage drawing.

For all of these processes, the even dispersion of the CNT within the polymer solution or melt is very important. However, in terms of achieving excellent axial mechanical properties, alignment and orientation of the polymer chains and the CNT in the composite is necessary. Fiber alignment is accomplished in post-processing such as drawing/annealing and is key to increasing crystallinity, tensile strength, and stiffness [137].

9.3.24 *FILTER APPLICATIONS*

Nonwovens composed of fibers made from glass, paper or polymers are highly porous membranes – the total porosity typically being of the order of 80 to 95% – which can be used to remove solid particles, dust particles, aerosols, fine fluid droplets from a stream either composed of a gas or a fluid. Water filtration is a topic that is of enormous importance worldwide. Air filtration is highly important for a broad range of industrial applications including power plants, and the same holds for fuel filtration – a must in modern car engines – or coalescence filtration of gasoline for airplanes. Typical high-efficiency filter requirements are that the filters should capture all fluid or solid particles, respectively, surpassing a specified size and that the capture probability should be as high as possible, say in the range of 99 or 99.9% [138–141].

9.3.24.1 Antimicrobial Air Filter

It is well known that heating, ventilating, and air conditioning (HVAC) air filters usually operated in dark, damp, and ambient temperature conditions, which is susceptible for bacterial, mold, and fungal attacks, resulting in unpredictable deterioration and bad odor. To solve this problem, functionalization of the surface of filtering media with antimicrobial agents for long-lasting durable antimicrobial functionality is of current interest. In 2007, Jeong and Youk [142] explored the electrospun polyurethane cationomer (PUCs) nanofiber mats with different amounts of quaternary ammonium groups in antimicrobial air filter. They found that PUCs exhibited very strong antimicrobial activities against Staphylococcus aureus and Escherichia coli. Ramakrishna and co-workers [143] induced the silver nanoparticles based on different electrospun polymer [cellulose acetate (CA); polyacrylonitrile (PAN); and polyvinylchloride (PVC)] nanofiber for antimicrobial functionality owing to the remarkable antimicrobial ability of silver ions and silver compounds.

9.3.24.2 Basic Processes Controlling Filter Efficiencies

It is helpful at this stage to recall some basic processes controlling filter efficiencies in general, that is, to a first approximation independent of the fiber diameters [144, 145]. What is known for conventional nonwovens composed of fibers with diameters well in the 10–100 micrometer range is that the filter efficiency is controlled by various types of capture processes happening within the nonwoven as the gas/fluids carrying particles pass through their pores. These basic processes are depicted in Figure 9.23.

The first process to be considered is the interception. Particles following the gas stream around the fiber as depicted in Figure 9.23a are intercepted by the fiber surfaces if the particles pass the fibers at a distance not larger than the particle diameter. It is obvious that larger particles tend to enhance the probability for such an interception.

The second process of importance is the impaction, as shown schematically in Figure 9.23b. Particles do not follow in this case the deflection of the gas stream due to the neighborhood of the fiber as a solid object but because of inertia effects follow the original path. This in turn causes the particle to impact on the fiber surface. Impaction tends to grow in importance as the flow velocity of the gas increases.

Finally, diffusion plays a role in controlling the capture efficiency. Here, particles carried by streamlines that pass the fiber at sufficient distance not to cause a direct interception nevertheless come into contact with the fiber surface because of diffusional motions, as depicted in Figure 9.23c. Diffusion tends to be of importance for smaller particles and low flow velocities.

Figure 9.24 gives a survey on the regimes in a flow-rate/particle-diameter diagram in which either the diffusion, the interception or finally the impaction dominate.

So, at low particle sizes diffusion more or less dominates the control of the filter efficiency, particularly for small flow velocities, whereas the

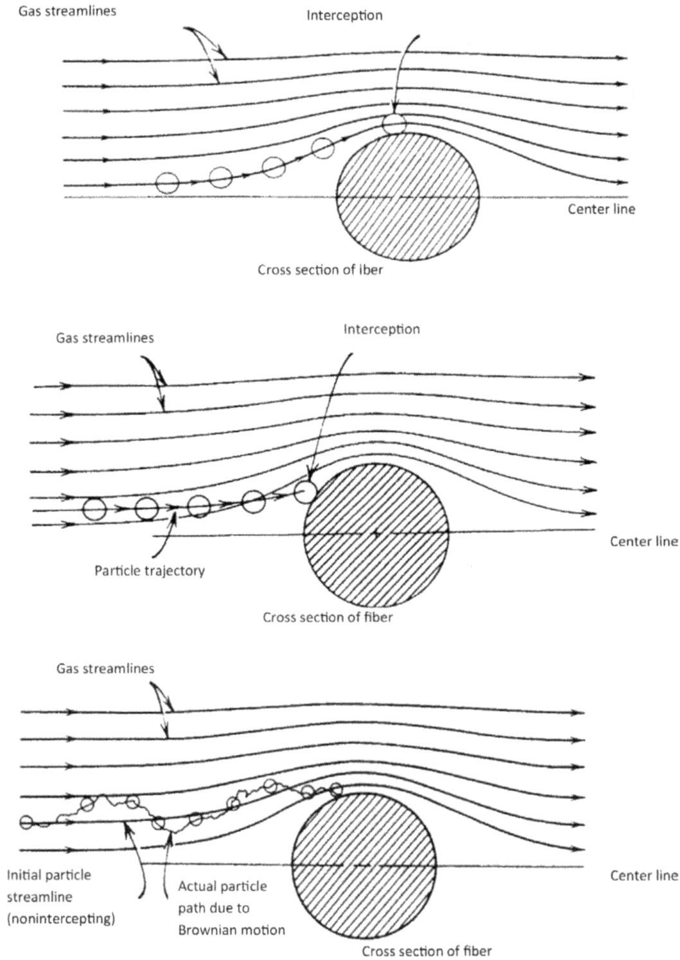

FIGURE 9.23 Molecular processes contributing to filtering effects: (a) interception, (b) impaction, (c) diffusion.

impaction dominates for large flow velocities and particle sizes, with the three processes contributing in different ratios at intermediate particle sizes and flow velocities.

In recent years, membrane separations have been applied in various industries, such as chemical, food, pharmaceutical, automobile-manufacturing, and metal-finishing industries. As the most popular membranes, polymeric membranes still have an inherent drawback—the

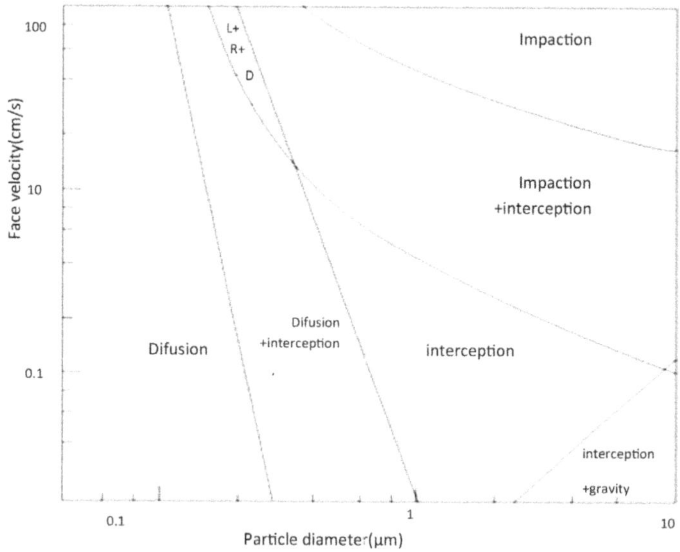

FIGURE 9.24 Survey on the regimes in a flow rate/particle diameter diagram in which either the diffusion, the interception or finally the impaction dominate diffusion.

tradeoff effect between permeability and selectivity, which means that more permeable membranes are generally less selective and vice versa. Hybrid membrane incorporating both organic and inorganic components is a convenient and efficient approach to avoid the tradeoff effect. Membranes commonly consist of a porous support layer with a thin dense layer on top that forms the actual membrane. Several researches are investigating the effects of incorporation of CNTs to develop mixed matrix membranes. In these membranes, the choice of both these components is a problem of materials selection, and also involves several fundamental issues, such as polymer-chain rigidity, free volume, and the altered interface – all of which influence transport through the membrane.

9.4 NANOTEXTILE AND TISSUE ENGINEERING FROM A BIOLOGICAL PERSPECTIVE

Nanofibers have yield potential applications in areas such as filtration, recovery of metal ions, drug release, dentistry, tissue engineering, catalysts and enzyme carriers, wound healing, protective clothing, cosmetics, biosensors, medical implants and energy storage. Improvement of catalytic efficiency of immobilized enzymes via materials engineering is demonstrated through the preparation of bioactive nanofibers. The nanofibers are produced by electrospinning, can be followed by the chemical attachment of a model enzyme. On the other hand, from a biological perspective, almost all human tissues and organs are deposited in some kind of nanofibrous form or structure.

9.4.1 CATALYST AND ENZYME CARRIERS

A carrier for catalyst in chemistry and biology is used to preserve high catalysis activity, increase the stability, and simplify the reaction process. An inert porous material with a large surface area and high permeability to reactants could be a promising candidate for efficient catalyst carriers. Using an electrospun nanofiber mat as catalyst carrier, the extremely large surface could provide a huge number of active sites, thus enhancing the catalytic capability. The well-interconnected small pores in the nanofiber mat warrant effective interactions between the reactant and catalyst, which is valuable for continuous flow chemical reactions or biological processes. The catalyst can also be grafted onto the electrospun nanofiber surface via coating or surface modification [1].

9.4.1.1 Catalysis

It is well known that nanostructured materials have opened new possibilities for creating and mastering nanoobjects for novel advanced catalytic materials. In general, catalysis is a molecular phenomenon and the reaction occurs on an active site. A crucial step in catalysis is how to remove and recycle the catalyst after the reaction. The immobilization of catalysts in materials with large surface area advances an interesting solution to this problem. Taking the large surface area and high porosities, electrospun nanofibers, as a novel catalysts or supports for catalysts, have been widely investigated in catalytic field.

9.4.1.2 Electrochemical Catalysts

Scientists investigated the electrochemical catalytical properties based on Pd/polyamide (Pd/PA6) electrospun nanofiber mats for the oxidation of ethanol in alkaline medium in which Pd/PA6 was directly used as electrocatalytic electrodes.

Simultaneously, other scientists also explored the electrochemical catalytic properties based on electrospun Pt and PtRh nanowires for dehydrogenative oxidation of cyclohexane to benzene (Figure 9.25). In contrast to the conventional Pt nanoparticle catalysts (e.g., carbon/Pt or Pt black), Pt and PtRh electrospun nanowires electrocatalysts exhibited higher catalytic

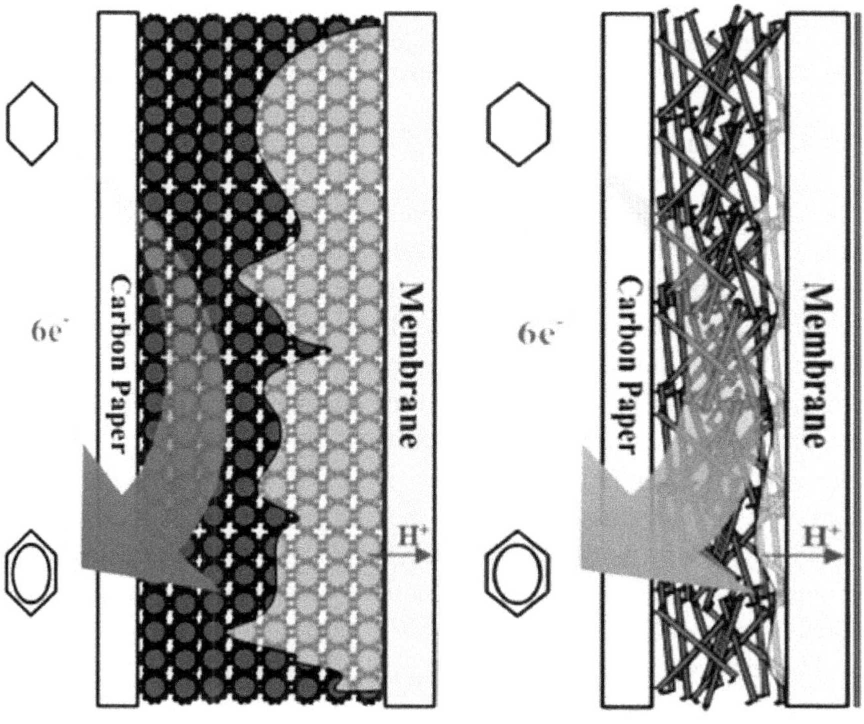

FIGURE 9.25 Schematic illustration for cyclohexane electrooxidation over the nanoparticles (a) and nanowire (b) catalysts.

activities with the same metal loading amount. Furthermore, PtRh nanowires performed the best catalytic activities with a maximum power density of ca. 23 $mWcm^{-2}$. Such higher electrocatalytic performances than Pt nanoparticles are attributed to the inherent physicochemical and electrical properties of 1D nanostructures.

1. Nanowire catalysts could provide facile pathways for the electron transfer by reducing the number of interfaces between the electrocatalysts, whereas the nanoparticles are likely to impose more impedance for electrons to transfer particle to particle.
2. Adding Rh to Pt to form the PtRh alloy can facilitate the adsorption/desorption properties of benzene and cyclohexane along with the modification of the C–H bond breaking ability.
3. The rougher morphology of PtRh nanowires comprise of small nanoparticles, which can provide high catalytic area.

9.4.1.3 Catalysis

The combination of nanofibers and catalysis seems, at first, to be rather odd. However, considering homogeneous catalysis as a first example it is well known that a huge problem is the removal and recycling of the catalytic agent after the reaction, either from the reaction solution or from the product. Complex separation methods involving in some cases several processing steps have been used for this purpose. The implantation of homogeneous or also heterogeneous catalysts into nanofibers poses an interesting solution for these problems. Now of course, the very nature of homogeneous reactions requires that the catalyst is molecularly dispersed in the same phase as the reaction compounds, so that these compounds come into intimate contact via diffusional motions, thus allowing for the reaction to proceed. The common phase is generally a solution or a melt.

However, considering nanofibers made from polymers in which the catalyst is molecularly dispersed it is well known that smaller but also larger molecules can perform surprisingly rapid diffusional processes in polymer matrices in the amorphous phase, in a partially crystalline phase, above and even below the glass-transition temperature. So, it is highly probable that reaction compounds that are dispersed in a solution or melt surrounding the nanofibers with catalysts dispersed in them can diffuse into the fiber matrix, make contact with the catalyst via diffusion.

Finally, the product molecules diffuse out of the fiber again. In fact, experiments to be discussed in the following have shown that this actually is the case.

One problem that has to be solved is to keep the catalyst within the fiber, despite allowing it to diffuse in the polymer matrix. By choosing the nature of the polymer carrier appropriately to induce specific interactions between carrier and catalysts, by attaching the catalyst via flexible spacers to the polymer backbone, one is able to achieve this goal.

The reaction mixture can circulate around the fibers, as is the case, for example, in the continuously working microreaction technique, or the fibers fixed on a carrier can be immersed repeatedly into a reaction vessel to catalyze the content of the vessel. In addition to the short diffusion distances within nanofibers, the specific pore structures and high surface areas of nanofiber nonwovens allow a rapid access of the reaction components to the catalysts and of the products back into the reaction mixture.

For homogeneous catalysis, systems consisting of core-shell nanofibers combined with proline and $Sc(OTf)_3$ ($TfO = CF_3SO_3$) catalysts were fabricated, for instance, by template methods described above (TUFT (tubes by fiber template process) method). In contrast to conventional catalysis in homogeneous solution or in microemulsions, for which the conversion is 80%, the fiber systems achieved complete conversion in the same or shorter reaction times. The fibers can be used several times without loss of activity. Furthermore, nanofibers were used as carriers for enzymes, where the enzymes were either chemically attached to the electrospun fibers or directly dispersed in the nanofibers during the electrospinning process. High catalyst activities were reported in this case as well. Current activity will certainly lead to a broad range of catalytic systems.

The use of polymer nanofibers in heterogeneous catalysis was analyzed for nanofibers loaded with monometallic or bimetallic nanoparticles (such as Rh, Pt, Pd, Rh/Pd, and Pd/Pt) has been reported in the literature. These catalyst systems can be applied in hydrogenation reactions, for example. To fabricate such fibrous catalyst systems, polymer nanofibers are typically electrospun from solutions containing metal salts (such as $Pd(OAc)_2$) as precursors. In the next step, the salts incorporated in the fibers are reduced, either purely thermally in air or in the presence of a reducing agent such as H_2 or hydrazine). The Pd nanoparticles formed have diameters in the range of 5–15 nm, depending on the fabrication method. The catalytic properties of these mono- or bimetallic nanofiber catalysts were investigated in several model hydrogenations, which demonstrated that the catalyst systems are highly effective.

9.4.1.4 Enzymes

Chemical reactions using enzymes as catalysts have high selectivity and require mild reaction conditions. For easy separation from the reaction solution, enzymes are normally immobilized with a carrier. The immobilization efficiency mainly depends on the porous structure and enzyme-matrix interaction. To immobilize enzyme on electrospun nanofibers, many approaches have been used, including grafting enzyme on fiber surface, physical adsorption, and incorporating enzyme into nanofiber via electrospinning followed by crosslinking reaction.

To graft enzymes on nanofiber surface, the polymer used should possess reactive groups for chemical bonding. In some studies, polymer blends containing at least one reactive polymer were used. The immobilized enzymes normally showed a slightly reduced activity in aqueous environment compared with the un-immobilized native counterpart, but the activity in non-aqueous solution was much higher. For example, α-chymotrypsin was used as a model enzyme to bond chemically on the surface of electrospun PS nanofibers. The enzyme was measured to cover over 27.4% monolayer of the nanofiber surface, and the apparent hydrolytic activity of the enzyme-loaded was 65% of the native enzyme, while the activity in non-aqueous solution was over 3 orders of magnitude higher than that of its native enzyme under the same condition. In another study using PAN nanofibers to immobilize lipase, the tensile strength of the nanofiber mat was improved after lipase immobilization, and the immobilized lipase retained >90% of its initial reactivity after being stored in buffer at 30°C for 20 days, whereas the free lipase lost 80% of its initial reactivity. Also the immobilized lipase still retained 70% of its specific activity after 10 repeated reaction cycles. In addition, the immobilized enzyme also showed improved pH and thermal stabilities. Ethylenediamine was used to modify PAN nanofiber mat to introduce active and hydrophilic groups, followed by a chitosan coating for improvement of biocompatibility.

Enzymes were incorporated into nanofibers via electrospinning, and subsequent crosslinking the enzymes incorporated effectively prevented their leaching. In the presence of PEO or PVA, casein and lipase were electrospun into ultra-thin fibers. After crosslinking with 4,4′-methylenebis(phenyl diisocyanate) (MDI), the fibers became insoluble, and the lipase encapsulated exhibited 6 times higher hydrolysis activity towards olive oil than that of the films cast from the same solution. The cross-linked enzymes in nanofibers showed very high activity and stability. For example, the immobilized α-chymotrypsin in a shaken buffer solution maintained the same activity for more than two weeks.

In addition to chemical bonding, the enzymes were also applied onto nanofibers simply via physical adsorption. Polyacrylonitriles-2-methacryloyloxyethyl phosphoryl choline (PANCMPC) nanofiber was reported to have high biocompatibility with enzymes because of the formation of phospholipid microenvironment on the nanofiber surface. Lipase on the nanofibers showed a high immobilization rate, strong specific activity and good activity retention.

9.4.1.5 Photocatalysis

The increasing industrial needs and growing urbanization have led to water scarcity issues around the globe and the wastewater produced has to be treated for re-utilization of clean water in daily activities. Among the various techniques, the heterogeneous photocatalysis system is an effective method for treating wastewater and photodegrading organic pollutants. The semiconductor metal oxides have been used as photocatalysts where upon irradiation of sunlight, create electron-hole pairs, which in turn produce radicals in different pathways as shown in Figure 9.26 [14].

The photocatalytic mechanism is as follows: upon irradiation semiconductor metal oxides eject an electron from the valance band to the conduction band, thereby leaving behind a hole in the valence band. The generated electrons and holes produce superoxide radicals to degrade the pollutants by reacting with chemisorbed oxygen on the catalyst surface and oxygen in the aqueous solution [15, 16].

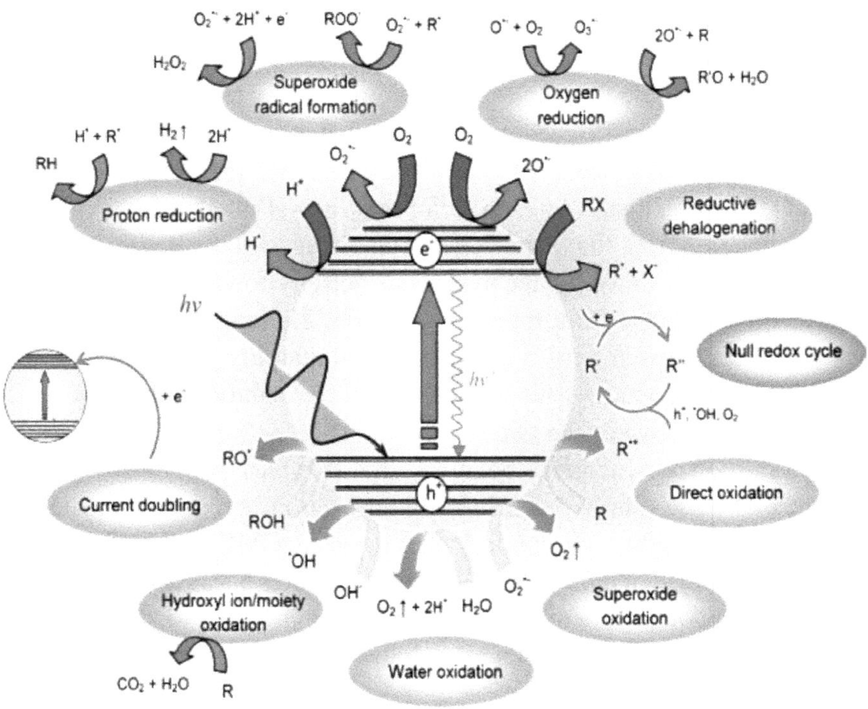

FIGURE 9.26 Possible photocatalytic mechanism of generating free radicals in the conduction band (CB) and valence band (VB) of semiconducting metal oxides.

Semiconducting metal-oxides such as TiO_2, ZnO, Fe_2O_3, WO_3, Bi_2WO_6, CuO and many more have been widely used as oxidative photocatalysts for effective removal of industrial pollutants and wastewater treatment [17, 18].

Scientists synthesized novel rice grain- shaped TiO_2 mesostructures by electrospinning and observed the phase change with increasing temperature from 500°C to 1000°C. A comparison study was performed on the photocatalytic activity of the TiO_2 rice grain and P-25. They observed an enhanced photocatalytic activity on alizarin red dye in rice grain-shaped TiO_2 which was due to its single crystalline nature and larger surface area than P-25. Scientists synthesized anatase TiO_2 nanofibers by utilizing a simple electrospinning technique and were able to grow high dense rutile TiO_2 nanorods along the fibers using hydrothermal treatment. The nanofibril-like morphology of TiO_2 nanorods/nanofibers with rutile and anatase phase was able to degrade the rhodamine-6G effectively under UV radiation. Researchers adopted the core/shell technique for synthesizing hollow mesoporous TiO_2 nanofibers with a larger surface area of around 118 m^2/g.

9.4.2 MEDICINAL APPLICATIONS FOR ELECTROSPUN NANOFIBERS

9.4.2.1 Nanotechnology and Medicinal Applications in General

It is obvious that the combination of nanoscience/nanotechnology with medicine makes a lot of sense for many reasons. One major reason is that the nanoscale is a characteristic biological scale, a scale related directly to life. DNA strains, globular proteins such as ferricins, viruses are all on this scale. For instance, the tobacco mosaic virus is actually a nanotube. The dimensions of bacteria and of cells tend to be already in the micrometer range, but important subunits of these objects such as the membranes of cells have dimensions in the nm scale.

So there are certainly good reasons for addressing various types of medicinal problems on the basis of nanoscience, of nanostructures. Scaffolds used for engineering tissues such as bone or muscle tissues may be composed of nanofibers mimicking the extracellular matrix (ECM), nanoscalar carriers for drugs to be carried to particular locations within the body to be released locoregionally rather than systemic are examples. Wound healing exploiting fibrillar membranes with a high porosity and pores with diameters in the nanometer scale, thus allowing transport of fluids, gases from and to the wound yet protecting it from bacterial infections, are further examples for the combination of nanoscience and medicine with the focus here on nanofibers.

A highly interesting example along this line certainly concerns inhalation therapy. The concept is to load specific drugs onto nanorods with a given length and diameter rather than onto spherical objects such as aerosols as already done today. The reasoning is that such spherical particles tend to become easily exhaled so that frequently only a minor part of them can become active in the lung. Furthermore, the access to the lung becomes limited with increasing volume of these particles.

However, rather long fibers are known to be able to penetrate deeply into the lung. The reason is that the aerodynamic radius controls this process with the aerodynamic radius of rods being controlled mainly by the diameter and only weakly by the length, as detailed later in more detail. So, inhalation therapy based on nanorods as accessible via nanofibers offers great benefits.

In the following, different areas where nanofibers and nanorods for that matter can contribute to problems encountered in medicine will be discussed.

9.4.2.2 An Overview

Regenerative medicine combines the principles of human biology, materials science, and engineering to restore, maintain or improve a damaged tissue function. Regenerative medicine is divided into cell therapy or "cell transplantation" and "tissue engineering."

The National Institute of Biomedical Imaging and Bioengineering (NIBIB) defines tissue engineering as "a rapidly growing area that seeks to create, repair and/or replace tissues and organs by using combinations of cells, biomaterials, and/or biologically active molecules."

Tissue engineering has emerged through a combination of many developments in biology, material science, engineering, manufacturing and medicine. Tissue engineering involves the design and fabrication of three-dimensional substitutes to mimic and restore the structural and functional properties of the original tissue. The term 'tissue engineering' is loosely defined and can be used to describe not only the formation of functional tissue by the use of cells cultured on a scaffold or delivered to a wound site, but also the induction of tissue regeneration by genes and proteins delivered in vivo.

Cell transplantation is performed when only cell replacement is required. However, in tissue engineering, the generated tissue should have similar properties to the native tissue in terms of biochemical activity, mechanical integrity and function. This necessitates providing a similar biological environment as that in the body for the cells to generate the desired tissue. Figure 9.27 summarizes the important steps in tissue engineering. First, cells are harvested from the patient and are expanded in cell culture medium. After sufficient expansion, the cells are seeded into a porous scaffold along with signaling molecules and growth factors that can promote cell growth and proliferation. The cell-seeded scaffold will be then placed into a bioreactor before being implanted into the patient's body. As it is evident in Figure 9.27 three major elements of tissue engineering include [28]:

(a) cells,
(b) scaffolds, and
(c) bioreactors.

A. Cells

Cells are the building block of all tissues. Therefore, choosing the right cell source with no contamination that is compatible with the recipient's immune system is the critical step in tissue engineering. Stem cells are employed as the main source of cells for tissue engineering and are taken from autologous, allogeneic or xenogeneic sources for different applications. Stem cells are divided into the following three groups: embryonic stem cells (ESCs), induced pluripotent stem cells, and adult stem cells. ESCs are isolated from the inner cell mass of pre-implantation embryos. These cells are considered pluripotent since they can differentiate into almost any of the specialized cell types.

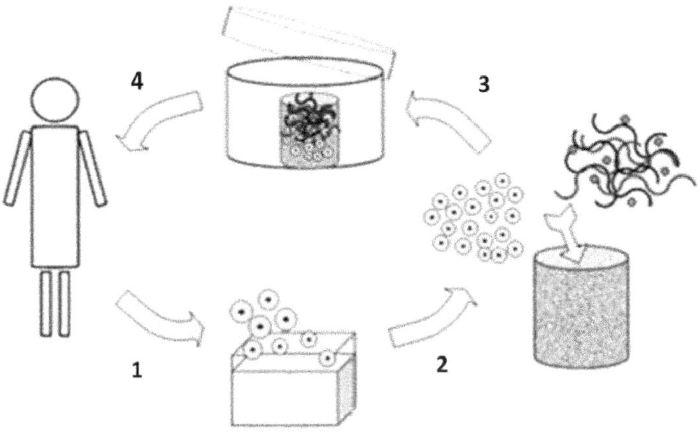

FIGURE 9.27 Schematic diagram summarizing the tissue engineering process.

Induced pluripotent stem cells are the adult cells that have been transformed into pluripotent stem cells through programming. Among adult stem cells, Mesenchymal stem cell is widely used as a multipotent source. Mesenchymal stem cell is derived from bone marrow stroma and can differentiate into a variety of cell types *in vitro*. Other sources of adult stem cells include the amniotic fluid and placental derived stem cells [29, 30].

B. Scaffolds

Scaffolds are temporary porous structures used to support cells by filling up the space otherwise occupied by the natural ECM and by providing a framework to organize the dissociated cells. A biocompatible and biodegradable material is chosen for tissue engineering scaffolds, which have sufficient porosity and pore-interconnectivity to promote cell migration and proliferation, and allow for nutrient and waste exchange. The rate of degradation should be tuned with the rate of cell growth and expansion, so that as the host cells expand and produce their own ECM, the temporary material degenerates with a similar rate. Moreover, the by-products should be confirmed to be non-toxic. Mechanical properties (Figure 9.28) should also match that of the native ECM [31, 32].

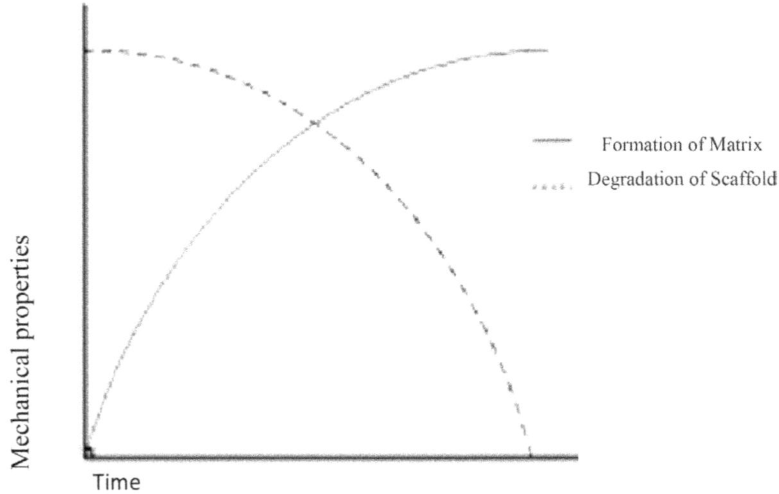

FIGURE 9.28 Mechanical properties of an ideal scaffold.

C. Bioreactors

In-vitro tissue engineering requires bioreactors in order to provide sufficient nutrients and oxygen to the cells while removing the toxic materials left by the proliferating cells. Moreover, the essential cell-specific mechanical stimuli are also provided by bioreactors. Each type of cell (cartilage, bone, myocardium, endothelial, etc.) has different requirements in terms of pH, oxygen tension, mechanical stimulation, temperature, etc. As a result, it is necessary to use cell specific bioreactors for generation of different tissues [33].

For the treatment of tissues or organs in malfunction in a human body, one of the challenges to the field of tissue engineering/biomaterials is the design of ideal scaffolds/synthetic matrices that can mimic the structure and biological functions of the natural ECM. Human cells can attach and organize well around fibers with diameters smaller than those of the cells. In this regard, nanoscale fibrous scaffolds can provide an optimal template for cells to seed, migrate, and grow. A successful regeneration of biological tissues and organs calls for the development of fibrous structures with fiber architectures beneficial for cell deposition and cell proliferation [34].

In general, tissue engineering involves the fabrication of three-dimensional scaffolds that can support cell in-growth and proliferation. In this context, the generation of scaffolds with tailored, biomimetic geometries (across multiple scales) has become an increasingly active area of research [35].

9.4.2.3 Tissue Engineering: Background Information

There is without any doubt a constantly increasing need for tissues and organs to replace those that have been damaged by sickness or accidents. It is also without any doubt that this need cannot be covered by allogeneic transplants.

The shortage of donor organs, immunological problems, and possibly contamination of the donor tissue limit the use of organ transplants. This is a good reason for having a closer look at the emerging science of tissue engineering.

Tissue engineering involves the cultivation of various types of tissues to replace such damaged tissues or organs. Cartilage, bone, skin tissue, muscle, blood vessels, lymphatic vessels, lung tissue, and heart tissue are among the target tissues. In vitro, in vivo as well as combination approaches are known. In vitro approaches taken in tissue engineering rely on the seeding of specific cells on highly porous membranes as scaffold. Both homologous and autologous cells have been used for this purpose. Autologous cells, that is, cells that are harvested from an individual for the purpose of being used on that same individual, have the benefit of avoiding an immunologic response in tissue engineering. The bodies of the patients will not reject the engineered tissue because it is their own tissue and the patients will not have to take immunosuppressive drugs.

The concept is that such scaffolds will *mimic* to a certain extent the extracellular matrix surrounding cells in living tissue. The extracellular matrix is known to have a broad range of tasks to accomplish. It embeds the cells of which the particular tissue is composed, it offers points of contacts to them, provides for the required mechanical properties of the tissue. So, the expectation is that cells seeded on adequate porous scaffolds experience an enhanced proliferation and growth, covering finally the whole scaffold. A further task is to define the three-dimensional shape of the tissue to be engineered. Ideally, such a scaffold may then be reimplanted into the living body provided that an appropriate selection of the nature of the seeded cells was done.

So, as an example, to replace muscles, muscle cells might be chosen for the seeded cells. Yet, frequently rather than choosing specific cell lineages stem, cells such as, for instance, mesenchymal stem cells, to be discussed below in more detail, are seeded for various reasons. The proliferation of such cells is, in this case, just one step, the next involving the differentiation of the cells along specific target cell lines depending of the target tissue. To induce such differentiation various types of biological and chemical signals have been developed. To enhance proliferation it will in general be necessary to include functional compounds such as growth factors, etc., into the scaffold membranes, as discussed below.

Another approach used currently less often (that is, an in vivo approach) in tissue engineering consists in implanting the original scaffold directly into the body to act as nucleation sites for self-healing via seeding of appropriate cells.

Because the tissue engineering technology is based essentially on the seeding of cells into three-dimensional matrices, the material properties of the matrix as well as its architecture will fundamentally influence the biological functionality of the engineered tissue. One has to keep in mind that the carrier matrix has to fulfill a diverse range of requirements with respect to biocompatibility, biodegradability, morphology, sterilizability, porosity, ability to incorporate and release drugs, and mechanical suitability. Also, the scaffold architecture, porosity and relevant pore sizes are very important. In general, a high surface area and an open and interconnected 3D pore system are required for scaffolds.

These factors affect cell binding, orientation, mobility, etc. The pores of scaffolds are, furthermore, very important for cell growth as nutrients diffuse through them. The minimum pore size required is decided by the diameter of cells and therefore varies from one cell type to another. Inappropriate pore size can lead to either no infiltration at all or nonadherence of the cells. Scaffolds with nanoscalar architectures have bigger surface areas, providing benefits for absorbing proteins and presenting more binding sites to cell membrane receptors.

Biological matrices are usually not available in sufficient amounts, and they can be afflicted with biological-infection problems. It is for this reason that during the last decades man-made scaffolds composed of a sizable number of different materials of synthetic or natural, that is biological, origins have been used to construct scaffolds characterized by various types of architectures.

These include powders, foams, gels, porous ceramics and many more. However, powders, foams, and membranes are often not open-pored enough to allow cell growth in the depth of the scaffold; consequently, the formation of a three-dimensional tissue structure is frequently restricted. Even loose gel structures (e.g., of polypeptides) may fail. Furthermore, smooth walls and interfaces, which occur naturally in many membranes and foams, are frequently unfavorable for the adsorption of many cell types.

From a biological viewpoint, almost all of the human tissues and organs are deposited in nanofibrous forms or structures. Examples include: bone, dentin, collagen, cartilage, and skin. All of them are characterized by well-organized hierarchical fibrous structures realigning in nanometer scale.

Nanofibers are defined as the fibers whose diameter ranges in the nanometer range. These have a special property of high surface area and increased porosity which makes it favorable for cell interaction and hence it makes its a potential platform for tissue engineering. The high surface area to volume ratio of the nanofibers combined with their microporous structure favors cell adhesion, proliferation, migration, and differentiation, all of which are highly desired properties for tissue engineering applications. There are mainly three techniques involved in synthesizing nanofibers namely electrospinning, self- assembly, and phase separation [36].

a) Phase Separation

In this technique water-polymer emulsion is formed which is thermodynamically unstable. At low gelation temperature, nanoscale fibers network is formed, whereas high gelation temperature leads to the formation of platelet-like structure. Uniform nanofiber can be produced as the cooling rate is increased. Polymer concentration has a significant effect on the nanofiber properties, as polymer concentration is increased porosity of fiber decreased and mechanical properties of fiber are increased. The final product obtained is mainly porous in nature but due to controlling the key parameters we can obtain a fibrous structure (Figure 9.29). The key parameters involved are as follows [36].

(a) type of polymers and their viscosity;
(b) type of solvent and its volatility;
(c) quenching temperature; and
(d) gelling type.

b) Self Assembly

It is a powerful approach for fabricating supra molecular architectures. Self-assembly of peptides and proteins is a promising route to the fabrication of a variety of molecular materials including nanoscale fibers and fiber network scaffolds (Figure 9.30). The main mechanism for a generic self-assembly is the intermolecular forces that bring the smaller unit together [36, 37].

c) Electrospinning

It is a term used to describe a class of fibers forming processes for which electrostatic forces are employed to control the production of the fiber.

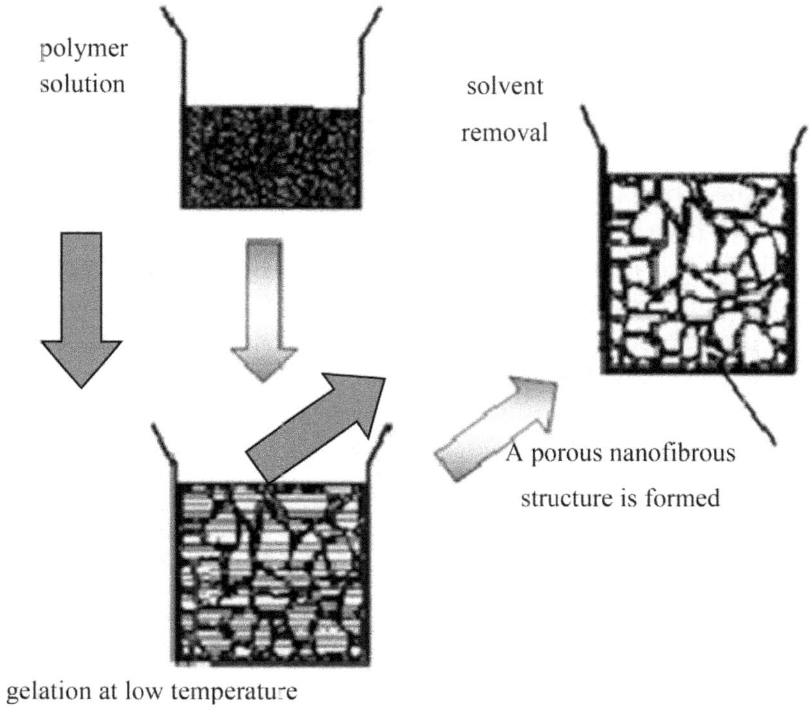

FIGURE 9.29 Nanofibrous structure production through phase separation.

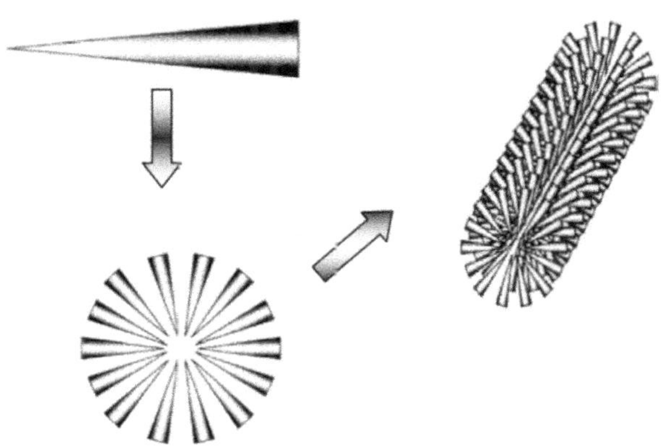

FIGURE 9.30 Schematic presentation of self-assembled nanofiber production.

Electrospinning readily leads to the formation of continuous fibers ranging from 0.01 to 10 μm. Electrospinning is a fiber forming processes by which electrostatic forces are employed to control the production of fibers. It is closely related to the more established technology of electrospraying, where the droplets are formed. "Spinning" in this context is a textile term that derives from the early use of spinning wheels to form yarns from natural fiber. In both electrospinning and electrospraying (Figure 9.31), the role of the electrostatic forces is to supplement or replace the conventional mechanical forces (e.g., hydrostatic, pneumatic) used to form the jet and to reduce the size of the fibers or droplets, hence the term "electrohydrodynamic jetting." Polymer nanofibers fabricated via electrospinning have been proposed for a number of soft tissue prostheses applications such as blood vessel, vascular, breast, etc. In addition, electrospun biocompatible polymer nanofibers can also be deposited as a thin porous film onto a hard tissue prosthetic device designed to be implanted into the human body. This method will be discussed in detail, later.

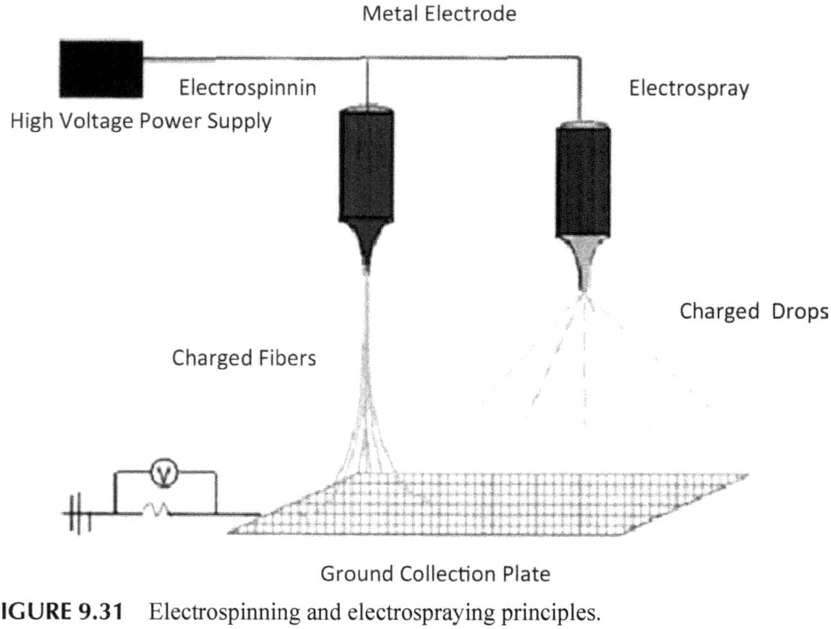

FIGURE 9.31 Electrospinning and electrospraying principles.

9.4.2.4 Nanofibers in Tissue Engineering Applications

A rapidly growing field of application of polymer nanofibers is their use in tissue engineering. The main areas of intensive research are nerve, blood vessel, skeletal muscle, cartilage, bone and skin tissue engineering.

9.4.2.4.1 Nerve Tissue Engineering

Application of electrospun polymeric nanofibers for nerve tissue regeneration is a very significant issue. The most important observation is the elongation and neurite growth of cells parallel to fiber direction, and the effect of fiber diameter is not so significant. They have found that aligned nanofibrous scaffolds are good scaffolds for neural tissue engineering [38].

9.4.2.4.2 Skin Tissue Engineering

Nanofibers exhibit higher cell attachment and spreading, especially when coated with collagen than microfibers. The results, which researchers were obtained, prove the potential of electrospun nanofibers in wound healing and regeneration of skin and oral mucosa [39].

9.4.2.4.3 Blood Vessel Tissue Engineering

A number of studies have shown that the biodegradable polymers mimic the natural ECM and show a defined structure replicating the in vivo-like vascular structures and can be ideal tools for blood vessel tissue engineering [40].

9.4.2.4.4 Skeletal Muscle Tissue Engineering

Skeletal muscle is responsible for maintenance of structural contours of the body and control of movements. Extreme temperature, sharp traumas or exposure to myotoxic agents are among the reasons of skeletal muscle injury. Tissue engineering is an attractive approach to overcome the problems related to autologous transfer of muscle tissue. It could also be a solution to donor shortage and reduction in surgery time. The studies demonstrate the absence of toxic residuals and satisfactory mechanical properties of the scaffold [41].

9.4.2.4.5 Cartilage Tissue Engineering

There are three forms of cartilage in the body that vary with respect to structure, chemical composition, mechanical property and phenotypic characteristics of the cells. These are hyaline cartilage, fibrocartilage and elastic cartilage. Cells capable of undergoing chondrogenic differentiation upon treatment with appropriate factors and a 3-D scaffold that provides a suitable environment for chondrogenic cell growth are the two main requirements for successful cartilage tissue engineering. In addition, there are some other conditions to fulfill. First, the matrix should support cartilage-specific matrix production; second, it should allow sufficient cell migration to achieve a good bonding to the adjacent host tissue and finally, the matrix should provide enough mechanical support in order to allow early mobilization of the treated joint [42].

9.4.2.4.6 Bone Tissue Engineering

Bone engineering has been studied for a long time to repair fractures and in the last decades, used in preparation of dental and orthopedic devices and bone substitutes. Bone tissue engineering is a more novel technique, which deals with bone restoration or augmentation. The matrix of bone is populated by osteogenic cells, derived of mesenchymal or stromal stem cells that differentiate into active osteoblasts. Several studies have demonstrated that it is possible to culture osteogenic cells on 3-D scaffolds and achieve the formation of bone. They designed a novel 3-D carrier composed of micro and nanofibers and have observed that cells used these nanofibers as bridges to connect to each other and to the microfibers. Furthermore, a higher ability for enhancement of cell attachment and a higher activity was observed in the nano/microfiber combined scaffolds compared to the microfibrous carrier. The fibrous scaffolds improved bone formation [43, 44].

9.4.3 TISSUE ENGINEERING: CELLS AND SCAFFOLDS

At this point in history tissue engineering is largely an Edisonian exercise in which the scaffold provides mechanical support while host-appropriate cells populate the structure and the deposit ECM components specific to the organ targeted for replacement. The current goal is a 'neotissue' that the body can "work with" and eventually adapt to carry out the full range of expected biological activities. The primary constituent of the various ECM's involved is typically collagen; the ratios of collagen type and hierarchical organization define the mechanical properties and organization of the evolving neotissue. In addition, the ECM provides cells with a broad range of chemical signals that regulate cell function [45–47]. Cells have been mainly cultured at the surface of the electrospun materials instead of in the bulk material. 2D monolayer culture models are easy and convenient to set up with good viability of cells in culture. Although, cells on electrospun surfaces have shown 3D matrix adhesion, considerations must be made at a 3D level to truly assess the potential of electrospun biomaterials for tissue engineering by providing cells with the 3D environment found in natural tissues [48, 49].

In recent years, there have been a large number of patients who suffered from the bone defects caused by tumor, trauma or other bone diseases. Generally, autogenetic and allogenetic bones are used as substitutes in treatment of bone defects. However, secondary surgery for procuring autogenetic bone from patient would bring donor site morbidity and allogenetic bone would cause infections or immune response [50, 51]. Thus, it is necessary to find new approach for the bone regeneration. As a promising approach, the tissue engineering develops the viable substitutes capable of repairing or regenerating the functions of the damaged tissue. For the bone tissue engineering, it requires a scaffold system to temporarily support the cells and direct their growth into the corresponding tissue in vivo [52, 53].

There are three basic tissue-engineering strategies that are used to restore, maintain, or improve tissue function, and they can be summarized as cell transplantation, scaffold-guided regeneration, and cell loaded scaffold implantation [54, 55].

1. Cell transplantation involves the removal of healthy cells from a biopsy or donor tissue and then injecting the healthy cells directly into the diseased or damaged tissue. However, this technique does not guarantee tissue formation and generally has less than 10% efficiency.

2. Scaffold-guided regeneration involves the use of a biodegradable scaffold implanted directly into the damaged area to promote tissue growth.

3. Cell-loaded scaffold implantation involves the isolation of cells from a patient and a biodegradable scaffold that is seeded with cells and then implanted into the defect location. Prior to implantation, the cells can be subjected to an in vitro environment that mimics the in vivo environment in which the cell/polymer constructs can develop into functional tissue. This in vitro environment is generally the result of a bioreactor, which provides growth factors and other nutrients while also providing mechanical stimuli to facilitate tissue growth. The first phase is the in vitro formation of a tissue construct, by placing the chosen cells and scaffold in a metabolically and mechanically supportive environment with growth media [35, 56]. The key processes occurring during the in vitro and in vivo phases of tissue formation and maturation are: (1) cell proliferation, sorting and differentiation; (2) extracellular matrix production and organization; (3) degradation of the scaffold; and (4) remodeling and potentially growth of the tissue [57] (Figure 9.32).

It is generally accepted that electrospinning has the potential to fabricate scaffolds as it results in a material with sufficient strength, nanostructure, biocompatibility, and economic attractiveness. Structures composed of the thin fibers generated by the electrospinning fall into this category as demonstrated by the widespread use of the process. Electrospinning produces non-woven meshes containing fibers ranging in diameter from tens of microns to tens of nanometers [58, 59].

A scaffold design based on nanofibers can successfully mimic the structure and components of ECM component in the body and therefore properties of other native tissues. Specifically, the ECM consists of a cross-linked network of collagen and elastin fibrils (mechanical framework), interspersed with glycosaminoglycans (biochemical interactions). In spite of its remarkable diversity due to the presence of various biomacromolecules and

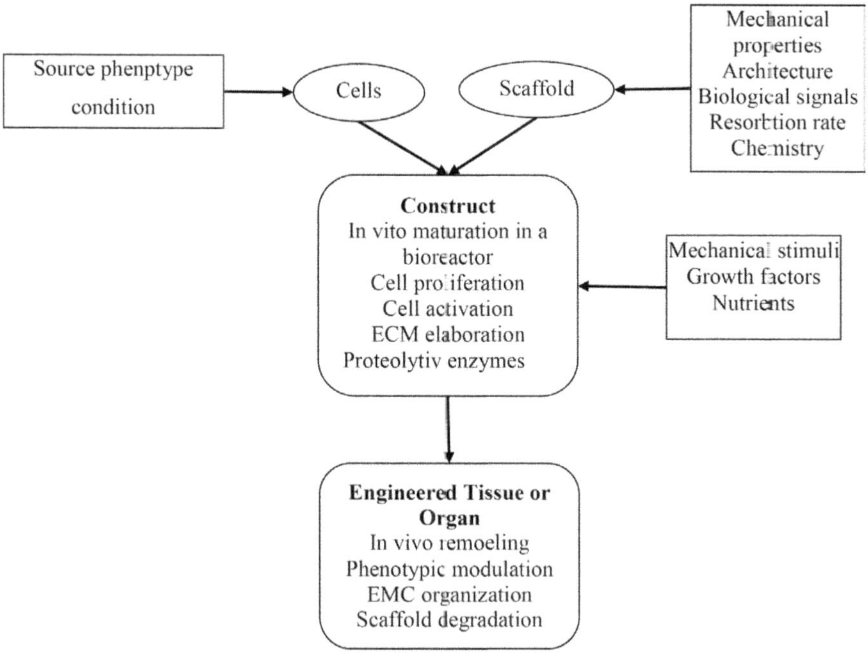

FIGURE 9.32 Tissue engineering paradigm

their organization, a key feature of native ECM is the nanoscale dimension of its internal components [60].

9.4.4 *EXTRACELLULAR MATRIX*

9.4.4.1 Collagen

Collagen types II, VI, IX, X, and XI are found in articular cartilage, although type II accounts for 90–95% of the collagen in the matrix. Type II collagen has a high amount of bound carbohydrate groups, allowing more interaction with water than some other types. Types IX and XI, along with type II, form fibrils that interweave to form a mesh. This organization provides tensile strength as well as physically entrapping other macromolecules. Although the exact function of types IX and XI are unknown, type IX has been observed to bind superficially to the fibers and extending into the inter-fiber space to interact with other type IX molecules, possibly acting to stabilize the mesh structure. Type X is found only near areas of the matrix that are calcified [61, 62].

9.4.4.2 Proteoglycans

Proteoglycans are composed of about 95% polysac-charide and about 5% protein. The protein core is associated with one or more varieties of glycosaminoglycan (GAG) chains. GAG chains are unbranched polysaccharides made from disaccharides of an amino sugar and another sugar. At least one component of the disaccharide has a negatively charged sulfate or carboxylate group, so the GAGs tend to repel each other and other anions while attracting cations and facilitating interaction with water. Hyaluronic acid, chondroitin sulfate, keratan sulfate, dermatan sulfate and heparan sulfate are some of the GAGs generally found in articular cartilage [63].

There are both large aggregating monomers and smaller proteoglycans present in articular cartilage. The aggregating proteoglycans, or aggregans, are composed of monomers with keratan sulfate and chondroitin sulfate GAGs attached to the protein core. In most aggregan molecules, link proteins connect many (up to 300) of these monomers to a hyaluronic acid chain. Aggregans fill most of the interfibrillar space of the ECM and are thought to be responsible for much of the resilience and stress distribution in articular cartilage through their ability to attract water. There are no chemical bonds between the proteoglycans and collagen fibers; aggregation prevents diffusion of the proteoglycans out of the matrix during joint loading [64, 65].

The smaller proteoglycans include decorin, biglycan and fibromodulin. They have shorter protein cores and fewer GAG chains than their larger counterparts. Unlike aggregans, these molecules do not affect physical properties of the tissue, but are thought to play a role in cell function and organization of the collagen matrix [64].

9.4.4.3 Noncollagenous Proteins

In contrast to proteoglycans, glycoproteins have only a small amount of oligosaccharide associated with the protein core. These polypeptides help to stabilize the ECM matrix and aid in chondrocyte-matrix interactions. Both anchoring CII and cartilage oligomeric protein anchor chondrocytes to the surrounding matrix. Other noncollagenous proteins commonly found in most tissues, such as fibronectin and tenascin, are also observed in articular cartilage and are believed to perform similar functions as the glycoproteins [66].

9.4.4.4 Tissue Fluid

Tissue fluid is an essential part of hyaline cartilage, comprising up to 80% of the wet weight of the tissue. In addition to water, the fluid contains gases, metabolites and a large amount of cations to balance the negatively charged GAG's in the ECM. It is the exchange of this fluid with the synovial fluid that provides nutrients and oxygen to the avascular cartilage. In addition, the entrapment of this fluid though interaction with ECM components provides the tissue with its ability to resist compression and return to normal shape after deformation [28].

9.4.5 SCAFFOLDS

It is evident that scaffolds mimicking the architecture of the extracellular should offer great advantages for tissue engineering. The extracellular matrix surrounds the cells in tissues and mechanically supports them, as discussed above. This matrix has a structure consisting of a three-dimensional fiber network, which is formed hierarchically by nanoscale multifilaments. An ideal scaffold should replicate the structure and function of the natural extracellular matrix as closely as possible, until the seeded cells have formed a new matrix.

The use of synthetic or natural nanofibers to build porous scaffolds, therefore, seems to be especially promising and electrospinning seems to be the fabrication method of choice for various reasons. Electrospinning first of all allows construction of nanofibers from a broad range of materials of synthetic and natural origin. The range of accessible diameters of the nanofibers is extremely broad covering the range from a few nm up to several micrometers. Secondly, the nonwovens composed of nanofibers and produced by electrospinning have a total porosity of up to 90%, which is highly favorable in view of the requirements defined above. By controlling the diameter of the nanofibers one is able to control directly the average pore sizes within the nonwovens.

When constructing the scaffold by electrospinning the material choice for nanofibers is important. One often chooses degradable polymers designed to degrade slowly in the body, disappearing as the cells begin to regenerate. The degradation rate must therefore match the regeneration rate of tissue in this case.

Biocompatible and biodegradable natural and synthetic polymers such as polyglycolides, polylactides, polycaprolactone (PCL), various copolymers, segmented polyurethanes, polyphosphazenes, collagens, gelatin, chitosans, silks, and alginates are used as the carrier materials. Mixtures of gelatin and chitosans or synthetic polymers like PCL and PEO (polyethyleneoxide) are also employed, as are PCL modified by grafting, and copolymers coated or grafted with gelatin. The material choice for the applications depends upon the type of scaffold required, nature of the tissues to be regenerated and their regeneration time. The correct material helps in fulfilling the requirement of specific mechanical properties and degradation times for the particular application.

The highly porous nature of the scaffolds is apparent in all cases and the fiber diameter and thus the pore dimensions can be controlled over a sizable range in all cases, as discussed above. Functional compounds such as growth factors can be introduced in large quantities of up to 50% and more into the fibers if required, by adding these compounds to the spinning solution.

A variety of cells (e.g., mesenchymal stem cells, endothelial cells, neural stem cells, keratinocytes, muscle cells, fibroblasts, and osteoblasts) have been seeded onto carrier matrices for the generation of target tissues (such as skin tissue, bone, cartilage, arteries, and nerve tissue). The diameters of the fibers used generally conform to the structural properties of the extracellular matrix and are of the order of 100 nm. However, in some cases, fibers with diameters of less than 100 nm or of the order of 1 nm were used. The observation is that the number of cells located on the scaffold increase in time due to proliferation processes, if the nature of the scaffold is chosen appropriately.

A frequent requirement for the growth of cells in tissue engineering is that the cells are not oriented randomly within the scaffold but are oriented planar or even uniaxial. Tissue engineering of bones or muscles is an example.

In several studies, the proliferation behavior of cells in such fiber structures was compared with that on films cast from the same polymer material. The results showed that the fiber architecture generally affects cell growth positively. For endothelial cells, however, it was reported that a smoother surface can be beneficial for cell adhesion and proliferation. Another conclusion made was that the biocompatibility of a material improves with decreasing fiber diameter. Porosity also seems to have a favorable influence on cell growth. For instance, it was observed that mesenchymal stem cells form branches to the pores on porous nanofibers.

Another important requirement is that the scaffolds are porous enough to allow cells to grow in their depths, while being provided with the necessary nutrients and growth factors. The degree of porosity and the average pore dimensions are significant factors for cell proliferation and the formation of three-dimensional tissues. Depending on the cell type, the optimal pore diameters are 20–100 nm; pore diameters larger than 100 nm are in general not required for optimal cell growth. It was also found that cells could easily migrate to a depth of about 100 nm, but encounter problems at greater depths.

One solution to this problem involves a layer-by-layer tissue-generation procedure. In this approach, cells are uniformly assembled into multilayered three-dimensional (3D) structure with the assistance of electrospun nanofibers. This approach offers lot of flexibility in terms of varying cell seeding density and cell type for each cell layer, the composition for each nanofiber layer, precise control of fiber layer thickness, fiber diameter, and fiber orientation. A further answer to this problem consists in introducing the cells directly during the preparation step of the scaffold via electrospinning. The concept is to combine the spinning process of the fibers with an electrospraying process of the cells.

Combination of electrospinning of fibers and electrospraying of cells Scaffolds based on nanofiber nonwovens offer a lot of further advantages. One important prerequisite for a scaffold is a sufficient mechanical compatibility. Cartilage, for example, is characterized by a Young's modulus of about 130 MPa, a maximum deformation stress of about 20 MPa, and a maximum deformation of 20–120%; the corresponding values for skin tissue are 15–150 MPa, 5–30 MPa, and 35–115%. These ranges of values can be achieved with electrospun nanofibers. For instance, for scaffolds composed of electrospun collagen fibers with diameters of about 100 nm, a Young's modulus of 170 MPa and maximum deformation stress of 3.3 MPa were found. However, the maximum elongation is usually less than 10%. Another important finding is that the fibers can impart mechanical stress to the collective of growing cells. It was reported that the production of extracellular tissue is greater if oriented rather than unoriented matrix fibers are employed. This production can be significantly increased by the application of a periodical mechanical deformation (typically 6%).

Mimicking functional gradients (one of the important characteristic features of living tissue), that is, gradients in composition, microstructure and porosity in scaffolds are also possible in a simple way by electrospinning. For example layer-by-layer electrospinning with composition gradients via controlled changes in the composition of the electrospinning solutions provide functional gradient scaffolds. The incorporation of bioactive agents in electrospun fibers will lead to advanced biofunctional tissue-engineering scaffolds. The biofunctionalization can alter the efficiency of these fibers for regenerating biological functional tissues. The bioactive agents can be easily incorporated onto fibers just by mixing them in electrospinning solution or by covalent attachment.

Scaffolds fabricated from electrospun nanofibers have definitely several advantages. However, considerable room for optimization remains with respect to architecture, surface properties, porosity, mechanical and biomechanical properties and functional gradient, and also with respect to the seeding of cells in the three-dimensional space and the supply of nutrients to the cells. It is often observed that the cells preferentially grow on the surfaces or that they initially adhere to the carrier fibers, but then detach after differentiation. Toxicity of the organic solvents used for electrospinning is another issue for in vivo applications. The solution is to use either water-soluble polymers for electrospinning with subsequent cross-linking after scaffold formation or to make use of water-based polymeric dispersions. One final remark: early investigations on the co-growth of different types of cells on scaffolds are very promising. For instance, the co-growth of fibroblasts, keratinocytes, and endothelial cells was reported; the astonishing result is that co-growth enhances cell growth. So, a lot can be expected from scaffolds for tissue engineering based on electrospun nanofibers.

Ideal scaffolds probably approximate the structural morphology of the natural collagen found in the target organ. The ideal scaffold must satisfy a number of often conflicting demands: (1) appropriate levels and sizes of porosity allowing for cell migration; (2) sufficient surface area and a variety of surface chemistries that encourage cell adhesion, growth, migration, and differentiation; and (3) a degradation rate that closely matches the regeneration rate of the desired natural tissue [35]. While a broad range of tissue engineering matrices has been fabricated, a few types of synthetic scaffolding show special promise. Synthetic or natural materials can be used that eliminate concerns regarding unfavorable immune responses or disease transmission [58, 59].

Synthetic tissues help stimulate living tissues to repair themselves in various parts of the human body, such as cartilage, blood vessels, bones and so forth, due to diseases or wear and tear. Victims whose skin are burned or scalded by fire or boiling water may also find an answer in synthetic tissues. The newest generation of synthetic implant materials, also called biomaterials, may even treat diseases such as Parkinson's, arthritis and osteoporosis. The uses of synthetic tissues are numerous. And there are several methods available to create them. One method is to make use of scaffold fabrication technology. Under this technology, synthetic tissues are cultured and placed on the scaffold that is shaped accordingly to, say a tendon or ligament, and then grafted onto the damaged part of the body. Once the new tissues grow over the damaged part of the organ or achieved sufficient structural integrity, the scaffolds would eventually degrade until only the tissues remain. It is also possible to use biocompatible materials that do not degrade, in which case, the scaffolds remain harmlessly in the body.

Synthetic polymers typically allow a greater ability to tailor mechanical properties and degradation rate. Clearly, the electrospinning process can eventually be developed to achieve successful utilization in vivo on a routine basis. Electrospinning offers the ability to fine-tune mechanical properties during the fabrication process, while also controlling the necessary biocompatibility and structure of the tissue engineered grafts [67]. The ability of the electrospinning technique to combine the advantages of synthetic and natural materials makes it particularly attractive, where a high mechanical durability, in terms of high burst strength and compliance (strain per unit load), is required. Advances in processing techniques, morphological characteristics and interesting, biologically relevant modifications are underway [68, 69].

However, there is one big drawback. This is the unstable dynamical behavior of the liquid jet, which is formed during the electrospinning process. This instability inhibits the fibres to be aligned in a regular way, which is crucial to satisfy the scaffolds specifications. Many researches have been done to study this jet behavior. When the instability of the jet can be controlled, electrospinning cannot only be adapted to produce high quality scaffolds for tissue engineering, but also for many other applications.

9.4.6 WOUND HEALING

An interesting application of electrospun nanofibers is the treatment of large wounds such as burns and abrasions [70–72]. It is found that these types of wounds heal particularly rapidly and without complications if they are covered by a thin web of nanofibers, in particular, of biodegradable polymers. Such nanowebs have suitable pore size to assure the exchange of liquids and gases with the environment, but have dimensions that prevent bacteria from entering. Mats of electro-spun nanofibers generally show very good adhesion to moist wounds. Furthermore, the large specific surface area of up to $100 m^2 g^{-1}$ is very favorable for the adsorption of liquids and the local release of drugs on the skin, making these materials suitable for application in hemostatic wound closure. Further, multifunctional bioactive nanofibrous wound healing dressings can be made available easily simply by blending with bioactive therapeutic agents (like antiseptics, antifungal, vasodilators, growth factors, etc.) or by coaxial electrospinning. Compared to conventional wound treatment, the advantage is also that scarring is prevented by the use of nanofibers.

The nanofibrillar structure of the nanoweb promotes skin growth, and if a suitable drug is integrated into the fibers, it can be released into the healing wound in a homogeneous and controlled manner. The charging of biodegradable nanofibers with antibiotics was realized with the drugs cefazolin and mefoxin. Generally, different drugs with antiseptic and antibiotic effects, as well as growth and clotting factors, are available for wound healing. Polyurethane (PU) is widely used as the nanoweb material because of its excellent barrier properties and oxygen permeability. Electrospun mats of PU nanofibers as wound dressings were successfully tested on pigs. Histological investigations showed that the rate of epithelialization during the healing of wounds treated with nanofiber mats is higher than that of the control group. Another promising and, in contrast to PU, biodegradable material is collagen. The wound healing properties of mats of electrospun fibers of type I collagen can be investigated on wounds in mice. It was found that especially in the early stages of the healing process better healing of the wounds was achieved with the nanofiber mats compared to conventional wound care. Blends of collagen or silk and PEO were also electrospun into fibers and used in wound dressings.

Numerous other biodegradable polymers that can be electrospun can be applied in wound healing, for example, PLA and block-copolymer derivatives, PCL, chitin, and chitosan. Using tetracycline hydrochloride as a model drug, it was shown that the release kinetics can be adjusted by varying the polymer used for the fabrication of the nanofibers. Poly[ethylene-co-(vinyl acetate)] (PEVA), PLA, and a 50:50 mixture of the two polymers were investigated. With PEVA, faster drug release was observed than with PLA or the blend. With PLA, burst release occurred, and the release properties of the blend are intermediate between those of the pure polymers. The morphology of the fibers and their interaction with the drug are critical factors. The concentration of the drug in the fibers also affects the release kinetics. The higher the concentration, the more pronounced the burst, evidently because of an enrichment of the drug on the surface.

Handheld electrospinning devices have been developed for the direct application of nanofibers onto wounds. In such a device, a high voltage is generated with the voltage supplied by standard batteries. The device has a modular construction, so that different polymer carriers and drugs can be applied, depending on the type of wound, by exchanging containers within the spinning device.

9.4.7 TRANSPORT AND RELEASE OF DRUGS

Nanostructured systems for the release of drugs (or functional compounds in general) are of great interest for a broad range of applications in medicine, including among others tumor therapy, inhalation and pain therapy [72–74].

Nanoparticles (composed of lipids or biodegradable polymers, for example) have been extensively investigated with respect to the transport and release of drugs. Such nanostructured carriers must fulfill diverse functions. For example, they should protect the drugs from decomposition in the bloodstream, and they should allow the controlled release of the drug over a chosen time period at a release rate that is as constant as possible. They should also be able to permeate certain membranes (e.g., the blood/brain barrier), and they should ensure that the drug is only released in the targeted tissue. It may also be necessary for the drug release to be triggered by a stimulus (either external or internal) and to continue the release only as long as necessary for the treatment. A variety of methods have been used for the fabrication of such nanoparticles, including spraying and sonification, as well as self-organization and phase-separation processes. Such nanoparticles are primarily used for systemic treatment. Experiments are currently being carried on the targeting and enrichment of particular tissues (vector targeting) by giving the nanoparticles specific surface structures (e.g., sugar molecules on the surface).

A very promising approach is based on the use of anisometric nanostructures that is, of nanorods, nanotubes, and nanofibers for the transport and release of drugs. In the focus of such an approach will, in general, be a locoregional therapy rather than a systemic therapy. In a locoregional therapy the drug carriers are localized at the site where the drug is supposed to be applied. Such anisometric carriers can be fabricated by electrospinning with simultaneous incorporation of the drugs via the spinning solution. Another approach envisions the preparation of core-shell objects via coaxial electrospinning where the drug is incorporated in the core region of the fibers with the shell being composed of a polymer.

Nanofibers with incorporated super paramagnetic Fe_3O nanoparticles serve as an example the carrier should be possible with the application of an external magnetic field.

An interesting property of super paramagnetic systems is that they can be heated by periodically modulated magnetic fields. This feature allows drug release to be induced by an external stimulus.

A broad set of in vitro experiments on the release kinetics of functional molecules has been performed among others by fluorescence microscopy. The experiments often have demonstrated that the release occurs as a burst, that is, in a process that is definitely nonlinear with respect to time. It was, however, found that the release kinetics, including the linearity of the release over time and the release time period, can be influenced by the use of core-shell fibers, in which the core immobilizes the drugs and the shell controls their diffusion out of the fibers.

In addition to low molecular weight drugs, macromolecules such as proteins, enzymes, growth factors and DNA are also of interest for incorporation in transport and release systems. Several experimental studies on this topic have been carried out. The incorporation of plasmidic DNA into PLA -b -PEG -b- PLA block copolymers and its subsequent release was investigated, and it was shown that the released DNA was still fully functional. Bovine serum albumin (BSA) and lysozyme were also electrospun into polymer nanofibers, and their activities after release were analyzed, again yielding positive results. In the case of BSA, is was shown that the use of core-shell fibers fabricated by the chemical vapor deposition (CVD) of poly (p-xylylene) PPX onto electrospun nanofibers affords almost linear release over time. Further investigations deal with the incorporation and release of growth factors for applications in tissue engineering. In the following, some specific applications of nanofibers in drug release are described in more detail.

9.4.8 APPLICATION IN TUMOR THERAPY

Nanofibers composed of biodegradable polymers were investigated with respect to their use in local chemotherapy via surgical implantation. A selection of approaches will be discussed in the following. The water-insoluble antitumor drug paclitaxel (as well as the antituberculosis drug rifampin) was electrospun into PLA nanofibers. In some cases, a cationic, anionic, or neutral surfactant was added, which influenced the degree of charging of the nanofibers. Analysis of the release kinetics in the presence of proteinase K revealed that the drug release is nearly ideally linear over time. The release is clearly a consequence of the degradation of the polymer by proteinase. Analogous release kinetics were found when the degree of charging was increased to 50%. Similar investigations were also carried out with the hydrophilic drug doxorubicin.

To obtain nanofibers with linear release kinetics for water-soluble drugs like doxorubicin, water–oil emulsions were electrospun, in which the drug was contained in the aqueous phase and a PLA-co-PGA copolymer (PGA: polyglycolic acid) in chloroform was contained in the oil phase. These electrospun fibers showed bimodal release behavior consisting of burst kinetics for drug release through diffusion from the fibers, followed by linear kinetics for drug release through enzymatic degradation of the fibers by the proteinase K. In many cases, this type of bimodal behavior may be desired. Furthermore, it was shown that the antitumor drug retained its activity after electrospinning and subsequent release. The drug taxol was also studied with respect to its release from nanofibers. These few examples show that nanofibers may in fact be used as drug carrier and release agents in tumor therapy.

9.4.9 INHALATION THERAPY

Finally, a unique application for anisometric drug carriers, inhalation therapy, will be discussed. The general goal is to administer various types of drugs via the lung. One key argument is that the surface of the lung is, in fact, very large, of the order of 150 m^2, so that this kind of administration should be very effective.

Indications for such treatments are tumors, metastases, pulmonary hypertension, and asthma. But these systems are also under consideration for the administration of insulin and other drugs through the lung.

Further advantages of anisometric over spherical particles as drug carriers for inhalation therapy are that a significantly larger percentage of anisometric particles remain in the lung after exhalation and that the placement of the drug carriers in the lung can be controlled very sensitively via the aerodynamic radius. To produce rod-shaped carriers with a given aerodynamic diameter, nanofibers were electrospun from appropriate carrier polymers that were subsequently cut to a given length either by mechanical means or by laser cutting.

Further progress in inhalation therapy will mainly depend first of all on finding biocompatible polymer systems that do not irritate the lung tissue and on the development dispensers able to dispense such rod-like particles.

Enzyme-loaded nanoparticles as well as some other nanodispersed biocatalysts are remarkable from several perspectives. It appeared that the use of nanofibers, typically, electrospun nanofibers, provide a large surface area for the attachment or entrapment of enzymes and enzyme reaction. In the case of porous nanofibers, they can reduce diffusion path of substrates from reaction medium to enzyme active sites, due to the reduced thickness.

Tissue engineering has provided a new medical treatment as an alternative to traditional transplantation methods. It is a promising area to repair or replace task of damaged tissues or organs. Natural polymers offer the advantage of being very analogous, often identical to macromolecular substances existing in the human body. Thus, the biological environment is prepared to distinguish and interact with natural polymers. Some of the natural polymers applied as scaffolds in the nervous system are collagen, gelatin hyaluronic acid, chitosan and elastin. Synthetic polymers can be tailored to develop a wide range of mechanical features and degradation rates. They can also be processed to minimize immune response.

Polymeric materials can lead to great development in tissue engineering of damaged nervous system, but there are still many questions to be answered before their application, such as type and characteristic of polymer and the complementary methods which are appropriate for specific neurological dysfunctions and further investigation is needed to promote them as ideal scaffolds for nervous tissue engineering.

KEYWORDS

- **collagen**
- **glycosaminoglycan**
- **polycaprolactone**
- **proteoglycans**
- **scaffolds**
- **tissue fluid**

REFERENCES

1. Wente, V. A., *Superfine Thermoplastic Fibers.* Industrial & Engineering Chemistry, 1956, 48(8), 1342–1346.
2. Buntin, R. R., Lohkamp, D. T. *Melt Blowing-One-Step WEB Process for New Nonwoven Products.* Tappi, 1973, 56(4), 74–77.
3. Zhou, F. L., Gong, R. H. *Manufacturing Technologies of Polymeric Nanofibers and Nanofiber Yarns.* Polymer International, 2008, 57(6), 837–845.
4. Angadjivand, S., Kinderman, R., Wu, T. *High Efficiency Synthetic Filter Medium,* 2000, Google Patents.
5. Zeleny, J., *The electrical discharge from Liquid Points, and a Hydrostatic Method of measuring the electric intensity at their surface.* Physical Review, 1914, 3(2), 69–91.
6. Formhals, A., *Process and Apparatus Fob Pbepabing,* 1934, Google Patents.
7. Taylor, G., *Disintegration of Water Drops in an Electric Field.* Proceedings of the Royal Society of London. Series, A. Mathematical and Physical Sciences, 1964, 280(1382), 383–397.
8. Gibson, P., H. S. Gibson, D. Rivin, *Transport Properties of Porous Membranes Based on Electrospun Nanofibers.* Colloids and Surfaces A: Physicochemical and Engineering Aspects, 2001, 187, 469–481.
9. Patarin, J., B. Lebeau, R. Zana, *Recent Advances in the Formation Mechanisms of Organized Mesoporous Materials.* Current Opinion in Colloid & Interface Science, 2002, 7(1), 107–115.
10. Weghmann, A. *Production of Electrostatic Spun Synthetic Microfibre Nonwovens and Applications in Filtration.* in *Proceedings of the 3rd World Filtration Congress, Filtration Society.* 1982, London.
11. Kruiēska, I., E. Klata, M. Chrzanowski, *New Textile Materials for Environmental Protection,* in *Intelligent Textiles for Personal Protection and Safety.* 2006, 41–53.
12. Yarin, A. L., E. Zussman, *Upward Needleless Electrospinning of Multiple Nanofibers.* Polymer, 2004, 45(9), 2977–2980.
13. Majeed, S., et al., *Multi-Walled Carbon Nanotubes (MWCNTs) Mixed Polyacrylonitrile (PAN) Ultrafiltration Membranes.* Journal of Membrane Science, 2012, 403, 101–109.
14. Macedonio, F., E. Drioli, *Pressure-Driven Membrane Operations and Membrane Distillation Technology Integration for Water Purification.* Desalination, 2008, 223(1), 396–409.
15. Merdaw, A. A., A. O. Sharif, G. A. W. Derwish, *Mass Transfer in Pressure-Driven Membrane Separation Processes, Part II.* Chemical Engineering Journal, 2011, 168(1), 229–240.
16. Van Der Bruggen, B., et al., *A Review of Pressure-Driven Membrane Processes in Wastewater Treatment and Drinking Water Production.* Environmental Progress, 2003, 22(1), 46–56.
17. Cui, Z. F., H. S. Muralidhara, *Membrane Technology: A Practical Guide to Membrane Technology and Applications in Food and Bioprocessing.* 2010, Elsevier. 288.
18. Shirazi, S., C. J. Lin, D. Chen, *Inorganic Fouling of Pressure-Driven Membrane Processes—A Critical Review.* Desalination, 2010, 250(1), 236–248.

19. Pendergast, M. M., E. M. V. Hoek, *A Review of Water Treatment Membrane Nanotechnologies.* Energy & Environmental Science, 2011, 4(6), 1946–1971.
20. Hilal, N., et al., *A comprehensive review of nanofiltration membranes: Treatment, pretreatment, modeling, and atomic force microscopy.* Desalination, 2004, 170(3), 281–308.
21. Srivastava, A., S. Srivastava, K. Kalaga *Carbon Nanotube Membrane Filters*, in *Springer Handbook of Nanomaterials*. 2013, Springer. 1099–1116.
22. Colombo, L., A. L. Fasolino, *Computer-Based Modeling of Novel Carbon Systems and Their Properties: Beyond Nanotubes.* Vol. 3. 2010, Springer. 258.
23. Polarz, S., B. Smarsly, *Nanoporous Materials.* Journal of Nanoscience and Nanotechnology, 2002, 2(6), 581–612.
24. Gray-Weale, A. A., et al., *Transition-state theory model for the diffusion coefficients of small penetrants in glassy polymers.* Macromolecules, 1997, 30(23), 7296–7306.
25. Rigby, D., R. Roe, *Molecular Dynamics Simulation of Polymer Liquid and Glass. I. Glass Transition.* The Journal of chemical physics, 1987, 87, 7285.
26. Freeman, B. D., Y. P. Yampolskii, I. Pinnau, *Materials Science of Membranes for Gas and Vapor Separation.* 2006, Wiley.com. 466.
27. Hofmann, D., et al., *Molecular Modeling Investigation of Free Volume Distributions in Stiff Chain Polymers with Conventional and Ultrahigh Free Volume: Comparison Between Molecular Modeling and Positron Lifetime Studies.* Macromolecules, 2003, 36(22), 8528–8538.
28. Greenfield, M. L., D. N. Theodorou, *Geometric Analysis of Diffusion Pathways in Glassy and Melt Atactic Polypropylene.* Macromolecules, 1993, 26(20), 5461–5472.
29. Baker, R. W., *Membrane Technology and Applications*. 2012, John Wiley & Sons. 592.
30. Strathmann, H., L. Giorno, E. Drioli, *Introduction to Membrane Science and Technology.* 2011, Wiley-VCH Verlag & Company. 544.
31. Chen, J. P., et al., *Membrane Separation: Basics and Applications*, in *Membrane and Desalination Technologies*, L. K. Wang, et al., Editors. 2008, Humana Press. 271–332.
32. Mortazavi, S., *Application of Membrane Separation Technology to Mitigation of Mine Effluent and Acidic Drainage*. 2008, Natural Resources Canada. 194.
33. Porter, M. C., *Handbook of Industrial Membrane Technology.* 1990, Noyes Publications. 604.
34. Naylor, T. V., *Polymer Membranes: Materials, Structures and Separation Performance*. 1996, Rapra Technology Limited. 136.
35. Freeman, B. D., *Introduction to Membrane Science and Technology. By Heinrich Strathmann.* Angewandte Chemie International Edition, 2012, 51(38), 9485–9485.
36. Kim, I., H. Yoon, K. M. Lee, *Formation of Integrally Skinned Asymmetric Polyetherimide Nanofiltration Membranes by Phase Inversion Process.* Journal of applied polymer science, 2002, 84(6), 1300–1307.
37. Khulbe, K. C., C. Y. Feng, T. Matsuura, *Synthetic Polymeric Membranes: Characterization by Atomic Force Microscopy*. 2007, Springer. 198.
38. Loeb, L. B., *The Kinetic Theory of Gases*. 2004, Courier Dover Publications. 678.
39. Koros, W. J., G. K. Fleming, *Membrane-Based Gas Separation.* Journal of Membrane Science, 1993, 83(1), 1–80.

40. Perry, J. D., K. Nagai, W. J. Koros, *Polymer membranes for hydrogen separations.* MRS bulletin, 2006, 31(10), 745–749.
41. Hiemenz, P. C., R. Rajagopalan, *Principles of Colloid and Surface Chemistry, revised and expanded.* Vol. 14. 1997, CRC Press.
42. McDowell-Boyer, L. M., J. R. Hunt, N. Sitar, *Particle transport through porous media.* Water Resources Research, 1986, 22(13), 1901–1921.
43. Auset, M., A. A. Keller, *Pore-scale processes that control dispersion of colloids in saturated porous media.* Water Resources Research, 2004, 40(3).
44. Bhave, R. R., *Inorganic membranes synthesis, characteristics, and applications.* Vol. 312. 1991, Springer.
45. Lin, V. S.-Y., et al., *A porous silicon-based optical interferometric biosensor.* Science, 1997, 278(5339), 840–843.
46. Hedrick, J., et al. *Templating nanoporosity in organosilicates using well-defined branched macromolecules.* in *Materials Research Society Symposium Proceedings.* 1998, Cambridge University Press.
47. Hubbell, J. A., R. Langer, *Tissue engineering* Chem. Eng. News 1995, 13, 42–45.
48. Schaefer, D. W., *Engineered porous materials* MRS Bulletin 1994, 19, 14–17.
49. Hentze, H. P., M. Antonietti, *Porous Polymers and Resins.* Handbook of Porous Solids: 1964–2013.
50. Endo, A., et al., *Synthesis of ordered microporous silica by the solvent evaporation method.* Journal of materials science, 2004, 39(3), 1117–1119.
51. Sing, K., et al., *Physical and biophysical chemistry division commission on colloid and surface chemistry including catalysis.* Pure and Applied Chemistry, 1985, 57(4), 603–619.
52. Kresge, C., et al., *Ordered mesoporous molecular sieves synthesized by a liquid-crystal template mechanism.* nature, 1992, 359(6397), 710–712.
53. Yang, P., et al., *Generalized syntheses of large-pore mesoporous metal oxides with semicrystalline frameworks.* nature, 1998, 396(6707), 152–155.
54. Jiao, F., K. M. Shaju, P. G. Bruce, *Synthesis of Nanowire and Mesoporous Low-Temperature LiCoO2 by a Post-Templating Reaction.* Angewandte Chemie International Edition, 2005, 44(40), 6550–6553.
55. Ryoo, R., et al., *Ordered mesoporous carbons.* Advanced Materials, 2001, 13(9), 677–681.
56. Beck, J., et al., *Chu, DH Olson, EW Sheppard, SB McCullen, JB Higgins and JL Schlenker.* J. Am. Chem. Soc, 1992, 114(10), 834.
57. Zhao, D., et al., *Triblock copolymer syntheses of mesoporous silica with periodic 50 to 300 angstrom pores.* Science, 1998, 279(5350), 548–552.
58. Joo, S. H., et al., *Ordered nanoporous arrays of carbon supporting high dispersions of platinum nanoparticles.* nature, 2001, 412(6843), 169–172.
59. Kruk, M., et al., *Synthesis and characterization of hexagonally ordered carbon nanopipes.* Chemistry of materials, 2003, 15(14), 2815–2823.
60. Rouquerol, J., et al., *Recommendations for the characterization of porous solids (Technical Report).* Pure and Applied Chemistry, 1994, 66(8), 1739–1758.
61. Schüth, F., K. S. W. Sing, J. Weitkamp, *Handbook of porous solids.* 2002, Wiley-Vch.
62. Maly, K. E., *Assembly of nanoporous organic materials from molecular building blocks.* Journal of Materials Chemistry, 2009, 19(13), 1781–1787.

63. Davis, M. E., *Ordered porous materials for emerging applications.* nature, 2002, 417(6891), 813–821.
64. Morris, R. E., P. S. Wheatley, *Gas storage in nanoporous materials.* Angewandte Chemie International Edition, 2008, 47(27), 4966–4981.
65. Cote, A. P., et al., *Porous, crystalline, covalent organic frameworks.* Science, 2005, 310(5751), 1166–1170.
66. El-Kaderi, H. M., et al., *Designed synthesis of 3D covalent organic frameworks.* Science, 2007, 316(5822), 268–272.
67. Jiang, J. X., et al., *Conjugated microporous poly (aryleneethynylene) networks.* Angewandte Chemie International Edition, 2007, 46(45), 8574–8578.
68. Ben, T., et al., *Targeted synthesis of a porous aromatic framework with high stability and exceptionally high surface area.* Angewandte Chemie, 2009, 121(50), 9621–9624.
69. Eddaoudi, M., et al., *Modular chemistry: secondary building units as a basis for the design of highly porous and robust metal-organic carboxylate frameworks.* Accounts of Chemical Research, 2001, 34(4), 319–330.
70. Lu, G. Q., X. S. Zhao, T. K. Wei, *Nanoporous materials: science and engineering.* Vol. 4. 2004, Imperial College Press.
71. Holister, P., C. R. Vas, T. Harper, *Nanocrystalline materials.* Technologie White Papers, 2003(4).
72. Smith, B. L., et al., *Molecular mechanistic origin of the toughness of natural adhesives, fibers and composites.* nature, 1999, 399(6738), 761–763.
73. Mann, S., G. A. Ozin, *Synthesis of inorganic materials with complex form.* nature, 1996, 382(6589), 313–318.
74. Mann, S., *Molecular tectonics in biomineralization and biomimetic materials chemistry.* nature, 1993, 365(6446), 499–505.
75. Busch, K., S. John, *Photonic band gap formation in certain self-organizing systems.* Physical Review E, 1998, 58(3), 3896.
76. Argyros, A., et al., *Electron tomography and computer visualization of a three-dimensional 'photonic' crystal in a butterfly wing-scale.* Micron, 2002, 33(5), 483–487.
77. Sailor, M. J., K. L. Kavanagh, *Porous silicon–what is responsible for the visible luminescence?* Advanced Materials, 1992, 4(6), 432–434.
78. Koshida, N., B. Gelloz, *Wet and dry porous silicon.* Current opinion in colloid & interface science, 1999, 4(4), 309–313.
79. Hentze, H.-P., M. Antonietti, *Template synthesis of porous organic polymers.* Current Opinion in Solid State and Materials Science, 2001, 5(4), 343–353.
80. Nakao, S.-I., *Determination of pore size and pore size distribution: 3. Filtration membranes.* Journal of Membrane Science, 1994, 96(1), 131–165.
81. Barrer, R. M., *Zeolites and their synthesis.* Zeolites, 1981, 1(3), 130–140.
82. Iijima, S., *Helical microtubules of graphitic carbon.* nature, 1991, 354(6348), 56–58.
83. Kaneko, K., K. Inouye, *Adsorption of water on FeOOH as studied by electrical conductivity measurements.* Bulletin of the Chemical Society of Japan, 1979, 52(2), 315–320.
84. Maeda, K., et al., *Control with polyethers of pore distribution of alumina by the sol-gel method.* Chem.& Ind., 1989(23), 807.
85. Matsuzaki, S., M. Taniguchi, M. Sano, *Polymerization of benzene occluded in graphite-alkali metal intercalation compounds.* Synthetic metals, 1986, 16(3), 343–348.

86. Enoki, T., H. Inokuchi, M. Sano, *ESR study of the hydrogen-potassium-graphite ternary intercalation compounds.* Physical Review B, 1988, 37(16), 9163.
87. Pinnavaia, T. J., *Intercalated clay catalysts.* Science, 1983, 220(4595), 365–371.
88. Yamanaka, S., et al., *High surface area solids obtained by intercalation of iron oxide pillars in montmorillonite.* Materials research bulletin, 1984, 19(2), 161–168.
89. Inagaki, S., Y. Fukushima, K. Kuroda, *Synthesis of highly ordered mesoporous materials from a layered polysilicate.* J. Chem. Soc., Chem. Commun., 1993(8), 680–682.
90. Vallano, P. T., V. T. Remcho, *Modeling interparticle and intraparticle (perfusive) electroosmotic flow in capillary electrochromatography.* Analytical chemistry, 2000, 72(18), 4255–4265.
91. Levenspiel, O., *Chemical reaction engineering*. Vol. 2. 1972, Wiley New York etc.
92. Li, Q., et al., *Interparticle and intraparticle mass transfer in chromatographic separation.* Bioseparation, 1995, 5(4), 189–202.
93. Setoyama, N., et al., *Surface characterization of microporous solids with helium adsorption and small angle x-ray scattering.* Langmuir, 1993, 9(10), 2612–2617.
94. Marsh, H., *Introduction to carbon science.* 1989.
95. Kaneko, K., *Determination of pore size and pore size distribution: 1. Adsorbents and catalysts.* Journal of Membrane Science, 1994, 96(1), 59–89.
96. Seaton, N., J. Walton, *A new analysis method for the determination of the pore size distribution of porous carbons from nitrogen adsorption measurements.* Carbon, 1989, 27(6), 853–861.
97. Dullien, F. A., *Porous media: fluid transport and pore structure*. 1991, Academic press.
98. Yang, W., et al., *Carbon Nanotubes for Biological and Biomedical Applications.* Nanotechnology, 2007, 18(41), 412001.
99. Bianco, A., et al., *Biomedical Applications of Functionalized Carbon Nanotubes.* Chemical Communications, 2005(5), 571–577.
100. Salvetat, J., et al., *Mechanical Properties of Carbon Nanotubes.* Applied Physics A, 1999, 69(3), 255–260.
101. Zhang, X., et al., *Ultrastrong, Stiff, and Lightweight Carbon-Nanotube Fibers.* Advanced Materials, 2007, 19(23), 4198–4201.
102. Arroyo, M., T. Belytschko, *Finite Crystal Elasticity of Carbon Nanotubes Based on the Exponential Cauchy-Born Rule.* Physical Review B, 2004, 69(11), 115415.
103. Wang, J., et al., *Energy and Mechanical Properties of Single-Walled Carbon Nanotubes Predicted Using the Higher Order Cauchy-Born rule.* Physical Review B, 2006, 73(11), 115428.
104. Zhang, Y., *Single-walled carbon nanotube modeling based on one-and two-dimensional Cosserat continua*, 2011, University of Nottingham.
105. Wang, S., *Functionalization of Carbon Nanotubes: Characterization, Modeling and Composite Applications*. 2006, Florida State University. 193.
106. Lau, K.-t., C. Gu, D. Hui, *A critical review on nanotube and nanotube/nanoclay related polymer composite materials.* Composites Part B: Engineering, 2006, 37(6), 425–436.
107. Choi, W., et al., *Carbon Nanotube-Guided Thermopower Waves.* Materials Today, 2010, 13(10), 22–33.
108. Sholl, D. S., J. Johnson, *Making High-Flux Membranes with Carbon Nanotubes.* Science, 2006, 312(5776), 1003–1004.

109. Zang, J., et al., *Self-Diffusion of Water and Simple Alcohols in Single-Walled Aluminosilicate Nanotubes.* ACS nano, 2009, 3(6), 1548–1556.
110. Talapatra, S., V. Krungleviciute, A. D. Migone, *Higher Coverage Gas Adsorption on the Surface of Carbon Nanotubes: Evidence for a Possible New Phase in the Second Layer.* Physical Review Letters, 2002, 89(24), 246106.
111. Pujari, S., et al., *Orientation Dynamics in Multiwalled Carbon Nanotube Dispersions Under Shear Flow.* The Journal of chemical physics, 2009, 130, 214903.
112. Singh, S., P. Kruse, *Carbon Nanotube Surface Science.* International Journal of Nanotechnology, 2008, 5(9), 900–929.
113. Baker, R. W., *Future Directions of Membrane Gas Separation Technology.* Industrial & Engineering Chemistry Research, 2002, 41(6), 1393–1411.
114. Erucar, I., S. Keskin, *Screening Metal–Organic Framework-Based Mixed-Matrix Membranes for CO_2/CH4 Separations.* Industrial & Engineering Chemistry Research, 2011, 50(22), 12606–12616.
115. Bethune, D. S., et al., *Cobalt-Catalysed Growth of Carbon Nanotubes with Single-Atomic-Layer Walls.* Nature 1993, 363, 605–607.
116. Iijima, S., T. Ichihashi, *Single-Shell Carbon Nanotubes of 1-nm Diameter.* Nature, 1993, 363, 603–605.
117. Treacy, M., T. Ebbesen, J. Gibson, *Exceptionally high Young's modulus observed for individual carbon nanotubes.* 1996.
118. Wong, E. W., P. E. Sheehan, C. Lieber, *Nanobeam Mechanics: Elasticity, Strength, and Toughness of Nanorods and Nanotubes.* Science, 1997, 277(5334), 1971–1975.
119. Thostenson, E. T., C. Li, T. W. Chou, *Nanocomposites in Context.* Composites Science and Technology, 2005, 65(3), 491–516.
120. Barski, M., P. Kędziora, M. Chwał, *Carbon Nanotube/Polymer Nanocomposites: A Brief Modeling Overview.* Key Engineering Materials, 2013, 542, 29–42.
121. Dresselhaus, M. S., G. Dresselhaus, P. C. Eklund, *Science of Fullerenes and Carbon nanotubes: their Properties and Applications.* 1996, Academic Press. 965.
122. Yakobson, B., R. E. Smalley, *Some Unusual New Molecules—Long, Hollow Fibers with Tantalizing Electronic and Mechanical Properties—have Joined Diamonds and Graphite in the Carbon Family.* Am Scientist, 1997, 85, 324–337.
123. Guo, Y., W. Guo, *Mechanical and Electrostatic Properties of Carbon Nanotubes under Tensile Loading and Electric Field.* Journal of Physics D: Applied Physics, 2003, 36(7), 805.
124. Berger, C., et al., *Electronic Confinement and Coherence in Patterned Epitaxial Graphene.* Science, 2006, 312(5777), 1191–1196.
125. Song, K., et al., *Structural Polymer-Based Carbon Nanotube Composite Fibers: Understanding the Processing–Structure–Performance Relationship.* Materials, 2013, 6(6), 2543–2577.
126. Park, O. K., et al., *Effect of Surface Treatment with Potassium Persulfate on Dispersion Stability of Multi-Walled Carbon Nanotubes.* Materials Letters, 2010, 64(6), 718–721.
127. Banerjee, S., T. Hemraj-Benny, S. S. Wong, *Covalent Surface Chemistry of Single-Walled Carbon Nanotubes.* Advanced Materials, 2005, 17(1), 17–29.
128. Balasubramanian, K., M. Burghard, *Chemically Functionalized Carbon Nanotubes.* Small, 2005, 1(2), 180–192.

129. Xu, Z. L., F. Alsalhy Qusay, *Polyethersulfone (PES) Hollow Fiber Ultrafiltration Membranes Prepared by PES/non-Solvent/NMP Solution.* Journal of Membrane Science, 2004, 233(1–2), 101–111.
130. Chung, T. S., J. J. Qin, J. Gu, *Effect of Shear Rate Within the Spinneret on Morphology, Separation Performance and Mechanical Properties of Ultrafiltration Polyethersulfone Hollow Fiber Membranes.* Chemical Engineering Science, 2000, 55(6), 1077–1091.
131. Choi, J. H., J. Jegal, W. N. Kim, *Modification of Performances of Various Membranes Using MWNTs as a Modifier.* Macromolecular Symposia, 2007, 249–250(1), 610–617.
132. Wang, Z., J. Ma, *The Role of Nonsolvent in-Diffusion Velocity in Determining Polymeric Membrane Morphology.* Desalination, 2012, 286(0), 69–79.
133. Vilatela, J. J., R. Khare, A. H. Windle, *The Hierarchical Structure and Properties of Multifunctional Carbon Nanotube Fibre Composites.* Carbon, 2012, 50(3), 1227–1234.
134. Benavides, R. E., S. C. Jana, D. H. Reneker, *Nanofibers from Scalable Gas Jet Process.* ACS Macro Letters, 2012, 1(8), 1032–1036.
135. Gupta, V. B., V. K. Kothari, *Manufactured Fiber Technology.* 1997, Springer. 661.
136. Wang, T., S. Kumar, *Electrospinning of Polyacrylonitrile Nanofibers.* Journal of applied polymer science, 2006, 102(2), 1023–1029.
137. Song, K., et al., *Lubrication of Poly (vinyl alcohol) Chain Orientation by Carbon nanochips in Composite Tapes.* Journal of applied polymer science, 2013, 127(4), 2977–2982.
138. Filatov, Y., A. Budyka, V. Kirichenko, *Electrospinning of micro-and nanofibers: fundamentals in separation and filtration processes.* 2007, Begell House Inc., Redding, CT.
139. Brown, R. C., *Air filtration: an integrated approach to the theory and applications of fibrous filters.* Vol. 650. 1993, Pergamon press Oxford.
140. Hinds, W. C., *Aerosol technology: properties, behavior, and measurement of airborne particles.* 1982.
141. Greiner, A., J. Wendorff, *Functional self-assembled nanofibers by electrospinning*, in *Self-Assembled Nanomaterials, I.* 2008, Springer. 107–171.
142. Jeong, E. H., J. Yang, J. H. Youk, *Preparation of polyurethane cationomer nanofiber mats for use in antimicrobial nanofilter applications.* Materials Letters, 2007, 61(18), 3991–3994.
143. Lala, N. L., et al., *Fabrication of nanofibers with antimicrobial functionality used as filters: protection against bacterial contaminants.* Biotechnology and Bioengineering, 2007, 97(6), 1357–1365.
144. Maze, B., et al., *A simulation of unsteady-state filtration via nanofiber media at reduced operating pressures.* Journal of Aerosol Science, 2007, 38(5), 550–571.
145. Payet, S., et al., *Penetration and pressure drop of a HEPA filter during loading with submicron liquid particles.* Journal of Aerosol Science, 1992, 23(7), 723–735.

PART II

BIOCHEMICAL SCIENCES

CHAPTER 10

A NOTE ON ADVANCED GENETIC ENGINEERING METHODOLOGY

A. I. BERESNEV,[1] S. V. KVACH,[1] G. G. SIVETS,[2] and A. I. ZINCHENKO[1]

[1]*Institute of Microbiology, National Academy of Sciences, 220141, Kuprevich Str., 2, Minsk, Belarus, Fax: +375(17)264-47-66; E-mail: zinch@mbio.bas-net.by*

[2]*Institute of Bioorganic Chemistry, National Academy of Sciences, 220141, Kuprevich Str., 5/2, Minsk, Belarus, Fax: +375 (17) 267-87-61; E-mail: gsivets@yahoo.com*

CONTENTS

ABSTRACT

The method raising efficiency of synthesis in recombinant *Escherichia coli* cells of pyrimidine nucleoside phosphorylase (PyrNPase) from

Thermusthermophilus by inserting silent mutations into initial fragment of gene encoding this enzyme has been described. It was proposed to increase temperature inducing PyrNPase synthesis by bacterial cells up to 42°C. Ability of mutant PyrNPase (mutPyrNPase) to catalyze phosphorolysis of 3′-fluoro-3′-deoxyuridine and 3′-fluoro-2,'3′-dideoxythymidine was demonstrated. Recombinant nucleoside phosphorylases were originally engaged for synthesis of purine fluoronucleosides such as 3′-fluoro-3′-deoxyadenosine, 3′-fluoro-3′-deoxy-2-aminoadenosine, 3′-fluoro-3′-deoxyguanosine, 3′-fluoro-2,'3′-dideoxy-2-fluoroadenosine, 3′-fluoro-2,'3′-dideoxy-2-chloroadenosine, 3′-fluoro-3′-deoxy-2-fluoroadenosine, 3′-fluoro-3′-deoxy-2-chloroadenosine.

10.1 INTRODUCTION

Progress of genetical engineering triggered development of recombinant microbial strains providing for hyperproduction of valuable enzymes – nucleoside phosphorylases involved in enzymatic synthesis of nucleosides. Mesophilic bacteria *Escherichia coli* are most frequently used for construction of new recombinant strains [1, 2]. A series of recent publications described production and properties of pyrimidine- and purine nucleoside phosphorylases generated by various thermophilic microorganisms, their activity towards pyrimidine and purine nucleosides, biotechnological potential to derive purine 2′-fluoronucleosides [3]. Thermal stability of these biocatalysts enables to run enzymatic processes at elevated temperatures resulting in increased substrate solubility, accelerated nucleoside-yielding procedure in the course of enzymatic transglycosylation reaction and its enhanced efficiency in comparison with chemical methods [4].

Bacteria *E. coli* are widely used for production of recombinant thermophilic pyrimidine- and purine nucleoside phosphorylases, yet low level of biocatalysts synthesis by bacterial cells should be noted. The main reason for low efficiency of geterologous expression of genes cloned in *E. coli* is formation of mRNA pins at initial transcript site obstructing ribosomal synthesis of polypeptide chain [5]. Earlier we engineered strain *E. coli* KNK-12/1 producing pyrimidine nucleoside phosphorylase (PyrNPase) of *Thermusthermophilus* and originally demonstrated that the resulting thermophilic PyrNPase displayed a unique capacity to catalyze reaction of 3′-fluoro-2,'3′-dideoxythymidine (3′-F-2,'3′-ddThd) phosphorolysis yielding α-D-pentofuranose-1-phosphate [6].

Enzymatic synthesis of 3′-fluoro-2,'3′-dideoxyguanosine (3′-F-2,' 3′-dGuo) – a potential chemical therapy agent for treatment of viral infections was accomplished using nucleoside phosphorylases and adenosine deaminase from *E. coli.*

Since PyrNPase of *T. thermophilus* is a thermostable protein, level of its synthesis in *E. coli* cells is relatively low, which limits enzyme application prospects for synthesis of modified purine nucleosides [3].

Previously constructed recombinant strain *E. coli* BM-D6 produced homologous purine nucleoside phosphorylase (PurNPase) acting (like PyrNPase) as a key catalyst transforming pyrimidine nucleosides into modified purine nucleosides via enzymatic transglycosylation reaction [7]. PurNPase catalyzes stereoselective reaction of intermediate α-D-pentofuranose-1-phosphate (product of pyrimidine nucleoside phosphorolysis mediated by PyrNPase) condensation with purine heterocyclic base leading to formation of modified purine nucleoside.

Aim of this research was to increase efficiency of producing *T. thermophilus* PyrNPase in cells of genetically engineered *E. coli* strain by optimizing the structure of the respective translated mRNA and to investigate enzymatic synthesis of purine 3′-fluoro-3′-deoxy- and 3′-fluoro-2,'3′-dideoxynucleosides possessing antiviral and cytostatic activities from the available pyrimidine nucleosides engaging tandem reactions in the presence of recombinant nucleoside phosphorylases [8].

10.2 MATERIALS AND METHODS

Strain *E. coli* DH5α (Invitrogen, USA) essential for generation of constructed recombinant vector, strain *E. coli* BL21(DE3) (Invitrogen, USA) used for expression of cloned genes, recombinant strain *E. coli* BM-D6 producing homologous PurNPase, recombinant strain *E. coli* KNK-12/1 – a source of *T. thermophilus* PyrNPase, strain *E. coli* pADD3 engineered to produce recombinant homologous adenosine deaminase were applied in this study.

Data of primary structure of gene encoding PyrNPase of *T. thermophilus* were taken from GenBank base of nucleotides sequences. Molecular optimization of mRNA nucleotide sequence related to initial site of transcribed gene composed of 35 nucleotides was carried out using RNAfold appendix of Vienna RNA software package (version 1.8.4). Selection of primers for PCR was based on the available data concerning primary structure of *T. thermophilus* PyrNPase. Restriction sites *Nde*I and *Sal*I linked to 5′-ends of the primers were anchored by six random nucleotides. Point mutations in initial segment of PyrNPase gene locus affecting secondary mRNA structure were induced by direct oligonucleotide primer. Reaction mixture for PCR assay (50 μL) included 67 mM Tris-HCl buffer (pH 8.3) comprising 17 mM $(NH_4)_2SO_4$, 3 mM $MgCl_2$, 0.1% Tween 20, 0.12 mg/mL BSA, 8% glycerol, four 2′-deoxynucleoside triphosphate (each in 0.2 mM concentration), 10 pmol of primers TGTATATCTCCTTCTTAAAGTTAAACAAAATTATTTC and TATGAACCCCGTAGCATTTATCCGGGAGAAGCGGGA, 150–200 ng of genomic DNA from *T. thermophilus* plus 1 U of *Pfu*-polymerase. Amplification procedure occurred as follows: 1 min at 98°C, (1 s at 98°C, 10 s at 60°C, 45 s at 72°C) – 25 cycles, 2 min at 72°C. PCR products were separated by electrophoresis in 1.5% agarose gel. The putative PyrNPase gene product was recovered and ligated into vector pET42a previously digested with *Nde*I and *Sal*I and dephosphorylated with alkaline phosphatase. A novel *E. coli* strain producing heterologous mutant PyrNPase (mutPyrNPase) was derived from this vector.

The bacteria were cultured at temperature 37°C and 42°C on Luria-Bertani medium supplemented with 50 μg/mL kanamycin. 0.5 mM IPTG was applied to induce synthesis of target enzymes in microbial cells. Process of PyrNPase and mutPyrNPase synthesis by *E. coli* was monitored by electrophoresis in polyacrylamide gel containing sodium dodecyl sulphate (SDS-PAGE). The percentage of enzymes of total cell proteins was calculated using TotalLab 120 software.

The grown bacterial cells were harvested by centrifugation, suspended in 10 mM potassium-phosphate buffer, pH 7.0 (PPB) and subjected to ultrasonic disintegration. Enzyme purification was achieved by heating bacterial sonic lysates comprising PyrNPase and mutPyrNPase from *T. thermophilus* at 80°C during 1 h with subsequent removal of denatured proteins by centrifugation.

PyrNPase and mutPyrNPase activities were evaluated spectrophotometrically via changes in absorbance (λ=300 nm) during phosphorolysis cleavage of thymidine. A unit of PyrNPase and mutPyrNPase activity was defined at the amount of enzyme sufficient to produce 1 μmole of thymine in 1 min under reaction conditions.

Process of 3′-fluoro-3′-deoxyuridine (3′-F-3′-dUrd) phosphorolysis was analyzed in the presence of mutPyrNPase: 2.5 U of enzyme activity were added to 3′-F-3′-dUrd in 2 mM PPB and the reaction was performed at 80°C with regular monitoring of phosphorolysis degree in the course of 96 h.

Enzymatic synthesis of purine 3′-fluoro-3′-deoxy- and 2,’3′-dideoxy-β-D-ribonucleosides was conducted from 3′-F-3′-dUrd and 3′-F-2,’3′-ddThd substrates produced by chemical methods described earlier. The reaction were carried out in 5 mM PPB at temperature 60°C during 5 days with the following variations in mixture composition: (1) 15 mM 3′-F-3′-dUrd, 10 mM adenine (Ade), 20 U/mL mutPyrNPase *T. thermophilus*, 200 U/mL PurNPase *E. coli*; (2) 15 mM 3′-F-3′-dUrd, 10 mM 2-aminoadenine ($2NH_{2-}$Ade), 20 U/mL mutPyrNPase *T. thermophilus*, 200 U/mL PurNPase *E. coli*; (3) 15 mM 3′-F-3′-dUrd, 10 mM $2NH_{2-}$Ade, 20 U/mL mutPyrNPase *T. thermophilus*, (after cooling to room temperature 200 U/mL of adenosine deaminase from *E. coli* was supplied to the reaction mixture and incubated at 25°C for 24 h); (4) 15 mM 3′-F-2′,3′-ddThd, 10 mM 2-fluoroadenine (2F-Ade), 20 U/mL mutPyrNPase *T. thermophilus*, 200 U/mL PurNPase *E. coli*; (5) 15 mM 3′-F-2′,3′-ddThd, 10 mM 2-chloroadenine (2Cl-Ade), 20 U/mL mutPyrNPase *T. thermophilus*, 200 U/mL PurNPase *E. coli*; (6) 15 mM 3′-F-3′-dUrd, 10 mM 2F-Ade, 20 U/mL mutPyrNPase *T. thermophilus*, 200 U/mL PurNPase *E. coli*; (7) 15 mM 3′-F-3′-dUrd, 10 mM 2Cl-Ade, 20 U/mL mutPyrNPase *T. thermophilus*, 200 U/mL PurNPase *E. coli*.

Product accumulation was controlled by thin-layer chromatography on Silica gel $60F_{254}$ plates (Merck, Germany). Chromatographic separation of components in reaction mixtures № 1, 2 and 3 was performed in the mixture of isopropanol–chloroform–25% aqueous ammonia in the ratio 10:10:1 (v/v), in the reaction mixture № 4 and 5 – with chloroform–ethanol in 4:1 ratio (v/v), in reaction mixtures № 6 and 7 – with n-butanole–25% aqueous ammonia in 7:1 ratio (v/v).

Location of substrates and products on the TLC-plate were registered in UV-light and the compounds were eluted in 10 mM PPB. Concentration of products in eluates was determined spectrophotometrically using standard coefficients of molar extinction. The absorbance spectra were recorded at spectrophotometer UV 1202 (Shimadzu, Japan).

The structure of newly synthesized nucleosides was confirmed by comparing their TLC and UV-spectroscopy data with reference specimens of purine-3′-fluoronucleosides obtained earlier by chemical methods and in some cases by ^{1}H NMR, UV and mass spectroscopy analyses after chromatography on silica gel plates.

The provided experiments values are confidence level of arithmetical means at 95% confidence Interval.

10.3 RESULTS AND DISCUSSION

It was stated above that one of the methods of mRNA structural optimization leading to increased efficiency of PyrNPase production by mesophilic bacteria *E. coli* is insertion of silent mutations into initial gene fragment encoding this enzyme. As a result it raises Gibbs free energy (G) of mRNA from modified gene locus and consequently reduces the chance of pin emergence. Using RNAfold application of Vienna RNA software package we found that for mRNA gene domain lacking silent mutations G equaled –11.6 kJ (Figure 10.1A). In this study 3 nucleotide substitutions spanning first 35 base pairs of PyrNPase gene locus enabled to increase G value to –5.9 kJ (Figure 10.1B).

The level of PyrNPase and mutPyrNPase synthesis by *E. coli* cells was compared at standard (37°C) and elevated temperature (42°C). Results of experiments are illustrated by Figure 10.2. Computer processing of

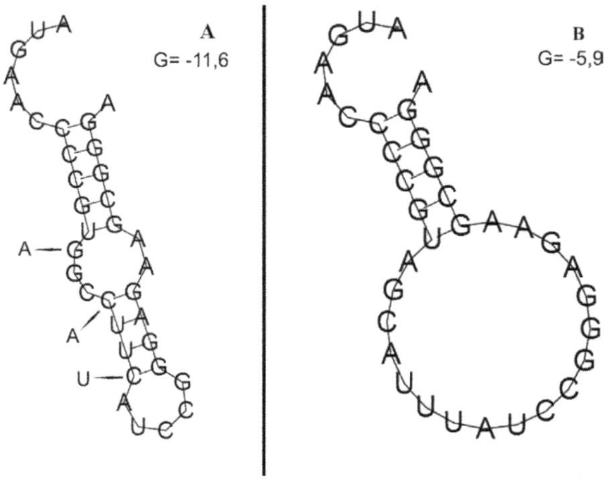

FIGURE 10.1 mRNA secondary structure of initial fragment of transcribed *T. thermophilus* PyrNPase gene prior to (A) and after (B) insertion of mutation.

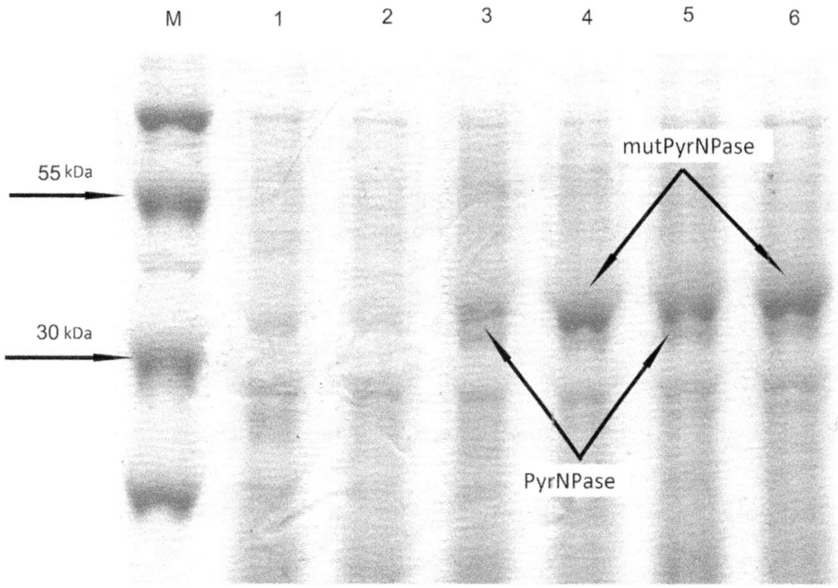

FIGURE 10.2 IPTG induction of synthesis in recombinant *E. coli* cells of PyrNPase and mutPyrNPase from *T. thermophilus* at various temperatures (M – marker proteins; 1 – lisate of non-induced cells producing PyrNPase; 2 – lisate of non-induced cells producing mutPyrNPase; 3 – lisate of induced cells producing PyrNPase at 37°C; 4 – lisate of induced cells producing mutPyrNPase at 37°C; 5 – lisate of induced cells producing PyrNPase at 42°C; 6 – lisate of induced cells producing mutPyrNPase at 42°C).

electrophoresis data demonstrated that the amount of synthesized bacterial mutPyrNPase exceeded that of PyrNPase by 1.7 times.

The information collected with the aid of TotalLab 120 was corroborated by measurements of enzyme activities of target proteins. The data are presented in Table 10.1. It was established that induction of enzyme synthesis by *E. coli* cells at 37°C and 42°C resulted in enhanced (1.7 times) mutPyrNPase activity expressed in U/g cells as compared to PyrNPase. At the next stage of research reaction of 3′-F-3′-dUrd phosphorolysis mediated by mutPyrNPase was examined. Experimental results are reflected in Figure 10.3. It was revealed that in the course of 4 day reaction engaging mutPyrNPase phosphorolysis degree of 3′-F-3′-dUrd reached 48% allowing to estimate it as a donor in tandem reaction of enzymatic transglycosylation yielding hardly available purine-3′-fluoronucleosides.

TABLE 10.1 Effect of Temperature on Induction of Synthesis of *T. thermophilus* PyrNPase and mutPyrNPase in *E. coli* Cells

Induction temperature, °C	PyrNPase		mutPyrNPase	
	U/mg cell mass	U/L cultural liquid	U/mg cell mass	U/L culturalliquid
37	4.5 ± 0.2	7–100 ± 50	7.5 ± 0.5	12.7 ± 0.9
42	8.2 ± 0.6	11 800 ± 130	12.7 ± 0.9	18 450 ± 220

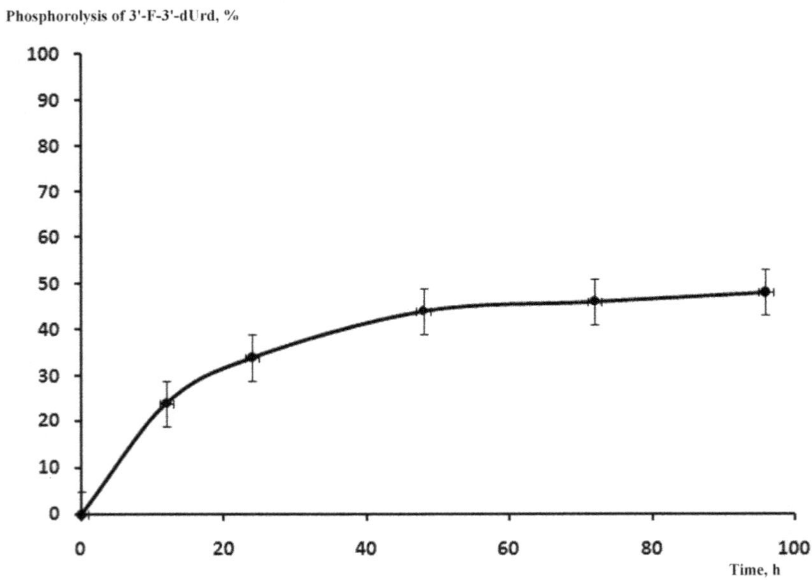

FIGURE 10.3 Dynamics of 3′-F-3′-dUrd phosphorolysis catalyzed by mutPyrNPase from *T. thermophilus*.

Enzymatic synthesis of novel and well-known purine 3′-fluoronucleosides previously produced exclusively by chemical methods was carried out from pyrimidine nucleosides (3′-F-3′-dUrd and 3′-F-2,’3′-ddThd) and purine heterocyclic bases (2Cl-Ade, 2F-Ade, $2NH_{2-}$Ade, Ade) in the presence of recombinant pyrimidine– and purine nucleoside phosphorylases as biocatalysts of phosphorolysis and stereoselective heterobase glycosylation with α-D-pentofuranose-1-phosphate, respectively. Completed reactions resulted in generation of modified purine nucleosides, like 3′-fluoro-3′-deoxyadenosine (3′-F-3′-dAdo), 3′-fluoro-3′-deoxy-2-aminoadenosine (3′-F-3′-d-$2NH_{2-}$Ado), 3′-fluoro-3′-deoxyguanosine (3′-F-3′-dGuo),

TABLE 10.2 Synthesis of Fluorinated Nucleosides

Number of reaction mixture	Product	Acceptor (10 mM)	Donor (15 mM)	Time, h	Product yield, mol.%
1	3′-F-3′-dAdo	Ade	3′-F-3′-dUrd	72	55±2
2	3′-F-3′-d-2NH_2-Ado*	2,6-DAP	3′-F-3′-dUrd	72	63±3
4	3′-F-2′,3′-dd-2F- Ado	2F-Ade	3′-F-2′,3′-ddThd	96	64±2
5	3′-F-2′,3′-dd-2Cl-Ado	2Cl-Ade	3′-F-2′,3′-ddThd	96	62±3
6	3′-F-3′-d-2F-Ado	2F-Ade	3′-F-3′-dUrd	96	53±2
7	3′-F-3′-d-2Cl-Ado	2Cl-Ade	3′-F-3′-dUrd	96	52±2

*3′-F-3′-d-2NH_2-Ado was transformed into 3′-F-3′-dGuoby introducing recombinant adenosine deaminase of *E. coli* into reaction mixture.

3′-fluoro-2,’3′-dideoxy-2-fluoroadenosine(3′-F-2,’3′-dd-2F-Ado),3′-fluoro-2,’3′-dideoxy-2-chloroadenosine (3′-F-2,’3′-dd-2Cl-Ado), 3′-fluoro-3′-deoxy-2-fluoroadenosine (3′-F-3′-d-2F-Ado), 3′-fluoro-3′-deoxy-2-chloroadenosine (3′-F-2,’3′-d-2Cl-Ado). The yield of end products (mol %) was calculated as the percentage of purine nitrogen bases fed into initial reaction mixture. Components of studied enzymatic reactions and yields of target nucleosides are provided in Table 10.2.

Summing up the obtained results, it should be noted that biotechnological synthesis of purine 3′-fluorinated nucleosides from corresponding pyrimidine analogs catalyzed by recombinant nucleoside phosphorylases appears extremely attractive in terms of producing new purine nucleoside derivatives and further elaboration of chemical-enzymatic processes generating biologically active compounds of this category. The proposed approach in some aspects is superior in efficiency than conventional chemical methods of producing similar modified purine nucleosides based, as a rule, on multistage scheme leading to output of anomeric nucleotide mixtures.

10.4 CONCLUSION

Using advanced genetic engineering methodology, a series of silent mutations was originally introduced into initial fragment of gene cloned in

E. coli and coding for PyrNPase of *T. thermophilus*. As a result 70% rise in the synthesis level of studied thermostable enzyme was achieved. We were to establish that pyrimidine fluoro-deoxynucleosides – 3′-F-3′-dUrd and 3′-F-2,'3′-ddThd may serve as substrates of phosphorolysis reaction catalyzed by thermostable PyrNPase of *T. thermophilus*. Moreover, recombinant nucleoside phosphorylases were never before involved in enzymatic synthesis of biologically significant purine 3′-fluoronucleosides, such as 3′-F-3′-dAdo, 3′-F-3′-d-2NH_{2}-Ado, 3′-F-3′-dGuo, 3′-F-2,'3′-dd-2F-Ado, 3′-F-2,'3′-dd-2Cl-Ado, 3′-F-3′-d-2F-Ado and 3′-F-2,'3′-d-2Cl-Ado.

KEYWORDS

- **modified nucleosides**
- **purine nucleoside phosphorylase**
- **pyrimidine nucleoside phosphorylase**
- **silent mutations**

REFERENCES

1. Mikhailopulo, I. A., Miroshnikov, A. I. (2010). ActaNatur. 2 (2), 36.
2. Mikhailopulo, I. A., Miroshnikov, A. I. (2011). Mendeleev Commun. 21, 57.
3. Szeker, K., Niemitalo, O., Casteleijn, M. G., Juffer, A. H., Neubauer, P., (2011). J. Biotechnol. 156, 268.
4. Szeker, K., Zhou, X., Schwab, T., Casanueva, A., Cowan, D., Mikhailopulo, I. A., Neubauer, P. (2012). J. Mol. Catal. B: Enzym. 84, 27.
5. Zhu, S., Ren, L., Wang, J., Zheng, G., Tang, P. (2012). Bioorg. Med. Chem. Lett. 22(5), 2102.
6. Berasneu, A. I., Kvach, S. V., Sivets, G. G., Zinchenko, A. I. (2013). Proceedings of the National Academy of Sciences of Belarus. Series of Biological Sciences.4, 71 (in Russian).
7. Burko, D. V., Eroshevskaya, L. A., Kvach, S. V., Shakhbazau, A. V., Kartel, N. A., Zinchenko, A. I. (2010). Biotechnology in Medicine, Foodstuffs, Biocatalysis, Environment and Biogeotechnology. Nova Science Publishers, Inc., New York, 1 p.
8. Qui, X. I., Xu, X. H., Qing, F. L. (2010). Tetrahedron. 66, 789.

CHAPTER 11

STABILITY FACTORS OF HERBACEOUS ECOSYSTEMS IN A BIOLOGICAL SENSE

R. A. Afanas'ev

Pryanishnikov All-Russian Scientific Research Institute of Agrochemistry, d. 31A, Pryanishnikova St., Moscow, 127550, Russia, Tel: +7-499-976-25-01; E-mail: rafail-afanasev@mail.ru

CONTENTS

ABSTRACT

In the evolutionary development herbaceous ecosystems elaborated the ways of protection from excess weight of wild large herd phytophages. Absence of this protection would lead not only to the disappearance of grassland vegetation but also to loss of soil due erosion processes. The author of this study believes that general protection measures include decrease of natural pasture productivity due to the vegetational change, the manifestation of forage herbs toxicity under adverse growth conditions and at destruction of the sod due to ungulates – the appearance of weeds which have poisonous and repellent properties and so not consumed by the animals; then there will the normal herbage restoration.

11.1 INTRODUCTION

Using a system approach the author gives a new interpretation of the known facts about grassland vegetation reaction to overload of pastures by animals. The reduction in pasture bioproductivity due to the vegetational change, toxicogenic defense reactions against eating of different plant species under adverse weather conditions and abundant appearance of inedible weeds in localities where with destroyed sod are evolutionary developed return reactions of herbaceous ecosystems to the demolition. From these positions we ought to consider mass appearance of weeds on cultivated land or while drastic meadow improving. Ecosystems perceive the violation of natural turf as impact of graminivorous animals and try restore status quo using their funds developed by the evolution. First of all these are meadow or field weeds.

11.2 EXPERIMENTAL PART

Materials of investigation were published messages of geobotany, phytosociology, ecology, meadland farming and other sciences associated with the study of vital activity of plant communities and also the results of own monitoring the state and dynamics of ley phytocenoses.

The methodological principles of generalization of the materials were the system approach [1, 2] and the actualism that is widely used in geology [3]. The system approach is to consider of a plant community and his biotope as a functional unit having the properties of self-regulation, protection from external actions through the development of appropriate responses. The method of the actualism is to recreate the history of development of the ecosystems based on the study of modern processes and conditions of their functioning as functional properties of plant communities have been developed in the process of long evolution. This fact is the basis for the retrospective assessment of the conditions and nature of these properties formation.

11.3 RESULTS AND DISCUSSION

According to modern data system is a dynamic organization of living and non-living elements in which there are external and internal circulations of substances determining its functional stability [4]. The most important condition for the stability of any system of the material world is its flexibility, the ability to direct the efforts of its parts (subsystems) and the whole system back to where danger, to avert this danger and save themselves. To some extent this property of the systems describes Le Chatelie's principle: "If a system at equilibrium experiences a change then as the result of ongoing processes the equilibrium shifts to partially counter-act the imposed change" [5, p.185].

Herbaceous ecosystems (biogeocenoses) are forms of existence of living matter, specifically plan cover of the earth. They meet all the requirements of a system as multi-component, self-regulatory "purposefully functioning structures capable of resolving problems in certain external conditions" [1, p.75]. So far herbaceous ecosystems have been studied mainly on specific aspects of their functioning without sufficient generalization of accumulated data. Meanwhile enormous factual material allows approaching its broad-scale understanding at the system level and gets "new system measurements, new genetic parameters of reality" [2, p.14].

It is the purpose of the present communication to asses the diverse properties of biogeocenoses from a system approach perspective and to identify cause-effect relationships in the dynamics of their floristic composition under the influence of external and internal factors.

Going directly to the statement of research materials, note that from the time of Leonardo da Vinci the main role in the formation of the soil cover along with other factors was given to higher plants including herbs [6]. Pedologists in particular L.O. Karpachevsky [7] believe that the formation of modern soil cover dates back to the Cretaceous period, when angiosperms are widely spread that is about 100 millions years ago, forming deciduous forests and meadows. Perhaps modern meadows differ from those that existed 100 millions years ago but their existence they are since the time and during this period meadow ecosystems developed adaptive response to external influences that threaten their existence. Thus, it is necessary to distinguish adaptation of separate types of plants and adaptation, characteristic for the entire community of these species, for example, phytocenoses as a whole. The system analysis allows you to select at least three categories of these adaptations that can be considered as factors of stability of herbaceous ecosystems. Though it may seem paradoxical, but all of these factors are aimed combating against the destruction of the soil cover, the preservation of its fertility, reducing the loss of plant nutrients into the environment. And more specifically, that they are directed against the negative impact on grassland cenoses primarily by large herbivores as well as natural phenomena causing the destruction of the sod.

The role of perennial grasses in the protection of soil from water and wind erosion is widely known [8]. However, until now, remained essentially out of sight scientists such function of herbaceous ecosystems as protection of soil from pasture erosion, for example, from destruction under the influence of large herbivorous animals. In natural conditions the wild herbivorous consume on average about 10% of the biomass of natural pastures [9], which corresponds to the general biological law of the energy pyramid [10]. However, when feeding or migration herd animals especially ungulates (buffaloes, bisons, tarpans, aurochs, antelopes, saigas, horses and other animals) could repeatedly be situations of heavy grazing, destruction of sod, which in combinations with weather anomalies – torrential rains, hurricanes – was supposed to lead to the destruction of the soil cover, death of ecosystems. In the process of biological evolution to survive could only such communities that due to the principle of natural selection have developed adequate defense remedies primarily of soil that was the basis of existence of the plant communities and also grassland landscape in general. Soil as bioinert body formed under the influence of vegetation, is unable to resist the active external actions, in particular the anthropogenesis [11]. Therefore, in the historical past, the soil could not develop appropriate protective mechanisms against the effects of major phytophages and the role of defenders of the soil from destruction in biogeocenoses was given to living beings – plants. Consider these mechanisms more with attraction of well-known facts.

The first is the reduction in the productivity of plant community with increased heavy grazing of aboveground biomass, which is designated as pasture digression. It is well known from the theory and practice of modern grassland agriculture [12]. Thus, according to I.V. Larin [13] and scientists to which he refers, with increasing systematic stocking in forest-meadow area first of all disappear high and semi high perennial grasses, forming the largest biomass: common timothy, tall oat grass, awnless brome, meadow fescue, red clover, meadow foxtail. These plants towering over the other attract animals in the first place. With frequent grazing such plants rapidly waste away and fall out of the grass cover giving way to low-growing and less productive grass – Kentucky bluegrass, fine bent grass, white clover and others which without encountering competition for light and nutrition from high grass begin to dominate in the grass cover.

At the further increase of stocking the stand composition changes more rapidly due to different eatability of plants. In these cases the plants, which are eaten most, also waste away and fall out of the grass cover. According to long-term observations, plants, grazed down 6–7 times during the summer, died or severely become sparse. On pastures remain inedible plants or plants, which are eaten not much: prostrate knotweed, silverweed, ladies'-mantle, dandelions, plantains and the like. The nutritional value of pastures for large herbivorous is falling, thus, to almost zero. The speed of pasture degradation depends on the stocking per unit of forage areas: the higher it is, the shorter the life cycle belonging to eatable species. Such dynamic change of grassland plants during high stocking from herbivores in all soil-climatic zones goes through procedure. The difference lies only in species composition of plant communities, replacing each other, and in the rate of substitution of one type by others. Rabotnov [14] pointed out that in England as a result of long unregulated grazing of sheep on pastures grew Nardus, bracken, heather, for example, they lost feed value.

The universality of the above-mentioned patterns, logically explainable by biological features of the different types of plants, from system approach should be seen as a defensive reaction of biogeocenoses from destruction of their grazing animals by decrease of stocking and in more general terms by reducing the population of herbivores in the region due to the lack of pasture forage. When the stocking of cattle on pastures decreases degraded pastures are recovered in the process of the so-called demutation [10]. However, an excessive stocking that may arise due to any reasons may not only lead to the degradation of pasture but also to pasture erosion, for example, to complete destruction of ecocenosis and destruction of the soil cover. So in modern conditions when excessive unregulated stocking of cattle in the mountains (the Caucasus, Altai, Buryatia) sod failure (formation of paths) may exceed 60% of pasture area and soil washout from deprived of vegetation places – more than half of its power. According some data [15], the annual reduction of soil profile in the Eastern regions of the Caucasus for this reason averaged 0.8 mm, for example, it decreased by 1 cm every 13 years. More striking manifestation of pasture erosion is observed on the Black Lands (Caspian lowland), where on winter pastures many years was converted cattle from different regions of the Northern Caucasus and Transcaucasia. As the result of excessive stocking here was almost completely destroyed not only the vegetation and soil, occurred desertification areas, up to the formation of drift sand.

But the decline, depression of pasture productivity is "the first line of defense" of grassy ecosystems. The second factor, or the mechanism of their sustainability, also directed against large herbivorous animals, triggering by the breach of the normal functioning of perennial grasses due to unfavorable weather conditions: long cold, drought, or vice versa waterlogging of the soil. In these conditions, ungulate excessive stocking on weakened plant communities may also lead to their death, sod destruction with subsequent erosion of soil cover. Protective function of herbaceous ecosystems in such situations is expressed in development in the organism of normal forage plants various substances, toxic for animals. This phenomenon is well known from literature [16, 17]. First of all, this accumulation in the herbs of nitrates by a violation of the synthesis of proteins and also retardation of the growth of plants due to adverse weather conditions: hailstorm and similar anomalies. The excessive consumption of such plants causes nitrate-nitrite toxicosis of animal. It is established that feed containing more than 0.07% $N\text{-}NO_3$, dangerous for animals and a doubling of the concentration can be fatal. In these conditions in plants of different families including Gramineae, Fabaceae, Cyperaceae hydrocyanic acid is formed by splitting of cyanogenic glycosides by relevant enzyme into glucose and hydrocyanic acid, which is a potent poison. Are marked many other toxicosis, caused by different glycosides, alkaloids, saponins, essential oils and other substances, resulting in elevated concentrations in plants at violation of the normal processes of growth and development. For example glycoside cumarin which is harmless in normal conditions and found in melilotus slow drying of wet habitats makes to dicumarin – highly toxic compound that causes the death of animals within 2–3 days. See the numerous cases of poisoning of herbivores due to eating plants affected by fungal diseases that infect weakened fodder plants. Characteristically, the most sensitive to the action of toxic compounds are the animals that came from other habitats (migrants), pregnant animal, young animals and the animals weakened due to starvation. Often there is the death of suckling calves because of the switch to the milk toxicants contained in the forage of the cows but not rendered visible harm to the health of the latter. The toxicity of many plants depends of the habitat, the phase of development and other factors that have been studied not enough. However, it is obviously that this property was developed by fodder grasses not accidentally and, at the system level, is the protection of biocenoses from herbivores with the deterioration of habitat conditions.

And, finally, on the last, the third obvious way to protect herbaceous ecosystems from destruction of the soil cover at the expense of weeds. It is in effect when the first two methods appear insufficient and animals, for whatever reasons, violate the integrity of the sod. It is the mass appearance of plants, usually called weeds. Among them are poisonous and indelible plants with distinct properties deter herbivores: toxicity, thorniness, hairiness, the presence of coarse stems, sharp smell and the like. To this group belong common thistle, plumeless thistle, sow thistle, black henbane, water hemlock, larkspur, aconite, white hellebore, datura – almost all the weeds of our gardens and arable lands – a total of more than 700 species, or about 15% of the floristic diversity of the natural pastures. The largest number of poisonous and noxious plants there are in the next families: Euphorbiaceae – 98% (29 species), Solanaceae – 97% (29 species), Equisetaceae – 81% (9 species), Ranunculaceae – 54% (117 species). Quite a lot of them are in Cruciferae (Brassicaceae) – 37% (60 species), Polygonaceae – 37% (39 species), Liliaceae – 26% (34 species). In the families Gramineae (Poaceae) and Fabaceae (Papilionaceae) there are 5% (25 and 28 species respectively), Cyperaceae – 1% (1 species). Thus, from more than 1000 species of Gramineae, Fabaceae and Cyperaceae only

54 species are dangerous for animals, whereas in the of miscellaneous herbs, which also include the weeds, there are about 700 species. Species diversity makes adequate herbage reactions to external influences depending on environmental conditions of their existence, including soil, intra- and interspecific, weather, phytosanitary, by a number of phytophages, and, as we can see, large herbivores.

In violation of the integrity of sod by ungulates animal's weeds quickly fill the gaps in the grass, preventing further appearance of animals on damaged areas. Striking the adequate responses of ecosystems to the negative impact of the animals: the greater the harm caused to the grass, the sharper repellent properties of weeds, which have experienced animals. In our time clearly this reaction can be traced about sheep enclosures, temporary stock stands of animals, where the degree of damage to turf decreases from the center of damage to the habitat periphery. The author of this study had to observe the emergence of a dense bed of a black henbane on the place of multiple milking in the valley of the small river in the Yaroslavl region, where the turf in the previous year was completely damaged on an area of about 100 m^2. On the periphery of the bed increased common thistles and musk thistles, and in the process of removal from the center of the bed their number and habitus respectively decreased. It is characteristic that in a year on the place of former bed was not one of the weeds; ring there was only green carpet of grass, although when seeding of black henbane millions of seeds of this plant were in the soil. And this is no accident. Williams [18] described in detail the change of tall weeds consisting of various weeds, to grasses with abandonment of arable land in fallow in the steppe zone. The fallow period with prevalence of tall weeds, which usually lasted one year replaced with the couch grass fallow (5–7 years), which in turn transformed into a solid fallow, consisting mainly of loose bunchgrasses (15–20 years), with a gradual transformation of the stipa steppe with the advantage of firm bunchgrasses. In conditions of formerly widespread in the South of Russia fallow farming system the restore of natural phytocenoses, appropriate to soil and climatic conditions of the steppe, lasted, so 20–30 years. On a similar scheme, but with a different floristic composition and duration of cenogenesis procenoses (intermediate cenoses) changed on the meadows of the humid zone in case of damage or destruction of the meadow sod by animals or technical means (with anthropogenesis). For example, on a small potato field (Tver region), previously fertilized by manure, procenoses of tall weeds a year after the end of treatment consisted of wormwood – *Artemisia vulgaris* L. (Figure 11.1), the following year, mainly from willow herb – *Chamerion angustifolium* L.

It weeds inhabit first of all arable land, because they are perceived by wildlife as areas with broken sod, requiring protection and restoration of natural vegetation – grass stand or forest depending on environmental conditions. If in modern conditions people define the stocking, in the nature it regulate themselves herbaceous communities, to be exact – herbaceous ecosystems, or ecosystems, including soil and local biota. These property ecosystems have developed in the process of long evolution (cenogenesis) and natural selection of systems: ecosystems failed to produce adequate protection, – have disappeared from the face of the earth as disappeared grass and soil on the Black Lands. But here in this guilty a man, against the arbitrariness of which nature was powerless. In natural condition, figuratively speaking, poison and antidote were developed simultaneously, and ecosystem natural selection happened, obviously, on the same principle as natural selection of individual species of plants or animals. Due to this in nature has been preserved dynamic equilibrium; the example of this equilibrium is the ecological balance in the system of predator-prey. The same

FIGURE 11.1 Wormwood (*Artemisia vulgaris*).

can be said about the dynamic equilibrium that exists between herbaceous communities, on the one hand, and grazing animals, on the other. In this case as a "predator" are herbivores (consumers), and in the role of "prey" – fodder grasses (producers). Upon termination of grazing occurs recovery (demutation) of productive phytocenoses that used in practice by providing pasture rest. Compilation of materials on the biological productivity of natural forage lands shows that it is inversely proportional to the intensity of grazing and is a regulator to stocking.

Interestingly, in phytocenological system of protection of soils from pasture erosion fit some insects. According Owen [19], in dry years in America locust destroyed up to 67% stock of phytomass on natural pastures, and the cattle had to be distilled in other places that he did not die of hunger. There was observed a direct link between stocking and number of locusts, the cause of which is the ecologist could not explain. In normal hydration or with a moderate stocking during the dry years the impact of the locusts was practically invisible or less significant. Similar phenomena have been observed previously in Russia, in particular in the Baraba steppe of Western Siberia [20]. From these facts it follows that the protection of herbaceous ecosystems. From these facts it follows that the protection of herbaceous ecosystems from stocking involved not only the vegetation, but also Orthoptera of this biotope, creating serious competition for large herbivores, reducing their numbers in case of increased threat to ecotope by the latter.

Appearing on the places of damaged sod, weeds treat juvenile undergrowth grass almost paternal care. Otherwise you will not say. First, it is the protection of seedlings of slow-growing perennial grasses from trampling by animals, second, from their grazing in young, immature age, thirdly, the accumulation of nutrients, especially nitrates, formed by mineralization of the destroyed sod, prevent them from losses due to leaching and denitrification, and, finally, the programmed destruction of weeds order to give the living place to the next vegetation formations, for example, cereal grass, passing it "inherited" nutrients, accumulated in the plant residues.

Although in the nature after community of tall weeds would grow usually wheatgrass as a dramatic example of nitrophyls, under the canopy of weeds with no less success you can grow types of forage grasses. This is evidenced by how scientific expertise and a wide practice of meadow grass cultivation. In particular, in the field experiment conducted in the state farm "Voronovo" of the Moscow region, on the fertile land of loam soil with coverless sowing in pure form or in mixtures of more than 10 types of cultivars of cereals and legumes, including oligotrophic, characteristic for the poor habitats (slough grass), in the year of planting white pig weed was growing abundantly. Weeding it manually on half of the area and leaving the other half before the end of vegetation did not reveal any significant difference in the condition of the herbage and yield of perennial grasses in any of subsequent years of research. These facts point to the specificity (commensalism) of relationships one-biennale weeds with perennial grasses developed under their canopy regardless of the floristic composition of each group: oligo-, meso- or eutrophes formed on the relevant fertility soils. However, fallow of tall weeds are able, apparently, to some extent, to control the floristic composition of the procenosis, which goes for a change. Research conducted in Timiryazev Moscow Agricultural Academy (V.A. Zvereva, unpublished data) on the effects of the extract from the seeds of sosnovsky cow parsnip (*Herackleum sosnowskyi*) on the germination of common valerian (*Valeriana officinalis*), St. John's wort (*Hypericum perforatum*), snow-on-the-mountain (*Euphorbia marginata*), green amaranth (*Amaranthus retrofiexus*) and harestail grass (*Lagurus ovatus*) showed that for two species of plants extract had inhibiting effect, for two other had stimulating effect and for one-additive. It should also be that the consistent successions (changes) of procenoses (intermediate cenoses) on the ruins of the sod from tall weeds to stable (climax) phytocenosis aimed at achievement of a definite purpose – to hold and accumulate in biogeocenoses formerly accumulated elements of mineral nutrition. This can be seen in the changing attitude of plants to have in the soil mobile nutrients, in particular nitrogen. The most demanding of them weeds such as mugwort (*Artemisia vulgaris*) white pig weed, common thistles, plumeless thistles, sow thistles, black henbane etc. High consumption of nitrogen weeds-eutrophes indicates at least such fact as the content of crude protein, not inferior legume grasses, from 18 to 28% (calculated on the dry weight).

From the system point of view, this change of plant groupings in place of the destroyed phytocenoses can be explained any otherwise than evolutionary developed way of herbaceous ecosystems to restore the "status quo" with the least loss of mobile plant nutrients from the destroyed sod formed by mineralization of its organic residues. One-biennale weeds, possessing powerful starting-growth and developing the greater weight, catch nitrogen and ash elements and when death passed them to subsequent herbaceous procenoses. Already loose bunchgrasses, which change rhizomatous grasses begin to inhibit the processes of mineralization of organic substances in soil that leads to a gradual accumulation of humus, and firm bunchgrasses with the prevalence of mycotrophic nutrition type complete the immobilization of nutrients transferring its main part in the organic form.

Significant role in the retention of nutrients in the soil at destruction of natural meadow turf by ungulates also plays a soil microflora. Now it is known [21], that in untilled soil at decomposition of plant residues most immobilization mineral soil nitrogen or nitrogen fertilizers [$N\text{-}NO_3$ and $N\text{-}NH_4$] by soil microorganisms occurs in the first 10–12 days, thus preventing its infiltration losses in the underlying soil and denitrification in the form of gaseous nitrogen forms. This process is accompanied by mineralization of organic substances in soil and plant residues with the release of mineral nitrogen, which is consumed by another group of soil microflora. With the advent of vegetation on the site of the destroyed natural turf role of soil microflora in the retention of nutrients from exogenous losses are gradually decreasing. According to Smelov [22], the number of microorganisms, mineralizing nitrogen of soil organic matter on the roots of loose bunchgrass – meadow fescue during 4 years of observations decreased from 9.2 to 1.4 milliards per 1 g of dry roots. Similar results were obtained with timothy. For the seventh year of life separate species of herbs accumulates from 34 to 47 tons of humus on 1 hectare transforming into immobile state 1.5–2 tons of nitrogen, from 0.5 to 1 ton of ash matter. From the results of our studies [23], it is also obvious that consort communications in biogeocenoses between herbs microflora and soil aim to the retention of nutrients in the soil including inhibition of nitrification, and more in sabulous in comparison with loamy. With equal doses of nitrogen fertilizers and almost the same removal of nitrogen with grass yield on irrigated pasture consisted mainly of cockfoot, and approximately equal to the content of mineral nitrogen in soil ratio nitrate form to ammonium one in the upper layer of sabulous on the average for vegetation period amounted 1–5, whereas in loamy – no more than 1–2.

Otherwise, the environment-forming role of biocenoses was reduced to a maximum retention of mineral nitrogen in its sphere adapting to the habitat nature (ecotope). Thus, even in the artificially created agrocenosis of cultural pastures under irrigation of sabulous where nitrates bigger risk of loss with infiltration waters, compared with loamy and where the best aeration, it would seem, must strengthen the processes of nitrification, the main fund of mineral nitrogen was presented ammonium form. Something of the kind also mentioned Korotkov [24] in his lysimeter studies with ^{15}N.

Characteristically, biological processes in the soil to some extent adapted by physical-chemical (bioinert) properties of these soils. According to our research, in the initial samples of poor sabulous and loamy soils the content of ammonia nitrogen was significantly higher in the loam at approximately equal to the content in both soil nitrate nitrogen (Figure 11.2). However, with sterilization samples and when washing them with a solution of ammonium nitrate followed by rinsing with distilled water it was found that ammonium stronger recorded sabulous soil than loamy (Figure 11.3). In other words, microbiological and physico-chemical processes in soils occur unidirectional – to hold

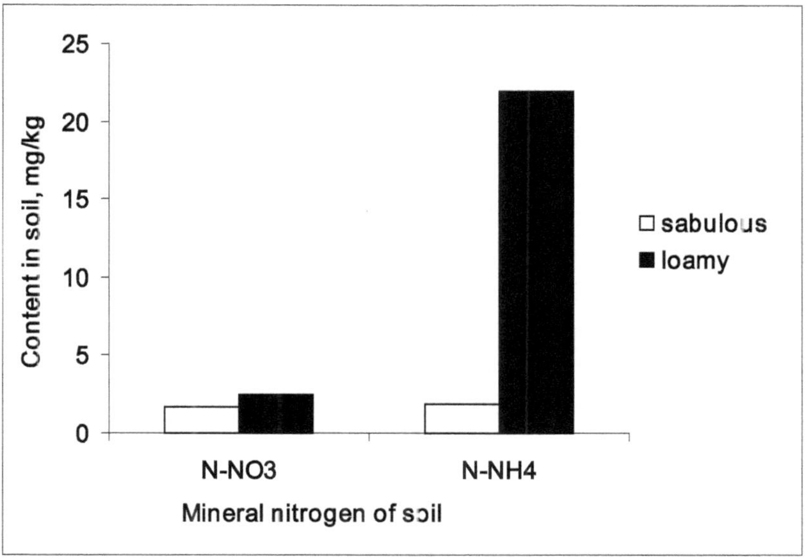

FIGURE 11.2 The content of mineral nitrogen in sabulous and loamy soils before application of nitrogen fertilizers.

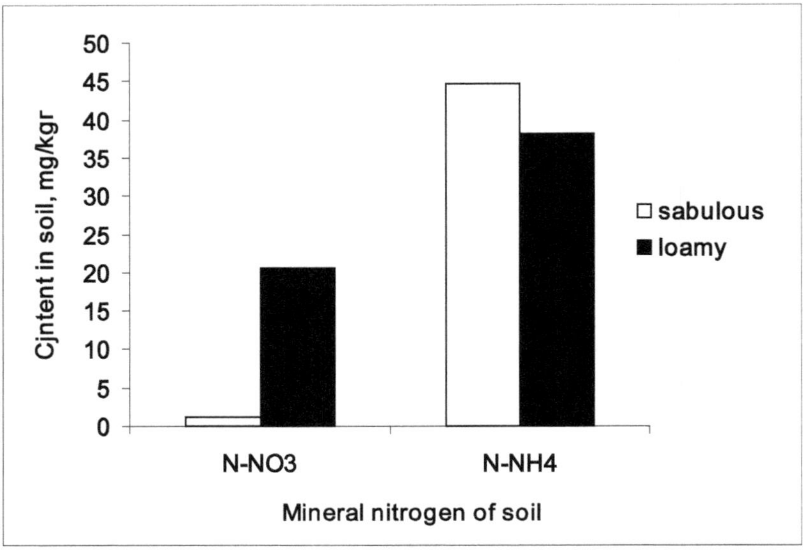

FIGURE 11.3 The content of mineral nitrogen in sabulous and loamy soils after application of nitrogen fertilizers.

nutrient elements in this context, nitrogenous substances, in strategic biogeocenoses and, above all, in their subsystems – soils, which are characterized as bioinert systems.

About that ecosystems, including soils, are self-regulatory systems, evidence and other facts. Kidin [21] in the experiments with ^{15}N established that with increasing doses of mineral fertilizers higher consumption of cultural plants increases the degree of immobilization of nitrogen by the soil microflora forming humus, on the one hand, and gaseous nitrogen losses of soil and fertilizer on physico-chemical and biological denitrification, on the other hand. It is known also, that the lack of soil nitrogen in natural plant communities appear accumulators of nitrogen – leguminous plants – clover, alfalfa, and other species, reducing then its abundance in phytocenoses in favor of grass species and diversity of herbs. It follows that the soils as if support within certain limits, the contents of the mobile nitrogen, the most sensitive than phosphate and potash, for the climax (relatively stable) plant communities.

On the whole, analysis of the results of research on retention in the system soil-plant of mineral forms of soil nitrogen shows that nitrate nitrogen is held mainly by soil microflora and plants: the soil microflora by placing it in an organic form (new growth of humus), plants due to the increase of biomass of weeds and the following plant associations, and also due to accumulation of nitrates (NO^-_3) in plant biomass. At that plants can accumulate in their biomass nitrates, except for the reproductive organs, without any functional limitations. In the soil environment, as shown above, nitrates can be lost due to denitrification and infiltration to groundwater. Ammonium nitrogen (NH^+_4) is held mainly by the soil, as in plants it cannot accumulate due manifestations of properties toxic to plants. It follows that between agents (subsystems) of the system soil-plant role of the main depot for nitrate nitrogen is given to the plant, for ammonium nitrogen – to soil.

In all these phenomena clearly one could trace antientropic trend of the herbaceous ecosystems in the soil, on the other hand, functioning, for example, accumulation and structuring of matter and energy (overground and underground mass of plants, biota and humus) in the number and proportion ensuring minimum loss into the environment. Soil cover of our planet was formed and stored due to this property of herbaceous (and forest) ecosystems. And not the last role in these processes played the ability of plant formation resist destructive effects on the soil by large herbivorous, their floristic biodiversity with severe soil-protecting functions.

11.4 CONCLUSION

In general biological sense, the sustainability of ecological systems is interpreted as the ability to resist the action or to return to the initial state after exposure. This chapter does not discuss factors of stability of herbaceous ecosystems associated with adaptive responses of plant communities on the change of soil, hydrological, meteorological and other conditions, which is the subject of synecology and is described by many geobotanists [14, 20, 25]. However, the subject of discussion in the work are the reactions of herbaceous ecosystems produced in the process of the evolutionary development of such ecosystems and aimed at soil conservation as a primary basis of their existence. Opened us at the system level, this aspect of the sustainability of natural ecosystems bases on known facts, directly or indirectly pointing to the specific, essentially passive, but adequate counteraction of phytocenoses to ungulates in danger of destroying the sod and soil cover. In general relations between herbivorous animals and phytocenoses obey the laws of biological systems "predator-prey" [10], where the role of "predator" belongs to herbivorous, and "prey" – phytocenoses. We found also the reason, functional predetermination of successive change of procenoses in the process of demutation (recovery) of destroyed vegetation which consists in evolutionary developed expediency of preservation and transfer of "inherited" moving nutrients produced when organic matter mineralization of former turf. When this was first shown soil-protecting role of weeds in nature, contrary to the established in the agriculture opinion of them as the plunderers of soil fertility [26]. We disclosed antientropic trend of successions herbaceous communities leading ultimately to the accumulation and structuration of mater and energy in ecosystems, ensuring their dynamic stability and, consequently, creation and preservation of soils in areas with grass vegetation.

From system positions seem unreasonable recommendations for implementation in forage production new fodder crops from plant species one way or another related to procenoses of tall weeds, for example, weeds with severe soil protection properties. An example is recommendation about cultivation for fodder mentioned above sosnovsky cow-parsnip [27] which did not lead to increased production of fodder bur caused the blockage this species many habitats suitable for its growth, in particular roadsides of roads and railways in the European part of Russia, other habitats with increased moisture and focal accumulation of nutrient elements.

KEYWORDS

- **animals**
- **ecosystems**
- **perennial grasses**
- **plants**
- **soil**
- **weeds**

REFERENCES

1. Sagatovsky, V. N. The experience of creation of a categorical apparatus of the system approach. "Philosophical Sciences" (in Rus.). 1976, № 3, p. 75 (in Russian).
2. Afanas'ev, V. G. System approach and society. Moscow: "Publishing House of Political Literature" (in Rus.), 1980, p. 14 (in Russian).
3. General biology with principles of historical geology. Moscow: "Higher School" (in Rus.). 1980, p. 4 (in Russian).
4. Chernikov, V. A., Aleksashin, A. V., Golubev, A. V. Agroecology. Moscow: "Kolos" ("Ear" in Rus.) Publishing House. 2000, p. 14 (in Russian).
5. Glinka, N. L. General chemistry, Leningrad: "Khimiya" ("Chemistry" in Rus.) Publishing House 1976, p. 185 (in Russian)
6. Krupenikov, I. A. The history of pedology. Moscow: "Nauka" ("Science" in Rus.) Publishing House 1981, 327 p. (in Russian).
7. Karpachevsky, L. O. Soil and pedosphere in space coordinates. "Agrarian Science" (in Rus.), 1995, № 4, 46–48 (in Russian).
8. Pavlovsky, E. S. Soil-protecting significance of natural forage lands. Natural forage resources of the USSR and their use. Moscow: Nauka ("Science" in Rus.) 1978, p. 74–78 (in Russian).
9. Rakitnikov, A. N. The use of natural forage resources as a factor of development of agriculture. Moscow: "Kolos" ("Ear" in Rus.) Publishing House. 1978, p. 35–47 (in Russian).
10. Reymers, N. F. Nature management. Moscow: "Mysl'" ("Thought" in Rus.) Publishing House, 1990, p. 152 (in Russian).
11. Sustainability of soils to natural and anthropogenic influences: Abstracts of All-Russian Conference. Moscow: Dokuchaev Scientific Research Institute of Pedology. 2002, 489 p. (in Russian).
12. Andreev, N. G. Meadland farming. Moscow: "Agropromizdat" ("Publishing House of Agricultural and Industrial Literature" in Rus.) Publishing House, 1985, 83–85 (in Russian).
13. Larin, I. V. Grassland science and pasture farming. Moscow-Leningrad: "Selhozgiz" ("Publishing House of Agricultural Literature" in Rus.) Publishing House, 1956, p. 63 (in Russian).
14. Rabotnov, T. A. Meadland farming. Moscow: Publishing House of Moscow State University. 1974, p. 349 (in Russian).
15. Erizhev, K. A. Mountain hay lands and pastures. Moscow: "Rodnik" ("Spring" in Rus.) Publishing House, 1998, 320 p. (in Russian).
16. Vilner, A. M. Forage poisonings. Moscow: "Kolos" ("Ear" in Rus.) Publishing House. 1974, 408 p. (in Russian).
17. Dimitrov, S. et al. Diagnostics of poisoning of animals. Moscow: "Agropromizdat" ("Publishing House of Agricultural and Industrial Literature" in Rus.) Publishing House. 1986, 284 p. (in Russian).
18. Williams, V. R. Collected papers. V.3. Moscow: "Selhozgiz" ("Publishing House of Agricultural Literature" in Rus.) Publishing House, 1949, 132–135 (in Russian).
19. Owen, O. S. The protection of natural resources. Moscow: "Kolos" ("Ear" in Rus.) Publishing House. 1977, 179–180 (in Russian).

20. Kurkin, K. A. Studies of meadow dynamics as systems. Moscow: "Nauka" ("Science" in Rus.) Publishing House, 1976, 287 p. (in Russian).
21. Kidin, V. V. Fundamentals of plant nutrition and application of fertilizers. Part 1. Moscow: Publishing House of Timiryazev Moscow Agricultural Academy. 2008, 415 p. (in Russian).
22. Smelov, S. P. Theoretical foundations of grassland science. Moscow: "Kolos" ("Ear" in Rus.) Publishing House.1966, p. 121–126 (in Russian).
23. Afanas'ev, R. A. Fertilizer of intensive irrigated pastures in the Nonchernozemic zone of the RSFSR. Summary of the doctoral thesis. Scriveri: Latvian Institute of Agriculture and Rural Economy. 1987, 44 p. (in Russian).
24. Korotkov, B. I. Regulation of water and nutrient regime of soils, organization and use of irrigated pastures: Irrigated pastures and hay lands in the Nonchernozemic zone. Moscow: "Rosselhozizdat" ("Publishing House of Russian Agricultural Literature" in Rus.) Publishing House, 1984, 9–12. (in Russian).
25. Sharashova, V. S. The sustainability of grassland ecosystems. Moscow: "Agropromizdat" ("Publishing House of Agricultural and Industrial Literature" in Rus.) Publishing House, 1989, 239 p. (in Russian).
26. Afanas'ev, R. A. Soil protection function of weed grasses in ecosystems. "Agricultural Biology" (in Rus.). 1983, №3, p. 11–15. (in Russian).
27. Vavilov, P. P. New forage plants. Field crops of USSR. Moscow: "Kolos" ("Ear" in Rus.) Publishing House. 1984, p. 59–63 (in Russian).

CHAPTER 12

A CASE STUDY ON COMPOSING PLANT-MICROBIAL ASSOCIATION FOR PHYTOREMEDIATION OF POLLUTED SOIL

A. A. FEDORENCHIK, N. V. MELNIKOVA, and Z. M. ALESCHENKOVA

Institute of Microbiology, National Academy of Sciences, Kuprevich str. 2, 220141 Minsk, Belarus, Fax: +375-17-267-47-66; E-mail: microbio@mbio.bas-net.by

CONTENTS

ABSTRACT

Plant-microbial association is made up by strain of alfalfa nodulating bacteria *Sinorhizobium meliloti* S3, strain of phosphate mobilizing bacteria *Serratia plymuthica* 57 and alfalfa cultivars (*Medicago sativa* L.) withstanding soil contamination with crude oil, diesel fuel and industrial oil.

Application of *S. meliloti* S3 + *S. plymuthica* 57 + *Medicago sativa* consortium intensifies crude oil (1%) disposal by 58.47% in 3 months.

12.1 INTRODUCTION

One of efficient methods for recultivation of polluted soil is phytoremediation engaging plant-microbial associations resistant to elevated contamination levels.

Many researchers reported beneficial effect of plant-microbial consortia on recovery of environmental media polluted with polycyclic aromatic hydrocarbons, synthetic surfactants, petroleum, chloroorganic, nitroaromatic, phosphoroaromatic compounds, other xenobiotics [1].

Elaboration of ecologically safe and economically justified methods aimed at restoring biological activity of contaminated and degraded agricultural land is extremely relevant for Belarus. Mechanisms of ecosystem autorecovery are rather complicated and time-consuming (over 10–25 years). Plants and associated symbiotic microorganisms are successfully used to raise efficiency of soil remediation. Favorable impact of plant species on soil regeneration is caused by their ability to uptake and transform toxic substances, to activate potential of microbial cenosis and, as a consequence, to intensify biochemical and chemical processes converting alien compounds in soil [2]. Plant survival in the polluted ground is promoted by contribution of rhizospheric microbial community, namely growth-stimulating activity and reduction of phytotoxicity due to degradation of the contaminants.

Application of natural adaptation mechanisms of plant-microbial associations lays the basis for modern biotechnologies aimed at remediation of impaired ecosystems. The shortage of land resources sets an urgent task to reclaim for agricultural management all types of contaminated and eroded soils. Nowadays plant species, including legume grasses are widely applied for recultivation of ecosystems polluted with oil and derived products. Alfalfa (*Medicago sativa L.*) may be used as a phytoremediating agent recovering soil biological activity. The problem of plant-microbial interaction directly boosting phytoremediation process is vital and it requires detailed investigation.

Objective of this research was to set up plant-microbial association providing for sustained growth and development of alfalfa and phytoremediation of soil contaminated with crude and industrial oil.

12.2 OBJECTS AND METHODS

The objects of studies were: strain of nodulating bacteria *S. meliloti* S3, strain of phosphate-mobilizing bacteria *S. plymuthica* 57, alfalfa (*Medicago sativa L.*), sod-podzol soil (humus – 2.38%, $pH_{(KCl)}$ – 6.0, P_2O_5 – 172 mg/kg soil, K_2O – 147 mg/kg soil, hydrolytic acidity – 1.82 mg/eq./100 g soil, total absorbed bases – 0.11 mg/eq./100 g soil), oil and derived products.

Activity of calcium phosphate dissolution by the tested bacterial strain was evaluated on liquid glucose-asparagine medium with phosphates [3]. Water-soluble phosphates wee determined according to Taussky and Shore, and at the end of experiment potentiometrically [4]. Amount of indolyl-3-acetic acid (IAA) produced by rhizobacteria was estimated by colorimetric method [5].

Sensitivity of rhizobacteria to crude oil and its derivatives was examined in submerged culture. 1% (v/v) pollutant was supplied into mannitol-yeast medium inoculated with test microbial culture in ratio 10% (v/v). Submerged fermentation was carried out in 100 mL Erlenmeyer flasks containing 50 mL of nutrient medium on laboratory shaker (agitation rate 200 rpm) at temperature 26±2°C during 48 h. Number of microorganisms was counted by dilution technique and plating of microbial suspension on agar nutrient medium. Residual oil levels in soil were measured gravimetrically [6]. Nitrogenase activity of rhizobia was assayed by acetylene method at gas chromatograph.

12.3 RESULTS AND DISCUSSION

Plant-microbial association is made up by strain of alfalfa nodulating bacteria *S. meliloti* S3, strain of phosphate mobilizing bacteria *S. plymuthica* 57 and alfalfa cultivars (*Medicago sativa* L.) withstanding soil contamination with crude oil, diesel fuel and industrial oil.

Strain of nodulating bacteria *S. meliloti* S3 shows nitrogen-fixing activity – 14.8 nM C_2H_4/hour and produces IAA in concentration 67 μg/mL. Treatment of alfalfa seeds by nodulating bacteria *S. meliloti* S3 resulted in 18% rise of green mass productivity. Studies on survival of strains *S. meliloti* S3 and *S. plymuthica* 57 during submerged mixed fermentation in media containing crude or industrial oil at 1% (w/v) concentration demonstrated that the pollutants did not inhibit growth and development of nitrogen-fixing and phosphate-mobilizing bacteria. Number of cells increased 1.3–1.7-fold, maximal cell titer recorded by 48 hours of fermentation equaled 7.1×10^9 and 5.8×10^9 CFU/mL, respectively.

Introduction of bacterial association *S. meliloti* S3 + *S. plymuthica* 57 into agrocenosis of *Medicago sativa* grown in soil contaminated with crude oil at 1% (w/w) concentration raised germination rate of alfalfa seeds by 9%, phytomass accumulation by 36%.

Plant-microbial association *S. meliloti* S3 + *S. plymuthica* 57 +*M. sativa* may be applied for remediation of soil polluted with petroleum, industrial oil and diesel fuel. The performed studies have shown that microbial treatment of alfalfa seeds intensified degradation of 1% crude oil in sod-podzol soil (Figure 12.1). Inoculation of alfalfa seeds with nodulating bacteria *S. meliloti* S3 accelerated oil decomposition (after 3 months) by 13.8% as compared to the control. Exposure of alfalfa seeds to phosphate-mobilizing bacteria *S. plymuthica* 57 and arbuscular mycorrhizal fungi (AMF) increased the rate of crude oil decay in soil by 28.4% and 22.7%, respectively. Maximal efficiency of petroleum disposal was achieved by plant-microbial association *S. meliloti* S3 + *S. plymuthica* 57 +*M. sativa* realizing the fastest process (58.47% up the control).

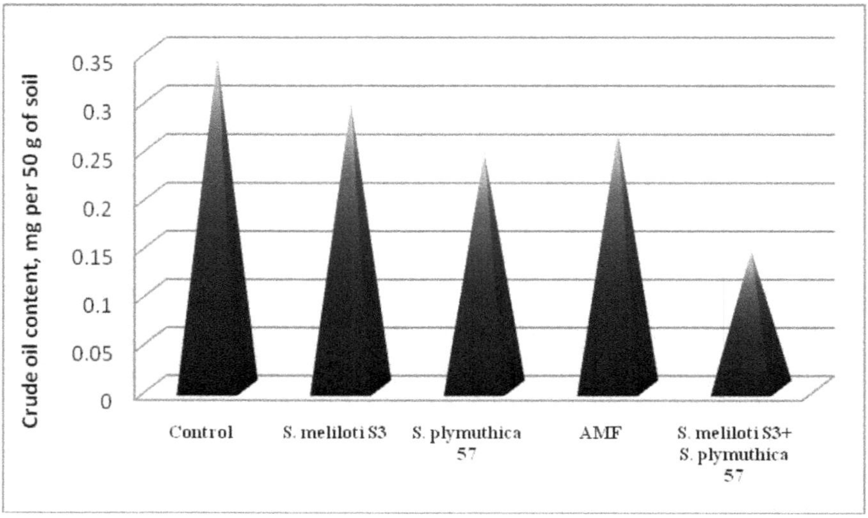

FIGURE 12.1 Effect of alfalfa seed microbial treatment on efficiency of crude oil decomposition by plant-microbial association.

Alfalfa is an excellent natural phytoextracting cultivar with respect to various pollutants. Increased share of leaves in overground part of the plant indicates efficiency of phytoevaporation as one of soil phytoremediation techniques. The foliage area of alfalfa plants in control variant equaled 43.3%. In experiments with soil introduction of industrial oil in concentrations 1 L/m^2 and 3 L/m^2 and seed treatment with nitrogen-fixing bacteria *S. meliloti S3* and phosphate-mobilizing cultures *S. plymutica 57* the leaf area rose to 49.2 and 52.5%, respectively.

Evaluation of solids content in alfalfa green mass revealed that in variants where seeds were processed with microbial association *S. meliloti S3 + S. plymutica 57* and soil was contaminated with industrial oil levels 1 and 3 L/m^3, the tested parameter dropped to 23.7 and 22.1%, respectively, in comparison with the control (24.9%). These variants were also distinguished by the highest foliation. Phytoevaporation is the capacity of plants to uptake oil hydrocarbons and maintain water balance, for example, draw from soil water and the pollutant. Reduced ratio of solids coupled to increased cell humidity is additional evidence of phytoevaporation efficiency in alfalfa crop. The most productive soil phytoremediation was accomplished by plant-microbial association *S. meliloti* S3 + *S. plymuthica* 57 + *M. sativa* in case of 1 L/m^2 industrial oil supply as confirmed by green mass of each plant and biomass harvest.

Falling solids ratio accompanied by larger leaf area of plants in variants with seed treatment by microbial association *S. meliloti* S3 + *S. plymuthica* 57 versus variants lacking microbial components unequivocally testifies to the involvement of bacterial tandem in oil degradation and soil remediation process.

Plant microbial association *S. meliloti* S3 + *S. plymuthica* 57 +*M. sativa* may be recommended for phytorecovery of soils contaminated with diesel fuel, crude and industrial oil. Application of the proposed plant-microbial consortium for phytoremediation of polluted soil could succeed only if hydrocarbon levels in soil don't exceed 1%. It should be noted that soil contamination is damaging to alfalfa seedlings, therefore crop plantings should be started with 1-month lag behind the pollution moment and seed doses should be increased by 20%.

Remediation of soil polluted with 1% spent industrial oil using alfalfa seed inoculation with microbial association enables to raise crop biomass productivity by 9.6 centners per hectare as compared to the control.

12.4 CONCLUSION

Application of symbiotic plant-microbial association is a promising method for phytoremediation of sod-podzol soil contaminated with crude oil and derived products.

KEYWORDS

- **alfalfa**
- **contamination**
- **crude and industrial oil**
- **degradation**
- **nodulating bacteria**
- **phosphate-mobilizing bacteria**
- **phytoremediation**

REFERENCES

1. Nazarov, A. V. Ilarionov, S. A. Biotechnology, 2005, 5, 54 (in Russian).
2. Diab, E. A. Res. J of Agric. Biol. Sci., 2008, 4(3), 604.
3. Palova, V. F. Agricultural Microbiology, 1982, 17(3), 321 (in Russian).
4. Umarov, M. M. Soil science, 1976, 11, 119 (in Russian).
5. Libbert, E. Physiologia Plantarum, 1969, 92, 51.
6. Lurje, A. A. Analytical Chemistry of Industrial Effluents. Chemistry, Moscow, 1984, 447 p. (in Russian).

CHAPTER 13

A CASE STUDY ON BIOLOGICAL ACTIVITY OF PERENNIAL GRASSES AND FIBER FLAX

GENRIETTA E. MERZLAYA and MICHAIL O. SMIRNOV

Pryanishnikov All-Russian Scientific Research Institute of Agrochemistry, d. 31A, Pryanishnikov St., Moscow, 127550, Russia, Phone: +7-499-976-25-01; E-mail: lab.organic@mail.ru, User53530@yandex.ru

CONTENTS

ABSTRACT

The use of fertilizers based on sewage sludge when optimizing their doses ensures improving the productivity of perennial grasses and fiber flax, increases the fertility of sod-podzolic soils, their biological activity, and does not cause heavy metal accumulation in soil and plant production.

13.1 INTRODUCTION

The most appropriate method for disposal of sewage sludge is recognized soil method with the cultivation of crops [1–7]. In Russia at wastewater processing annually you can receive more than 3.5 million tons of dry matter sludge. However, the use of sludge as fertilizer is often constrained due to adverse physical properties of their natural mass, the presence in them due to immature purification technologies increased concentrations of heavy metals in comparison with the standards, as well as due to the lack of reliable results on the effects of fertilizer on the basis of sludge in the system soil-plant. In this regard it is interesting accomplish studies to determine the effectiveness of sewage sludge and fertilizers of them in modern technologies of crop cultivation. Taking into account the above studies we have been conducted agro-ecological assessment of the actions of fertilizers based on sewage sludge coming from municipal treatment facilities.

13.2 EXPERIMENTAL PART

Fields experiments with fertilizers produced on the basis of sewage sludge were established in the Moscow and the Vologda regions. In the Moscow region, we accomplished micro field experiment on sod-podzolic clay loam soil when used as fertilizer composts from sewage sludge in city Moscow in different periods of storage and wood waste in the amount of 10% (calculated on the dry matter).

Compost 1 was prepared from fermented sewage sludge, coming directly from the filter-presses Kuryanovskaya aeration station, compost 2 – from the sludge of the same station after 10 years of posting on the sludge beds. For comparison, in the scheme of experiment we put options with bedding manure of cattle. All organic fertilizers were applied in two doses: 10 and 35 t/ha (calculated on the dry matter). It is important to note that the chemical composition of the sewage sludge of the Moscow aeration stations is subject to significant changes in time, which is explained by the improvement of the purification technologies and reduction of their discharge into drains mainly due to declines in many industries while market relations. In terms of experiment compost, which was prepared from sludge of long storage, was more polluted with heavy metals than compost of sludge, coming directly from the filter-presses.

Composts from sludge of various storage terms used in the experiment were characterized by high fertilizing value, contained organic matter – 48–52%, total nitrogen – 2–2.1% and had a neutral pH (Table 13.1).

From manure composts differed with less content of organic matter, nitrogen, potassium, but surpassed its phosphorus. Compost based on sludge

TABLE 13.1 Chemical Composition of Organic Fertilizers

Indicator	Bedding manure	Compost 1	Compost 2	The permissible content for sludge of groups [7]	
				1	2
Humidity, %	79.8	71.0	53.7		
Dry matter, %	20.2	29.0	46.3		
pH	7.0	7.4	7.2		
Content in dry substance:					
Organic matter, %	70.0	52.0	48.0		
Ash, %	29.8	48.0	52.0		
N %	2.7	2.0	2.1		
$N_{ammonium}$, %	0.06	0.01	0.01		
$N_{nitrate}$, %	0.02	0.02	0.04		
P_2O_5, %	2.4	5.3	5. 5		
K_2O, %	2.1	0.2	0.2		
C, %	35.1	26.0	24.0		
C:N	13	13	11		
Cu, mg/kg	36	425	1452	750	1500
Pb, mg/kg	6	50	167	250	500
Cd, mg/kg	2	8	42	15	30
Ni, mg/kg	16	104	353	200	400
Zn, mg/kg	160	1743	4589	1750	3500
Cr, mg/kg	60	147	774	500	1000
As, mg/kg	5	11	31	10	20

*Note 1 – Compost 1 was prepared from fermented sewage sludge, coming directly from the filter-presses Kuryanovskaya aeration station; Note 2 – Compost 2 was prepared from the sludge of the same station after 10 years of posting on the sludge beds.

from the filter-presses on the content of heavy metals met the standards [7]. At the same time compost from long-term storage on the sludge beds was contaminated zinc and cadmium, the contents of which, respectively, at 31 and 49% exceeded the permissible concentrations. The total amount of heavy metals in the compost was 2 times higher than in the compost based on sludge from the filter presses and 10 times higher than bedding manure.

Soil is sodpodzolic clay loam. In the original soil layer 0–20 cm contained 0.8% of organic carbon, 118 mg/kg of mobile P_2O_5, and 119 mg/kg of K_2O at pH 4.6.

When experimenting in 2000 we sowed cocksfoot – *Dactylis glomerata L.* – variety VIC 61 under the cover of spring barley "Zazersky 85." The experiment was carried out under the scheme: 1 – control without fertilizers; 2 – compost 1, 10 t/ha; 3 – compost 1, 35 t/ha; 4 – compost 2, 10 t/ha; 5 – compost 2, 35 t/ha; 6 – manure, 10 t/ha; 7 – manure, 35 t/ha. All organic fertilizers in both doses were applied into soil in 2000, in the following years we studied their aftereffects.

In the Vologda region, the studies were carried out in field experiment on sod-podzol middle loam soil, realized on three fields, which were introduced sequentially in 2010, 2011 and 2012. We studied the effect of compost from Vologda sewage sludge after biological treatment and peat in the ratio 1:1, which was applied for fiber flax. In the scheme of experiment were included variants with increasing doses of compost (2, 4, 6 t/ha dry matter), variant with mineral fertilizers – NPK (Nitrogen-phosphorus-kalium), EQ. (Equalent) – 4 t/ha of compost, variant – compost 2 t/ha + NPK, EQ. –2 t/ha of compost, as well as variant with granulated organic-mineral fertilizer, which was prepared of sewage sludge with the addition of nitrogen and potash fertilizers at the 5% of the active substance to the dry weight.

The compost contained (on dry matter) 2% of total nitrogen, 0.9% of phosphorus (P_2O_5), 0.3% of potassium (K_2O), and 67% of organic matter with pH 6.3. The content of heavy metals in compost was low (in mg/kg: Cd – 0.9, Zn – 147; Cr – 11.5; Pb – 11; Cu – 42; Hg – 0,097; Ni – 140. These compost indicators meet the standards of the Russian Federation.

Arable layer of sod-podzol middle loam soil before fertilizer application was characterized by a weak acid reaction (pH 5.1–5.3), contained 1.5–1.9% of organic matter, 230–290 mg/kg of mobile phosphorus (P_2O_5), 94–113 mg/kg of potassium (K_2O). The content of heavy metals in soil was low. Research in experiments was performed by standard methods [8]. Heavy metal content was determined in accordance with the methodological guidelines [9]. To determine carbon dioxide emissions from soil we used the method of infrared gasometry. Mathematical processing of experimental data was performed by the method of dispersion analysis [10] with the use of computer program STRAZ.

13.3 RESULTS AND DISCUSSION

According to results of researches, in the conditions of Moscow region the use of composting of sewage sludge in high doses compared with the control unlike variants with low doses significantly increased the yield of perennial grasses (Least significant difference (LSD_{05}) = 48 g of fodder units/ha), as evidenced by the data received in microfield experiment on the average for 10 years (Figure 13.1). At the same time bedding manure high doses during this period exceeded grass yield both composts in the same doses. It should be noted significant aftereffect of organic fertilizers in high doses in the next four years (35t/ha – calculated on the dry matter). While the aftereffect of

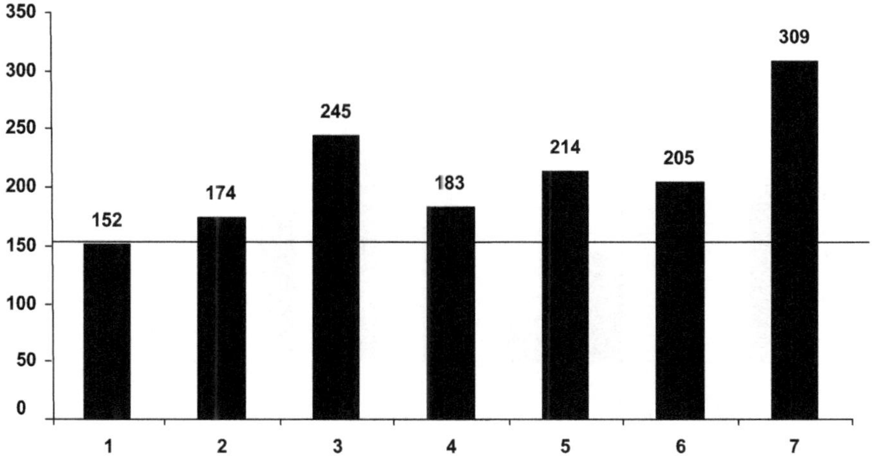

FIGURE 13.1 The yield of perennial grasses on average for 10 years (2000–2009), grams of fodder units with 1 m² (1 – control; 2 – compost 1, 10 t/ha; 3 – compost 1, 35t/ha; 4 – compost 2, 10 t/ha; 5 – compost 2, 35t/ha; 6 – manure, 10 t/ha; 7 – manure, 35 t/ha).

compost from sewage sludge of long-term storage manifested in 2010 and 2013, and the aftereffect of bedding manure – in 2010 and 2012.

The aftereffects of low doses of organic fertilizers (10 t/ha of dry matter) were shorter: the aftereffects of manure continued the first 2 years, the aftereffects of composts – 1 year. Despite many years of using herbage for hay, in the tenth year of experiment (2009), variants of high doses of compost and manure up to 32–57% remained valuable fodder culture planted in 2000 – cocksfoot (Figures 13.2 and 13.3). In control, where fertilizers were not applied, the content of cocksfoot in herbage was 51%, and invaded group of motley grass took 48%, for example, agrocenoses with application of high doses of manure and compost from long-term storage on the sludge beds according to the botanical composition approached control. It made little difference from the control on the content of the studied components in the variants on the 13th year of experiment. In this case, most cocksfoot (up 13%) remained in the use of high doses of manure – 35 t/ha of dry matter.

Under the influence of manure and composts changed agrochemical soil properties (Table 13.2), 2000 was the year of effect of fertilizer, and 2001, 2005, 2007, 2010, 2011 – years of aftereffect. In the year of application of

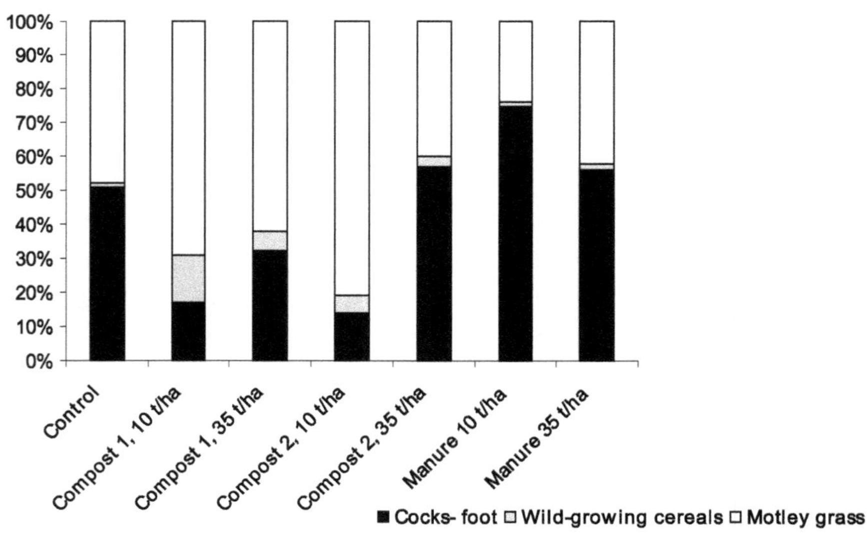

FIGURE 13.2 Botanical composition of cocksfoot herbage, 10th year of experiment.

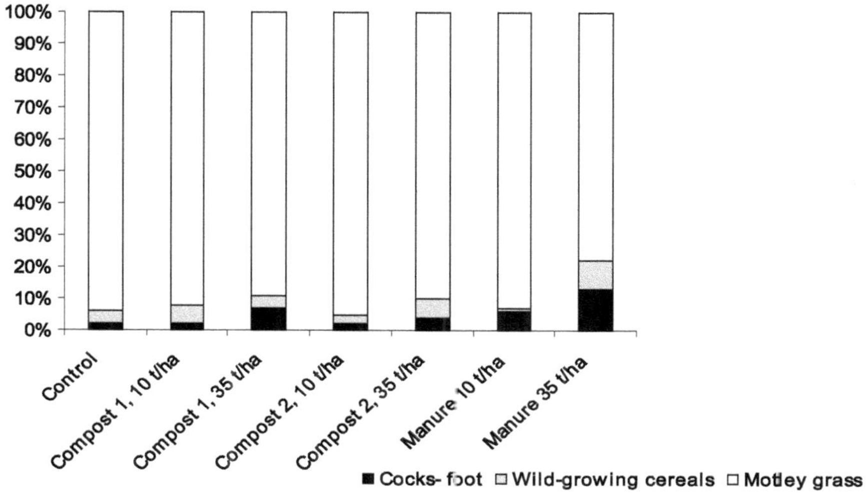

FIGURE 13.3 Botanical composition of cocksfoot herbage, 13th year of experiment.

TABLE 13.2 Agrochemical Properties of Soil

Type of fertilizer	2000 effect	The years of aftereffect				
		2001	2005	2007	2010	2011
		pH				
Control	3.8	3.9	3.9	4.4	4.2	4.4
Compost 1, 10 t/ha	3.8	4.1	4.4	4.2	4.3	4.5
Compost 1, 35 t/ha	4.2	4.5	4.8	4.5	4.5	4.6
Compost 2, 10 t/ha	3.9	4.0	4.3	4.5	4.2	4.5
Compost 2, 35 t/ha	4.1	4.5	4.5	4.5	4.4	4.5
Manure 10 t/ha	4.5	4.6	4.7	4.5	4.5	4.5
Manure 35 t/ha	4.5	4.6	4.7	4.5	4.5	4.6
		Humus, % C				
Control	0.75	0.72	0.69	0.93	0.99	0.95
Compost 1, 10 t/ha	0.75	0.69	0.64	0.87	0.92	0.92
Compost 1, 35 t/ha	0.79	0.89	0.98	1.01	0.95	1.06
Compost 2, 10 t/ha	0.71	0.69	0.69	0.83	0.86	0.90
Compost 2, 35 t/ha	0.79	0.89	0.81	1.06	0.89	0.93
Manure 10 t/ha	0.77	0.86	0.92	0.92	0.91	0.92
Manure 35 t/ha	0.88	0.95	0.98	0.98	1.02	0.97

TABLE 13.2 Continued

Type of fertilizer	2000 effect	The years of aftereffect				
		2001	2005	2007	2010	2011
		Mobile phosphorus (P_2O_5), mg/kg				
Control	110	110	111	114	105	99
Compost 1, 10 t/ha	180	125	110	205	152	145
Compost 1, 35 t/ha	320	300	310	380	370	322
Compost 2, 10 t/ha	160	140	90	180	148	141
Compost 2, 35 t/ha	220	240	260	415	277	285
Manure 10 t/ha	130	110	90	141	95	123
Manure 35 t/ha	270	220	180	247	168	183
		Mobile potassium (K_2O), mg/kg				
Control	96	96	55		92	69
Compost 1, 10 t/ha	101	99	39		98	68
Compost 1, 35 t/ha	90	102	39		98	71
Compost 2, 10 t/ha	95	96	35		99	57
Compost 2, 35 t/ha	99	100	42		100	70
Manure 10 t/ha	109	115	50		100	66
Manure 35 t/ha	120	109	98		118	75

all organic fertilizers pH in high doses markedly improved. This pattern remained during the next 5 years. Then to 7–11th year aftereffect noted the alignment of the pH values for different variants. In general you should specify that the manure and composts from both types of sludge produced a positive effect on the acid properties of the soil due to a neutral or slightly alkaline pH of the fertilizers and high content of calcium hear, which was capable with mineralization of organic matter enter into soil solution.

The content of organic carbon in the soil, compared to control was increased from applying high doses of all organic fertilizers used as a year of action, and in the aftereffect years. On the tenth-eleventh aftereffect years in comparison with the year of their introduction in all variants achieved a positive balance of organic substance in the soil. The same regularity was observed in the control variant. In general, we can say that the composting of sewage sludge in agrocenoses of perennial cereal grasses with their long mowing use improved humus status of sod-podzolic soils.

In the analysis of phosphate regime of the soil we marked its optimization under the influence of all kinds of organic fertilizers applied in high (35 t/ha) and low (10 t/ha) doses (except 2005). Content of mobile phosphorus in the soil increased more intensively, as a rule, when applying high doses of compost based on sludge from the filter presses. Application of compost containing potassium in lower doses practically had no effect on change of soil potash regime.

Much attention during the experiment was paid to the study of soil biological activity. Research results in 2008 jointly with the Department of ecology of Russian State Agricultural University – Timiryazev Moscow Agricultural Academy showed that composts on the basis of both kinds of sludge did not negatively impact on the microbial destruction of organic matter in the soil (in the layer 0–19 cm).

The total number of microorganisms on BEA (Beef-extract agar) and SAA (Starch-and-ammonia agar) in variants with organic fertilizers was close to the number on the control (11.6 million cells) and varied from 8.3 to 13.3 million cells in 1 g of dry soil. Higher values of this indicator were observed in the variants of long-term storage compost and manure used in high doses – 35 t/ha of dry matter.

In determining the activity of the carbon dioxide release from soils in the second year aftereffect of fertilizers we showed statistically significant changes of this indicator in relation to the control in versions with application of high doses of manure and both types of compost. And in variants with high doses of composts carbon dioxide emissions was more than when making a similar dose of manure. According the correlation analysis conducted on the basis of the experimental data; we established a connection CO_2 emissions with humus ($r = 0.87$) and with pH ($r = 0.67$) and grass yield (r = 0.79). Studying the activity of the carbon dioxide release from the soil at the end of the experiment (2012) showed (Figure 13.4) that in the variants with the organic fertilizers at the 11th year of their aftereffect the activity has declined sharply compared with the first term of definition (2001). However, high doses composts from sludge in relation to low doses contributed to the increase of CO_2 emission.

It is important to note that if you apply all studied fertilizers total heavy metal content in the soil does not exceed the standards of the Russian Federation. The heavy metal content in perennial grasses depended on the

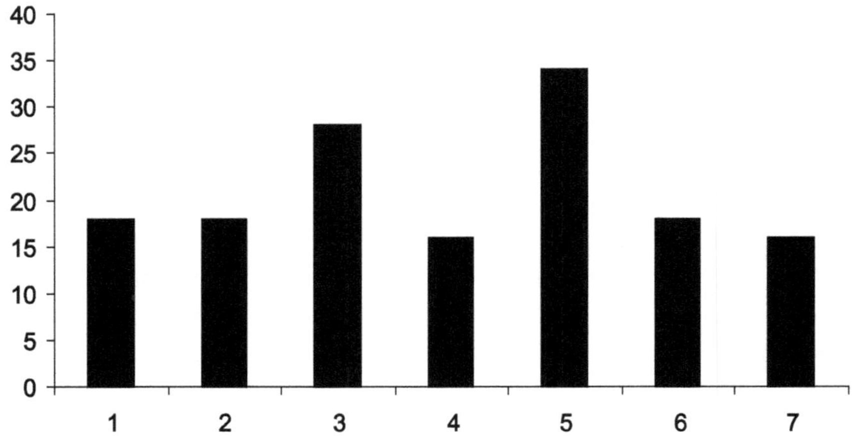

FIGURE 13.4 Carbon dioxide emission from soil (mcg C-CO_2/day) depending on the types and doses of manure, 2012 (1 – control, 2 – compost 1, 10 t/ha; 3 – compost 1, 35t ha; 4– compost 2, 10 t/ha; 5– compost 2, 35t/ha; 6 – manure, 10 t/ha; 7 – manure, 35 t/ha).

type and dose of composts, increasing by most indices at application of compost based on long-term storage of sludge with increasing dose, but not beyond acceptable levels.

Thus, on the basis of long-term studies in microfield experiment on sodpodzolic loamy soil we established that the application of organic fertilizers produced by fermentation of sewage sludge with wood waste with the optimization of their doses increased the biological activity and soil fertility, preserved the biodiversity of cenoses for long time, contributed to productive longevity of perennials. The aftereffect of compost based on sewage sludge in doses of 10 tones dry matter /ha was noted in the course of one year, bedding manure – for two years. The composts of sludge and manure when making in high doses – 35 t/ha of dry matter were characterized by a long aftereffect, which could be traced in the course of 10 years or more.

When studying the effect of fertilizers from sewage sludge of city Vologda on the yield of flax straw it was found that on average for 3 years experiment (2010–2011) compost application in a low dose of 2 t/ha was not been effective. At the same time increasing the dose of compost in 2 times provided a reliable response in flax straw yield. However, further increasing the dose of compost (3 times) was found to be inappropriate (Figure 13.5).

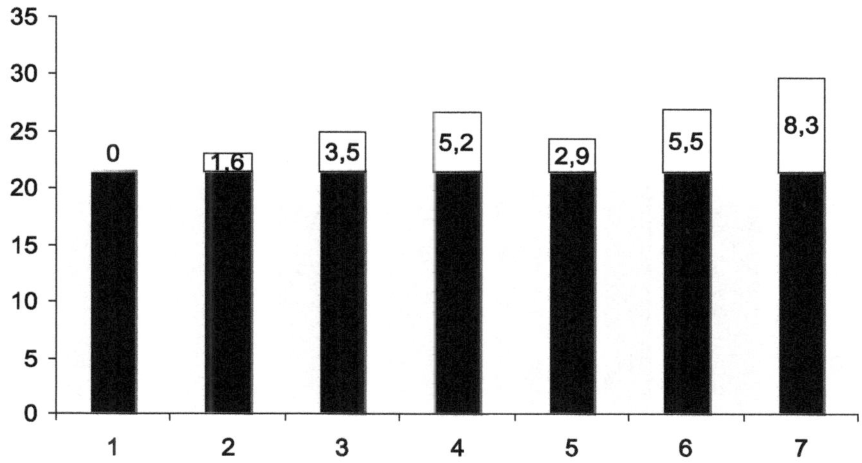

FIGURE 13.5 The yield and increase in yield of flax straw, c/ha.

The highest yield of flax straw 27–29.8 c/ha and statistically meaningful yield response in relation to the control without fertilizers at the level of 26–39% were achieved applying granulated organic-mineral fertilizers at a dose of 4 t/ha and combination compost with complete mineral fertilizer in variant 2 t/ha compost + NPK, EQ. t/ha compost.

The use only mineral fertilizers in doses equivalent to the sum of NPK 4 t/ha compost, though gave a true increase in the harvest of flax straw to control, but yielded almost all variants of experiment with organic fertilizers (the exception was only the lowest dose of compost – 2 t/ha).

On the yield of flax seeds, with humidity of 12% on average for 3 years (2010–2012) the most significant influence exerted the compost in a doze 6 t/ha and granulated organic mineral fertilizer that provided by 3.3 and 3.4 c/ha of seed or 43–48% above control, as well as organic mineral variant (2 t/ha compost + NPK EQ. 2 t/ha compost) where seed yield was 3–3.1 c/ha, that is by 30–35% exceeded the control (Figure 13.6).

According to the results of chemical analysis (Figures 13.7 and 13.8), the use of fertilizers based on sewage sludge in all the analyzed variants has not led to the accumulation of heavy metals in plant products. The contents of lead, cadmium, mercury and arsenic in flax straw and flax seeds when fertilizing sewage sludge were in permissible limits.

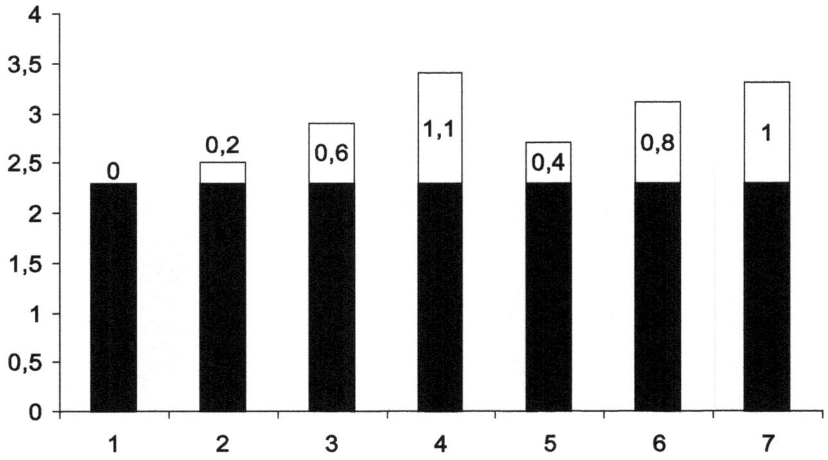

FIGURE 13.6 The yield and increase in yield of flaxseed, c/ha.

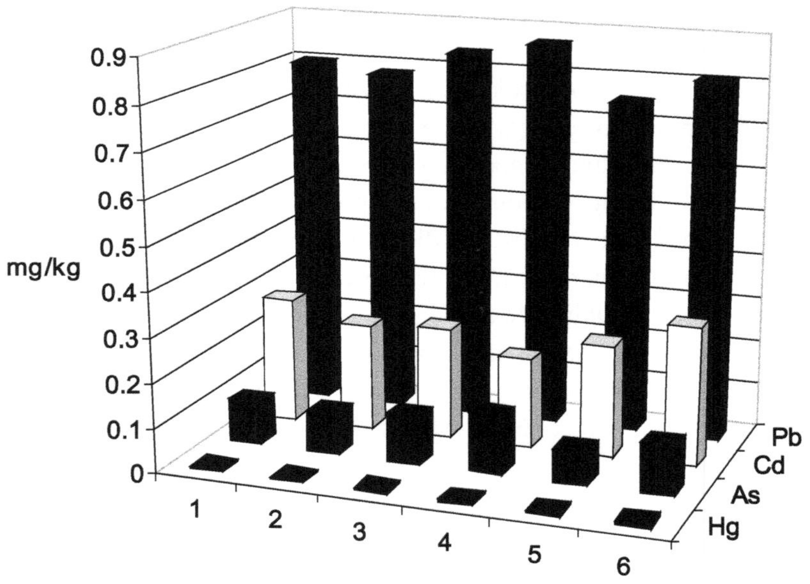

FIGURE 13.7 The content of heavy metals and arsenic in the straw of flax (1 – control, 2 – compost, 4 t/ha, 3 – NPK, 4 – compost + NPK, 5 – organic-mineral fertilizer, 6 – sewage sludge 4 t/ha).

We have not established appreciable influence of fertilizers produced from sewage sludge, on the accumulation of heavy metals and arsenic in the soil when compared with the control (Table 13.3).

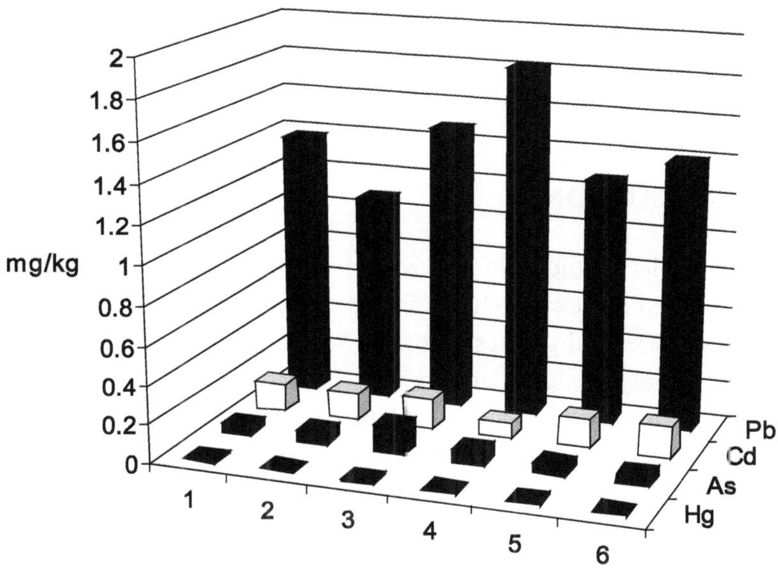

FIGURE 13.8 The content of heavy metals and arsenic in the seeds of flax (1 – control, 2 compost, 4 t/ha, 3 – NPK, 4 – compost + NPK, 5 – organic-mineral fertilizer, 6 – sewage sludge 4 t/ha).

TABLE 13.3 Influence of Fertilizers on the Total Contents of Heavy Metals and Arsenic in the Soil, mg/kg

Type of fertilizer	Cu	Zn	Pb	Cd	Ni	Cr	Hg	As
Control	5.5	24.1	5.9	0.35	10.0	8.5	0.026	2.23
Compost 2 t/ha	5.0	22.9	5.1	0.37	9.2	8.6	0.024	2.07
Compost 4 t/ha	5.1	22.9	5.5	0.35	9.3	8	0.023	2.36
Compost 6 t/ha	5.1	22.8	5.1	0.38	8.9	8.2	0.022	1.95
NPK, EQ. 4 t/ha of compost	6.0	25.0	5.7	0.43	9.0	7.9	0.027	2.09
Compost 2 t/ha + NPK, EQ. 2 t/ha of compost	5.8	24.2	5.9	0.39	9.7	8.2	0.027	1.97
Granulated organic mineral fertilizer, 4 t/ha	5.9	24.1	5.1	0.42	8.6	7.8	0.026	1.94
CC/RAC*	33–132	55–220	32–130	0.5–2.0	20–80		2.1	2–10

Note: CC – Critical concentration of every element; RAC – Roughly allowable concentration of every element [7]; EQ. – in a quantity equivalent to 4 (the 6th line) or 2 t/ha (the 7th line) of compost.

The content of all the investigated heavy metals in the soil of experiment variants with fertilizer application, according to 2012, was below the limits of CC/RAC, established in Russia.

13.4 CONCLUSION

Thus, the results of studies in field experiments have shown that the use of fertilizers based on sewage sludge in sod-podzolic soils reliably increased the yield of perennial grasses and flax. When optimizing doses of fertilizers from sewage sludge plant products (hay of perennial grass, straw and seeds of the flax) met Russian standards of content of heavy metals and arsenic. While the levels of heavy metals and arsenic in the soil were low and did not adversely affect on its ecological status.

KEYWORDS

- **fertilizers based on sewage sludge**
- **heavy metals**
- **sod-podzolic soils**
- **soil**

REFERENCES

1. Pryanishnikov, D. N. The selected works. Vol. 3. General issues of agriculture and chemicalization. Moscow: "Kolos" ("Ear" in Rus.) Publishing House, 1965, 639 p. (in Russian).
2. Resources of organic fertilizers in agriculture of Russia (Information and analytical reference book) (Ed. Eskov, A. I.). Vladimir: Russian Research Institute of Organic Fertilizers and Peat, 2006, 200 p. (in Russian).
3. Sanitary regulations and norms 2.1.7.573–96. Hygienic requirements for the use of wastewater and their sludge for irrigation and fertilization. Moscow: Department of Public Services of Russian Federation, 1997, 54 p. (in Russian).
4. Pahnenko, E. N. Sewage sludge and other non-traditional organic fertilizers. Moscow: Laboratory of Sciences, 2007, 311 p. (in Russian).

5. Ladonin, V. F., Merzlaya, G. E., Afanasev, R. A. Strategy of use of sewage sludge and composts on their basis in agriculture. Ed. Milashchenko, N. Z. Moscow: Agroconsult. 2002, 140 p. (in Russian).
6. Sychev, V. G., Merzlaya, G. E., Petrova, G. V. Ecologo-agrochemical properties and efficiency of vermi- and biocomposts. Moscow: Pryanishnikov Agrochemistry Research All-Russian Institute. 2007, 276 p. (in Russian).
7. State Standard of the Russian Federation R 17.4.3.07–2001, The nature conservancy. The soil. Requirements to the properties of sewage sludge when used as fertilizer. Moscow: Information-publishing center of the Russian Ministry of health. 2001, (in Russian).
8. Program and methodology of research in geographic network of field experiments on integrated application of chemicals in agriculture, Moscow: Pryanishnikov Agrochemistry Research All-Russian Institute. 1990, 187 p. (in Russian).
9. Collection of methods for the determination of heavy metals in soils, in hothouse soil and crop products. Eds. Ovcharenko, M. M., Kuznetsov, N. V. Moscow: Department of Agriculture and Food of Russian Federation, 1998, 97 p. (in Russian).
10. Dospehov, B. A. Methodology of field experiment. Moscow: "Kolos" ("Ear" in Rus.), 1979, 416 p. (in Russian).

CHAPTER 14

A CASE STUDY IN PRECISION AGROTECHNOLOGIES

RAFAIL A. AFANAS'EV

Pryanishnikov All-Russian Scientific Research Institute of Agrochemistry, d. 31A, Pryanishnikova St., Moscow, 127550, Russia, Tel: +7-499-976-25-01; E-mail: rafail-afanasev@mail.ru

CONTENTS

ABSTRACT

The chapter states regularities of the within-field variation of soil fertility, which are important for variable rate fertilizer application under conditions of precision agrotechnologies inclusive the technologies limiting agroeconomic efficiency. As is well known the usual (traditional) fertilizer practice stipulates their application taking into account-averaged indices of soil fertility: mobile plant food elements (N, P, K, etc.) content in the plow layer. At the same time a part of the plants gets excess of mineral nutrition, and other part – its deficiency. That results in shortage of agricultural products, its deterioration and also the pollution of the environment and the soil with the excesses of agrochemicals in overfertilized plots. In the last decades traditional technologies give place to high-precision agrotechnologies with differentiated fertilizer application taking into account within-field heterogeneity of soil fertility. There are several constraints for widespread adoption of high-precision agrotechnologies inclusive an underestimation of the character of within-field variability of soil quality. Our investigations reveal eight regularities of the within-field variation of agrochemical indices, which characterize soil fertility in arable soils. These regularities would allow more seriously estimate the efficiency of variable rate fertilizer application under conditions of precision agrotechnologies.

14.1 INTRODUCTION

The main procedures of precision agriculture include differentiated agrochemical application when taking into account within-field variability of fertility and crop status. The agrochemicals are mineral and organic fertilizers, amendments and other inputs. The usual (traditional) fertilizer practice utilizes average rate of fertilizer usage for an individual field. As a rule both procedures use soil analysis when calculating optimal rate of fertilizer. Users employing the traditional technology calculate the optimal rate by averaging results of soil analysis of the whole field; the alternative technology prescribes averaging the results for every within-field contour. Since now there are not visual contours of within-field boundaries all the processes of differentiated application of agrochemicals use satellite navigation system (GPS – Global Positioning System). There are many ways of admeasurement of these contours, which are differentiated one from another by fertility level. These ways are small-grid sampling and the agrochemical analysis of these samples throughout an entire field with later geostatistic data processing; preliminary yield, electroconductivity or landscape scanning; remote (aerospace) sensing of earth surface. Sampling and analyzing are carried out within the bounds of these contours formerly a priori distinguished and created by some way for calculating differentiated fertilizer and amendment rate and their application realized off-line.

For differentiated nitrogen additional fertilizing of vegetating crops they use the photometric methods of nitrogenous nutrition diagnostics by biomass green color or determine elasticity of herbage by crop-meter. The diagnostic devices of these technologies are coordinated with fertilizer applicator, which works on-line. This is the general scheme of preparation and carrying out of procedures for differentiated application of agrochemicals under conditions of precision agriculture.

Accordance with the usual standpoint of ordinary agrochemists they hope for increasing returns under conditions of differentiated application as compared with application by averaged rate because of different fertility of individual plots. Due to this additional expenditures caused by placing of hard- and dataware of high-precision agrotechnologies might be compensated. However, the experience of many agriculturalists shows necessity elaborating new ways of decision the problem. In particularly investigations performed in the USA (United States of America) (state Idaho) [1] over a 30-year period with conventional and variable rate nitrogen fertilizer application (data obtained from a seed potato operation) indicated discouraging result: variable rate nitrogen application was found to be unprofitable for the field when compared uniform nitrogen application since the total costs associated with variable rate fertilizer application outweighed the benefits obtained from maintaining the optimal plant available nitrogen levels. This is not unique information concerning the theme. Besides that there are a certain traditional character of agriculture and a sluggishness of land-users thinking. As a result in the last decade there is the decline of interest in the differentiated application of fertilizers. The decline could be explained by the cyclical nature of the development of new technologies [2].

Our investigations show that for practical use of differentiated application of fertilizers it is important more accurately take into account features of within-field variability of soil fertility and mineral nutrition including the features limiting prospective efficiency.

14.2 EXPERIMENTAL PART

The main methods of the investigations were statistical and graph-analytical analysis and the generalizations of the agrochemical characteristics of the plow layer of sod-podzolic clay loam soil. Site: Testing area of the Central Experiment Station of the Pryanishnikov All-Russian Agrochemistry Research Institute (Moscow region). Testing area was a part of cropping rotation field that comprised about 4 ha (200 × 200 m^2). The part included 400–10 × 10-m^2 plots on which 400 composite soil samples were taken and analyzed. Taking soil samples and their agrochemical analysis were carried out in accordance with the methods used in the agrochemical service and Russian scientific institutes. Humus (organic carbon) content was analyzed by Tiurins method (oxidation of soil organic matter in $K_2Cr_2O_7$ + H_2SO_4), mobile potassium and mobile phosphorus – by Kirsanov's method (0.2 normal HCl-soil extract), easy-mobile potassium and easy-mobile phosphorus -in low saline $CaCl_{2-}$extract, easy-hydrolysable nitrogen – in 0.5 normal H_2SO_4, N-NO_3–in H_2O-extract, pH – in saline suspension (KCl 1 normal) [3].

14.3 RESULTS AND DISCUSSION

We revealed eight regularities concerned within-field variability of agrochemical indices. The regularities would be important for technologies of precision agriculture. In the first place it was found that the soil reaction as well as humus and mobile phosphorus and potassium content more or less corresponded to a normal distribution.

In fact plots with the relatively average values of the easy-hydrolysable nitrogen, mobile phosphorus and potassium content occupied more than half of the area, while the number of plots with bulk and minority of the content were noticeably less (Figures 14.1–14.3). Thus, 350 plots contained 101–250 mg/kg mobile P_20_5 in their soil, while plots with lower and higher values of these indices occupied minimum square. Similar results characterized territorial distribution of sites with different easy-hydrolysable nitrogen and mobile potassium.

The second feature of soil spatial structure as used here is that the maximum characteristic of agrochemical indices variability (easy-hydrolysable nitrogen, mobile potassium and mobile phosphorus) was found in plots that had relatively smaller and larger values of the content while decreasing the variability in average interval (Figures 14.4–14.6). When investigating the plots which had the average values of mobile phosphorus

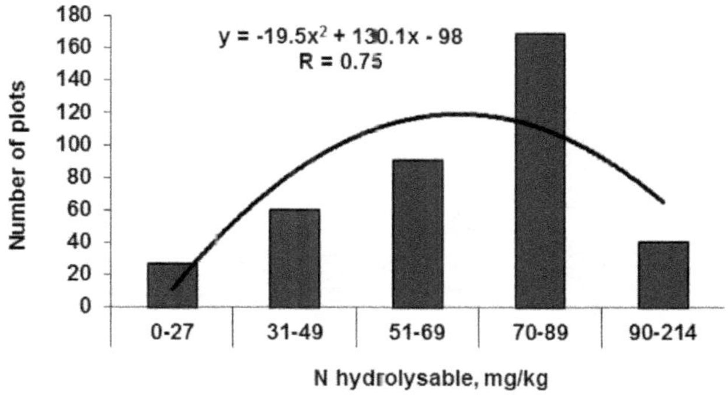

FIGURE 14.1 Curve of the distribution of the number of 400 plots versus N hydrolysable content in the CES testing area soil.

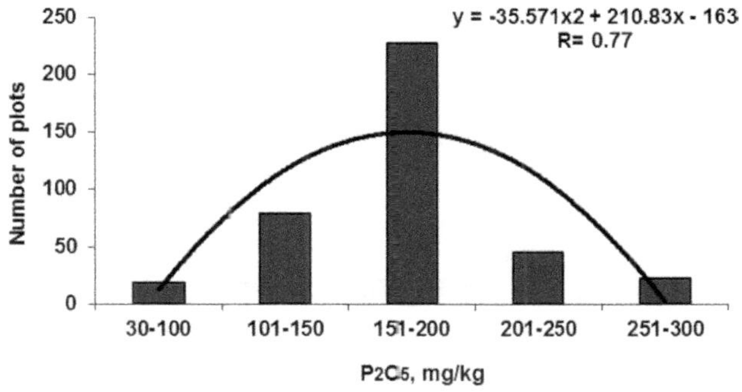

FIGURE 14.2 Curve of the distribution of the number of 400 plots versus mobile phosphorus (P_2O_5) content in the CES testing area soil.

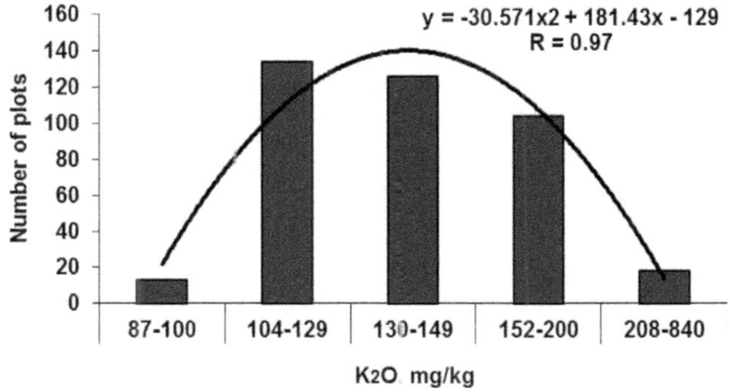

FIGURE 14.3 Curve of the distribution of the number of 400 plots versus mobile potassium (K_2O) content in the CES testing area soil.

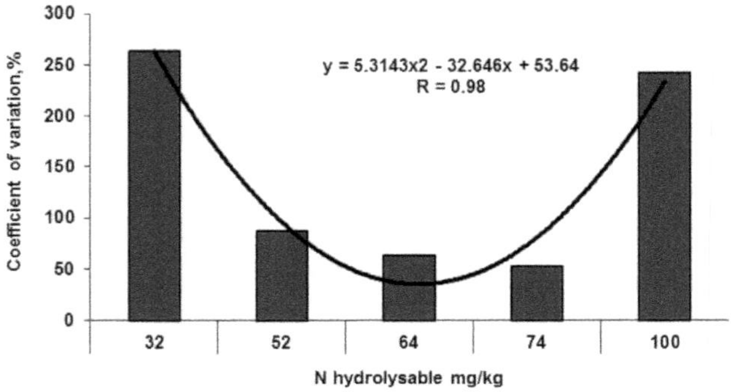

FIGURE 14.4 Variability of N hydrolysable content in soils of the 400 plots of agricultural test area.

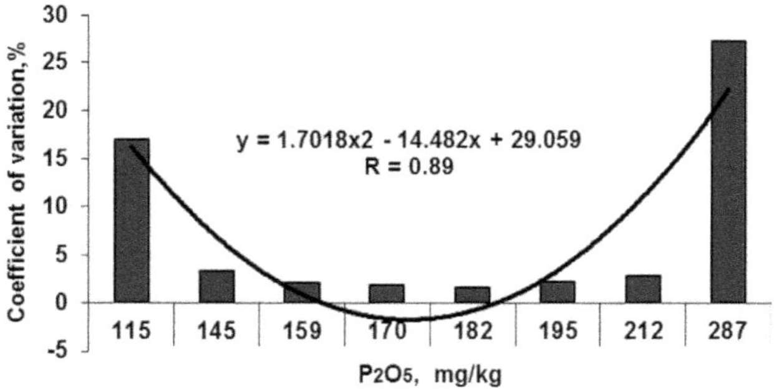

FIGURE 14.5 Variability of mobile phosphorus content in soils of the 400 plots of agricultural test site.

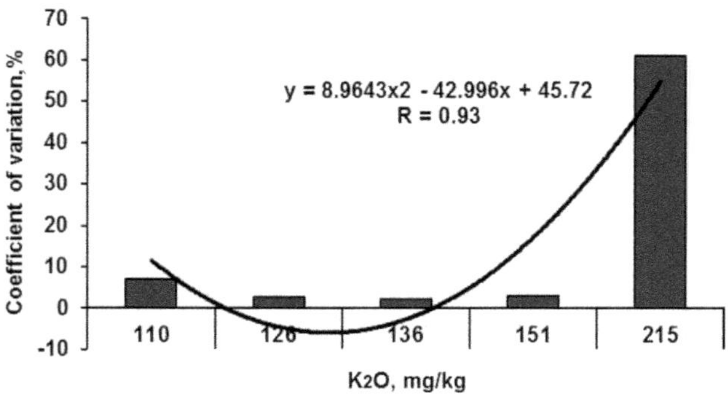

FIGURE 14.6 Variability of mobile potassium content in soils of the 400 plots of agricultural test site.

content we found that coefficient of variability of these indices was smaller than 5%, while it exceeded 15–25% for the plots with marginal indices. Coefficient of correlation between the theoretical calculations and the facts (R) was 0.89.

In accordance with the second regularity agroeconomic efficiency of variable rate fertilizer application in specific case will depend on the ratio of plots with marginal values of agrochemical characteristics to plots with their average characteristics: the efficiency increased when the ratio was more. But number of plots with marginal characteristics of the soil, according to the first regularity, was at the edges of the normal distribution curve, for example, their total area was far smaller than the area of the plots with average agrochemical characteristics for a given field. So the efficiency of variable rate agrochemicals applications determinates by these regularities.

The third regularity: intrafield variability of soil fertility can surpass inter-field variability. This was because an increase of the land area which was used for averaging-out of agrochemical indices reduced inter-contour variability and increased within-contour variability and vice versa. The regularity was found in the testing area of the Central Experiment Station of the Pryanishnikov All-Russian Agrochemistry Research Institute (Table 14.1).

According Table 14.1, variability of within-group humus as well as mobile P and K considerably (ten times as large) increased with increasing elementary plot area from 0.1 to 4 ha. It is necessary to take into

TABLE 14.1 Dependence of Variability of Soil Agrochemical Indices on the Area of the Plots Being Averaged

Number of plots	Plot area, ha	Coefficients of variation of agrochemical characteristics, V%		
		Humus	P_2O_5	K_2O
40	0.1	0.8	2.1	2.4
20	0.2	1.6	3.5	4.6
8	0.5	3.2	7.3	10.0
4	1.0	5.6	12.6	17.1
2	2.0	10.0	20.8	27.6
1	4.0	18.6	30.7	41.2

account this regularity when developing high agrotechnologies finding a compromise between the expediency of distinguishing fertility contours with minimum intracontour variability of soil fertility for increasing fertilizer efficiency, on the one hand, and the minimum number of such field contours for reducing the cost of taking and analyzing soil samples, on the other. The most acceptable is selection by methods of geostatistics in one plot as a rule no more than 5–6 contours with different level of soil fertility. Thus, it is necessary use every agrochemical index according to general area of a field. The fourth regularity of spatial heterogeneity of the soil cover consists in the smooth, gradual transition from the maximum values of the agrochemical indices to smaller and vice versa from smaller to maximum (Figure 14.7).

Line of trend that describes the change of mobile phosphorus content of 200m transect of the agricultural test site in limits 170–276 mg/kg P_2O_5 approximates the facts with coefficient of correlation $R = 0.88$. Consideration of this characteristic of the soil structure is very important from a practical viewpoint since when designing and creating machines for variable rate agrochemicals application it allows providing for a comparatively smooth change of their rates during travel over the field. This facilitates both designing and operating processes of fertilizer and amendment application under production conditions.

The fifth regularity of within-field variability of agrochemical indices in this context is the dependence of the level of soil fertility level on the meso- and microtopography of fields, which to a considerable extent causes all the above-mentioned regularities. Soils of one topographic location influence surrounding soils by leaching, nutrient transfer and

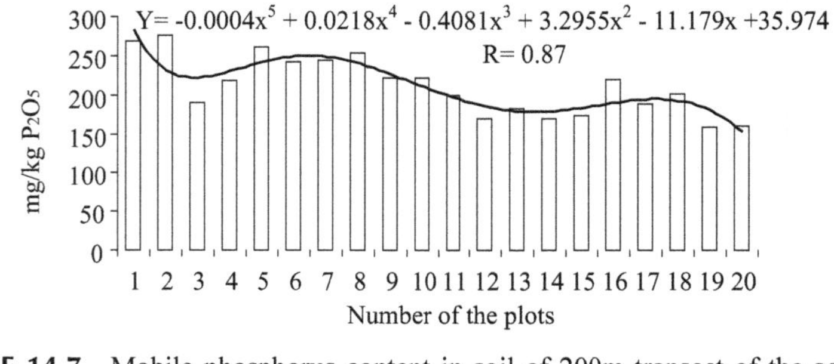

FIGURE 14.7 Mobile phosphorus content in soil of 200m transect of the agricultural test site.

deposition of chemical components. Under conditions of the testing area $N\text{-}NO_3$ content in soil of elevations prevails over $N\text{-}NH_4$ content, whereas in depressions ammonium-nitrogen content exceeds nitrate level due to greater anaerobiosis and reduction of nitrates migrating with subsurface and intrasol flow into depressions.

Greater mobile phosphorus and potassium content was also discovered in depressions (Figure 14.8).

This regularity allows the use of the results of topographic survey for preliminary revealing of fertility contours as elementary plots for soil sampling and agrochemical analyzing for the purpose of decreasing agrochemical soil analyzing costs compared to traditional grid sampling.

The sixth regularity of spatial heterogeneity of soil fertility represents noncoincidence of boundaries of various agrochemicals contours among themselves. The regularity was confirmed by the result: the coefficients of correlation between various agrochemical indices of intrafield contours were insignificant (Table 14.2).

Although there are similar tendencies of the spatial distribution of soil agrochemical characteristics between its separate parts in the limits of a field due peculiarity of the landscape corresponding to the sixth regularity nevertheless considerable differences of fertility of different intrafield parts are obvious. Therefore, for each within-field territorial contour different rates and ratios of nutrients in fertilizer corresponding to the agrochemical characteristics of this contour are necessary. Taking into account this thesis of the technology of variable rate fertilizer application it is

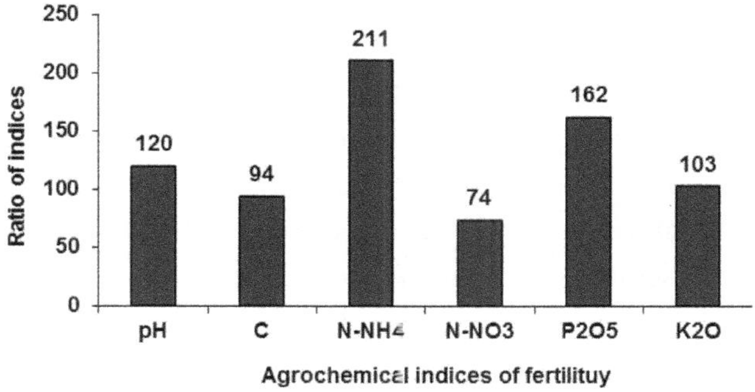

FIGURE 14.8 Ratio of soil agrochemical indices of depressions of the test site vice versa elevations.

TABLE 14.2 Coefficients of Correlation of Soil Agrochemical Indices at the Agricultural Test Site

Index	Humus, %	pH	P_2O_5		K_2O
			mg/kg	mg/L	mg/kg
pH	0.17				
P_2O_5, mg/kg	0. 9	0.23			
P_2O_5, mg/L	0.35	0.37	0.45		
K_2O, mg/kg	0	0.12	0.48	0.46	
K_2O, mg/L	0.07	0	0.3	0.3	0.75

required to use machines with simultaneous application of different types of fertilizers or repeated passes of machines adapted for applying one type of fertilizer.

It is necessary to refer greater values of variability of easy-mobile plant nutrients as compared with the values of less mobile nutrients to next (the seventh) regularity. As it is seen in the Figure 14.9; coefficients of variation of easy-mobile N, P, and K are half or twice as much again as the values of corresponding mobile nutrients. But for all that crop yield to a greater extent depends on an easy-mobile nutrients (in particular on phosphorus) than on a less mobile nutrients (Figure 14.10). Taking this regularity into account is very important for increase of agroeconomic efficiency of variable rate fertilizer application because it allows more punctually to take into consideration plant-nutrient need. In other words during the agrochemical analysis of soils side by side with traditional indices of its nutrient (mobile N, P, K) supply it is important to determine easy-mobile nutrients and to use these values while calculating the optimal fertilizer rates. This would increase fertilizer application efficiency and avoid losses of the fertilizers.

At the same time one ought to take into consideration the eighth regularity that was revealed by our investigation: although crop yield depends on easy mobile nutrients greater than it depends on less mobile forms this yield varies to a lesser degree than proper mobile forms. In particular as it as seen in the Figure 14.11, the values of coefficients of variation of easy-mobile phosphorus ($V = 52\%$) and potassium ($V = 62\%$) in soils of the 400 plots of agricultural test site to a marked degree surpass coefficients of variation of annual grasses hay yield ($V = 41\%$).

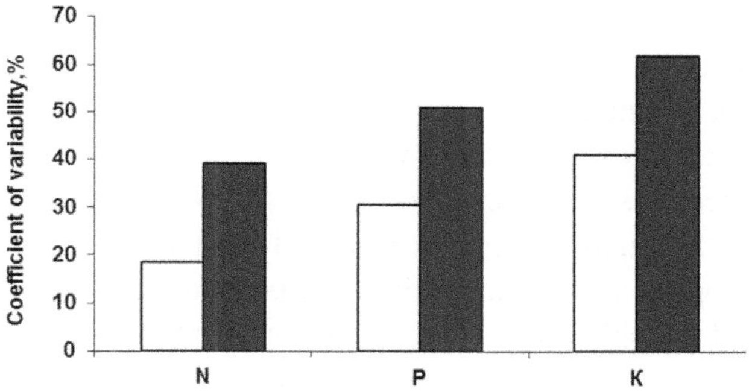

FIGURE 14.9 Variability of mobile (the left row of columns) and easy-mobile (the right row) NPK in soils of the agricultural test site.

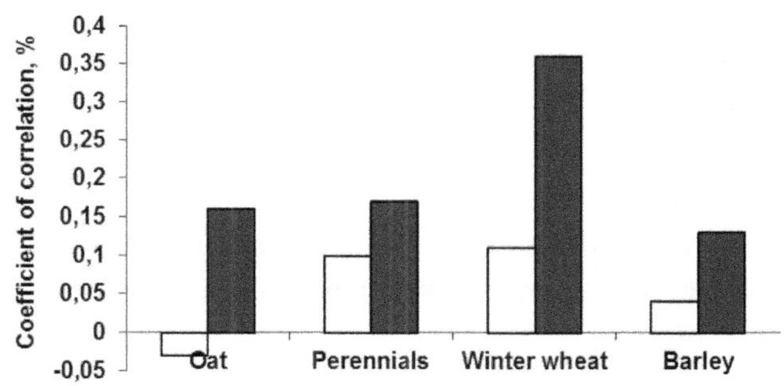

FIGURE 14.10 Influence of mobile (the left row of columns) and easy-mobile (the right row) phosphorus in soils of the agricultural test site on the yield of field crop.

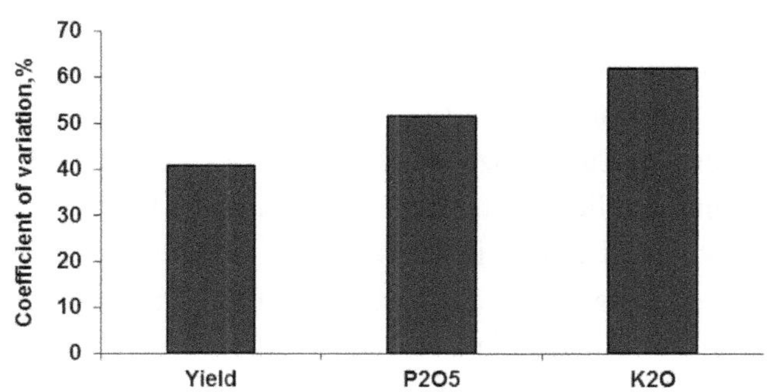

FIGURE 14.11 Variability of annual grasses hay yield and easy-mobile phosphorus and potassium in soils of the agricultural test site.

With lack of nutrition plants increase root development, they can increase solubility of hard-to-reach soil nutrients, biological activity of soil. The lack of nutrition also activates the functioning of soil microflora. Particularly it is known that activity of soil phosphatase sent off by microorganisms increases when content of mobile soil phosphorus decreases and vice versa [4, 5]. Many phosphororganic soil substances (phytin, phospholipids, nucleoproteids) are drawn into exchange of microorganisms with subsequent dephosphorylation and transition of phosphorus into soluble compounds. There are microorganisms in soil that capable transform hard-to-reach compounds of potassium whose exometabolites promote the transition of potassium into soil solution [6]. The activity of nitrogen fixers also increases with decrease of easy-mobile nitrogen content in soil. It is important take into account this regularity when estimating the efficiency of variable rate fertilizer application because to a certain degree differences between variants of traditional and differentiated fertilizer application become smooth through ecological flexibility of plants and also the flexibility of soil microflora which responds to dynamics of soil nutrients.

Of course the ability has their limits. For example studies of differentiated application of nitrogen fertilizer in Tyumen region of the Russian Federation where soils were leached Chernozemic, and crops – spring wheat, showed that application of limited rates of nitrogen counting on grain yield 2 t/ha variable rate application of nitrogen fertilizer didn't advantage first of all economically as compared with application by averaged rate. The plants at the expense of soil nitrogen equalized to a certain extent the difference in low rates of fertilizers applied in different parts of a field. However, with increasing of rates of nitrogen counting on grain yield 3–4 t/ha difference in nitrogen supply of plants because of its variable rate application resulted in essential economic efficiency [7]. So it is possible to come to a conclusion: if the differences between intrafield agrochemical indices or differentiated fertilizer rates are low one cannot expect essential efficiency of their application because of physiologic flexibility of plants.

The regularities of intrafield variability of soil fertility, demonstrated by the example of the testing area of the Central Experiment Station of the Pryanishnikov All-Russian Agrochemistry Research Institute, are characteristic for other types and subtypes of soils where we carried out similar investigations [8]. That indicates their similarity and expedience of utilization when developing high-precision agro technologies of variable rate fertilizer application under different soil-climatic conditions.

14.4 CONCLUSION

1. The regularities of intrafield variability of agrochemical indices which are general for all zonal soil types and subtypes may become theoretical grounds for more rational use of agrochemicals when developing technologies of variable rate fertilizer application. We found several distinguishing features of intrafield variability of soil fertility, which is necessary to take into account when trying increase efficiency of variable rate fertilizer and amendment application. These features are the dependence of agrochemical properties of soil on the meso- and microtopography of fields, smooth shape of conjugacy of soil contours with maximum and minimum values of agrochemical indices, territorial noncoincidence of contours regarding main agrochemical indices, greater values of variability of easy-mobile plant nutrients as compared with the values of less mobile nutrients, dependence of variability on the area of distinguished intrafield contours.
2. Intensity of intrafield heterogeneity of soil fertility has especial significance for choice of technology of fertilizer application (traditional or differentiated). There are two regularities which cause the intensity: firstly – as a rule the ratio of plots with marginal values of agrochemical characteristics to plots with their average characteristics is low, secondly – soils of marginal plots have the maximum value of agrochemical variability. Basically these regularities of soil intrafield fertility by and large reduce the efficiency of differentiated application of fertilizers.
3. Leveling of the difference in nutrient supply of plants in different parts of a field which is caused to a certain extent by ecologic flexibility of plants also influences on reduction of efficiency and so expediency of differentiation of fertilizer rate. It is necessary to define levels and criteria of soil fertility of agricultural lands for purpose of isolation of areas which are very promising for technologies of differentiated agrochemical application; bearing in mind

that the technologies require essential supplemental expenditures as compared with traditional fertilizer application by averaged rate.

KEYWORDS

- **precision agriculture**
- **regularities**
- **soil**
- **variability**
- **within-field contours**

REFERENCES

1. Watcins, K. B., Lu Yao-Chi, Wen-yuan, Huang. Economic and environmental feasibility of variable nitrogen fertilizer application with carry-over effects. Journal of Global Positioning System. Vol. 23, № 2, 401–426.
2. McGuire, J. Technology coordinator spatial Ag systems. The Ohio Geospatial Technologies Conference for Agriculture and Natural Resources Applications. March 24–26, 2003, Columbus. Ohio. 2003, 1–42.
3. Agrochemical methods of soil research. Moscow: "Nauka" ("Science" in Rus.) Publishing House, 1975, 656 p. (in Russian).
4. Rumyantseva, I. V., Devyatova, T. A., Afanas'ev, R. A., Merzlaya, G. E. Proceedings of Voronezh State University Scientific Session. Voronezh: Voronezh State University, 2011, 45–59 (in Russian).
5. Sychev, V. G., Listova, M. P., Derzhavin, L. M. Phosphate regime of soils for agricultural purposes. Bulletin of Geographic Network of Experiments with Fertilizers. Moscow: Pryanishnikov Agrochemistry Research All-Russian Institute, 2011, Issue 11, 64 p. (in Russian).
6. Merzlaya, G. E., Verkhovtseva, N. V., Seliverstova, O. M., Makshakova, O. V., Voloshin, S. P. Interconnection of the microbiological indices of derno-podzolic soil on application of fertilizer over long period of time. Problems of Agrochemistry and Ecology. 2012, № 2, 18–25 (in Russian).
7. Abramov, N. V., Abramov, O. N., Semizorov, S. A. Cherstobitov, S. V. Precision agriculture as a part of resource-saving technologies of crop cultivation. Geoinformatic technologies in agriculture. Proceedings of the International Conference. Orenburg: Publishing Centre of Orenburg State Agrarian University. 2013, 30–40 (in Russian).
8. Afanas'ev, R. A. Regularities of Intrafield Variability of Soil Fertility Indices. Journal of the Russian Agricultural Sciences. 2012, 38, 36–39.

CHAPTER 15

THE EFFECT OF PRE-INOCULATION OF SEEDS BY CELLS OF BACTERIA

Z. M. KURAMSHINA,[1] J. V. SMIRNOVA,[1] and R. M. KHAIRULLIN[2]

[1]*Sterlitamak branch of the Bashkir State University, 49 Lenin Ave., 453109 Sterlitamak, Russia, Phone/Fax: +7 (347) 343-38-69; E-mail: kuramshina_zilya@.mail.ru*

[2]*The Institute of Biochemistry and Genetics, Ufa Scientific Centre of The Russian Academy of Sciences, 71 October Ave., 450054 Ufa, Russia, Phone/Fax: +7 (347) 235-60-88; E-mail: krm62@mail.ru*

CONTENTS

ABSTRACT

The effect of pre-inoculation of seeds by cells of bacteria *Bacillus subtilis* (strains 26D and 11VM) was regarded on *Secale cereale* growth in conditions of the cadmium ions. It is shown that the bacterization of seeds increases plant resistance to the stress caused by the action of metal ions. To assess the toxic effect of cadmium and the protective effect of bacteria tolerance and relative indexes of plants were calculated. The absolute tolerance index of plants pretreated bacteria was higher than that of uninoculated plants in the conditions of cadmium exposure.

15.1 INTRODUCTION

Cadmium is one of the most dangerous trace elements, since it has a high cumulative effect, it is not biodegradable and can lead to serious physiological, biochemical disorders in the organism [1–3].

Due to the increasing environmental pollution of the cadmium ions the considerable attention is paid to seek different methods to reduce metal toxicity to plants. As against known agricultural methods (liming, increasing humates, manure, mobile forms of phosphorus), modern biotechnological methods are highly effective, environmentally friendly and cheap. The results of some studies [4–7] indicate that bacteria, such as associative rhizosphere, can play a significant role in plant resistance to the stress caused by toxic heavy metals.

In the agricultural practice for the protection of plants against diseases caused by phytopathogenic fungi can be used some antagonistic bacteria. One of its endophytic strain of *B. subtilis* 26D that is a base of biofungicide "Phytosporin-M." There is evidence [8-10] of the protective effect of *B. subtilis* 26D in relation to plants under various stresses. However, little work is devoted to the studying of the influence of endophytic bacteria on the plant growth under of the exposure to heavy metals.

The purpose of the present study was to investigate the influence of pre-inoculated seeds of *Secale cereale* by cells of *B. subtilis* strains 26D and 11VM on the plant growth upon the cadmium ions influence.

15.2 MATERIALS AND METHODS

The investigations were carried out on rye plants (*Secale cereale*, Chulpan-7 cultivar). The seeds were surface sterilized in 96% ethanol 3 min, then washed three times in distilled water.

Strains of *B. subtilis* 26D (collection ARRIAM St.-Pb. Pushkin, №128) and 11VM (ARRIAM №519) were given to the staff laboratory of biotechnology of the Bashkir State Agrarian University. In experiments the daily culture growing on meat-peptone agar at 37°C is used. The bacterial cells were washed with 0.01 M KCl solution and then the cell suspension was adjusted to the required concentration spectrophotometrically. 1 g seeds were inoculated with 0.02 mL bacteria cell suspensions; a titer of 1 billion/mL.

The inoculated and controlled seeds were grown in Petri dishes (d = 140 mm, h = 24 mm) on a filter paper moistened with distilled water or a solution of $Cd(NO_3)_2 \times 4H_2O$. The salt solution was prepared in the terms of the metal ion content.

The tolerance index of plants was determined as the ratio of the mean length of the shoot (root) of the plant grown on a solution with a certain concentration of metal ions, to the mean length of the shoot (root) control plants [11]:

$$I(\%) = \frac{\Delta L_{on}}{\Delta L_k} x100\%$$

For the plants that were treated by bacteria two tolerance indexes were calculated: relative and absolute ones. To calculate the relative tolerance index the plants, obtained from seeds treated by bacteria and grown on medium without cadmium, were considered as control ones. While calculating the absolute tolerance index the plants, obtained from the seeds treated with water, and germinated in distilled water were considered as control seedlings.

15.3 RESULTS AND DISCUSSION

The seed treatment by bacteria *B. subtilis* has a positive impact on the plant growth, but strains showed different activity in relation to plants. Thus, when the seeds were germinated in Petri dishes, we observed the stimulation of shoots 12–15% relative to control (untreated), roots – by 4–15%, respectively, for strains 26D and 11VM (Table 15.1).

It is known that researched *B. subtilis* strains are endophytic, which live in plant tissues without causing significant harm to the plant or to

TABLE 15.1 Effect of Cadmium Ions on the Length of Shoots and Roots of Rye Plants*

Concentration	Variant		
	Control	+ *B. sub.* 26D	+ *B. sub.* 11VM
Cd 0 mg/L	6.6±0.2 / 29.6±0.2	7.4±0.5 / 30.9±0.5	7.7±0.4 / 34.3±0.3
Cd 1 mg/L	8.3±0.4 / 33.6±0.5	8.5±0.4 / 35.8±0.4	8.5±0.3 / 36.7±0.3
Cd 10 mg/L	4.7±0.3 / 18.1±1.0	5.7±0.3 / 28.2±1.1	6.2±0.3 / 29.5±1.5
Cd 20 mg/L	4.4±0.4 / 14.0±0.4	5.1±0.3 / 17.2±0.3	6.2±0.5 / 17.7±0.6
Cd 40 mg/L	4.2±0.3 / 13.8±0.9	3.8±0.3 / 14.3±0.6	3.8±0.3 / 15.5±0.5

*Above the line – the length of shoots, under the line – the length of the roots.

benefit [12, 13]. It has been shown that the stimulation of the plant growth may be due to the increase of the indoleacetic acid level in plant tissues and bacterial cells expression of cytokinin-like activity [14–16].

Cadmium, being highly toxic element, significantly inhibits the growth of plants, wherein the root system is the most sensitive to the metal [17]. However, it is known that heavy metals, including cadmium, in very low concentrations do not cause significant changes and are even able to have a positive impact on the plants: decrease lipid peroxidation, stimulate growth, increase of the content of pigments [18].

The increase in the content of metal in the solution was resulted in a decrease of the growth rate of plants. At the concentration of 10 mg/L the growth of shoots of control plants was inhibited by 28%. Roots are noticeably responsive to the presence of cadmium and its growth was inhibited by 39%.

Further increase of the metal content in the solution resulted in the strong inhibition of the plant growth: at the concentration of 40 mg/L shoots the growth was inhibited by 36% and roots growth – by 53% (Table 15.1).

Anti-stress activity in relation to plants is well manifested for both strains (Table 15.2). In the presence of metal ions, higher roots of the

TABLE 15.2 Relative (RTI) and Absolute (ATI) Tolerance Index of Rye Plants to the Action of Cadmium Ions*

Concentration	Tolerance index, %				
	Control	+ *B. sub.* 26D		+ *B. sub.* 11VM	
	RTI	RTI	ATI	RTI	ATI
Cd 0 mg/L	100 / 100	100 / 100	111.8 / 104.6	100 / 100	117.3 / 116.0
Cd 1 mg/L	126.4 / 113.7	114.9 / 115.8	128.5 / 121.1	109.4 / 107.1	117.3 / 124.3
Cd 10 mg/L	71.1 / 61.4	77.1 / 91.4	86.2 / 95.5	80.6 / 85.9	94.5 / 99.7
Cd 20 mg/L	65.7 / 47.4	68.8 / 55.7	77.0 / 58.3	80.1 / 51.6	93.9 / 59.8
Cd 40 mg/L	63.2 / 46.7	52.0 / 46.2	58.2 / 48.3	48.8 / 45.1	57.3 / 52.4

*Above the line – tolerance index of shoots, under the line – tolerance index of the roots

relative tolerance index (RTI) were observed when seeds were treated by cells of *B. subtilis* 26D. Interestingly, that the shoot RTI values of the bacteria pre-treated plants in the case of the cadmium concentration of 1 mg/L were slightly lower than that of untreated plants.

The absolute tolerance index (ATI) of roots was always higher that the control value in the case of seeds treatment with both bacteria strains (Table 15.2). The similar rule was shown in the analysis of ATI of shoots of plants that were pre-treatment with *B. subtilis* 26D bacteria.

It is known that cadmium ions in toxic concentrations significantly inhibited the growth and development of plants, also they have a negative impact on photosynthesis, respiration, water metabolism and mineral nutrition [19]. In our experiments the treatment of plants by bacilli increased the tolerance index of rye plants. The growth-stimulating effect of the bacteria, probably due to the fact that the bacilli can produce phytohormone-like substances [14, 15], and also increase the content of nutrients in the plant-available form [20] and inhibit the development of pathogenic microorganisms [14].

Some authors explain the phytotoxicity of heavy metals at their ability to induce cells reactive oxygen species, leading to oxidative stress [21]. From this viewpoint, due to the ability of a plant organism under the influence of the metal to keep activity of antioxidant enzymes (catalase, peroxidase), the resistant plants are at a certain level [18]. We have shown earlier that white mustard plants, the seeds of which were pre-treated by the bacteria when exposed to high concentrations of cadmium, had higher growth and activity of catalase and peroxidase at the same time there was a simultaneous decrease in the intensity of lipid peroxidation [22]. We assume that the anti-stress effect of bacilli in the rye plants when exposed to cadmium ions is also related to the preservation of the activity of antioxidant enzymes.

15.4 CONCLUSION

Thus, the processing plant by endophytic strain *B. subtilis* weakened the toxic effects of cadmium ions. Each strain showed itself individually. The relative and absolute tolerance indexes of plants in almost all variants were higher when seeds were pretreated with the bacteria. On the ground of the investigations of some authors and our experimental data, we suggest that the endophytic bacteria stimulate the plant growth and increase immunity, keep up the activity of catalase and peroxidase, reduce the intensity of lipid peroxidation, stabilize cell membranes with the toxic effects of cadmium. The results are of considerable interest while developing techniques for the crop production on soils contaminated with heavy metal salts.

KEYWORDS

- **cadmium**
- **endophytic strains of *Bacillus subtilis***
- **rye**
- **tolerance index (TI)**

REFERENCES

1. Kvesitadze, G. I., Khatisashvili, G. A., Evstigeneeva, Z. G. Anthropogenic toxicants metabolism in higher plants. Nauka. 2005, 199 p.
2. Seregin, I. V., Ivanov, V. B. Physiological aspects of cadmium and lead toxic effects on higher plants. Russian journal of plant physiology. 2001, V. 48, 523–544.
3. Kabata-Pendias, A., Pendias, H. Trace elements in soil and plants. CRC Press LLC, Boca Raton, FL, 1992, 453 p.
4. Brown, M. E. Seed and root bacterization. Annual Review of Phytopathology. 1974, V. 12, 181–197.
5. Belimov, A. A., Kunakova, A. M., Safronova, V. I., Kozhemyakov, A. P., Stepanok, V. V., Yudkin, L.Yu., Alekseev Yu.V. Employment of rhizobacteria for the inoculation of barley plants cultivated in soil contaminated with lead and cadmium. Microbiology. 2004, T. 73. № 1, 99–106.
6. Burd, G. I., Dixon, D. G., Glick, B. R. Plant growth promoting that decrease heavy metal toxicity in plants. Canadian Journal of Microbiology. 2000, V. 46, 247–255.
7. Siunova, T. V., Kochetkov, V. V., Boronin, A. M. Effect of rhizosphere bacteria on nickel accumulation by barley plants. Agrochemicals. 2006, № 10, 80–84.
8. Mannanov, R. N., Sattarova, R. K. Antibiotics produced by *Bacillus* bacteria. Chemistry of Natural Compounds. 2001, V. 37. № 2, 117–123.
9. Mubinov, I. G. Wheat reaction to the action of the cells of the endophytic strain 26D Bacillus subtilis - basis biofungicide fitosporin: Author. Diss. Cand. Biol. Sciences. Ufa, 2007, 22 p.
10. Khairullin, R. M., Nedorezkov, V. D., Mubinov, I. G., Zakharova, R.Sh. Increasing of wheat firmness to abiotic stresses of endophytic strain Bacillus subtilis. Herald OSU. 2007, N 2, 129–134.
11. Wilkins, D. A. The measurement of tolerance to edaphic factors by means of root growth. New Phytol. 1978, № 80, 623–633.
12. Nedorezkov, V. D. Biological rationale for the use of endophytic bacteria in the protection of wheat against diseases in the Southern Urals: Author. Diss. Dr. agricultural Sciences. Saint-Petersburg, 2003, 41 p.
13. Kado, C. I. The Prokaryotes. Springer-Verlag: New-York, 1992, V. 2, p. 352.
14. Arkhipova, T. N., Melent'ev, A. I., Martynenko, E. V., Kudoyarova, G. R., Veselov, S.Yu. Comparison of effects of bacterial strains differing in their ability to synthesize cytokinins on growth and cytokinin content in wheat plants. Russian Journal of Plant Physiology. 2006, T. 53. № 4. C. 507–513.
15. Melentev, A. I. Aerobic spore-forming bacterium Bacillus Cohn in agroecosystems. Nauka. 2007, 147 p.
16. Smirnov, V. V., Reznik, S. R., Vasilevska, I. A. Spore-forming aerobic bacteria – producers of biologically active substances. Naukova Dumka, 1982, 278 p.
17. Cagno, R., Guidi, L., Stefani, A., Soldatini, G. F. Effects of cadmium on growth of *Helianthus annuus* seedlings: physiological aspects. New Phytologist. 1999, V. 144, 65–71.
18. Titov, A. F., Talanov, V. V., Kaznina, N. M., Laidinen, G. F. Plant resistance to heavy metals. Petrozavodsk: Karelian Research Centre, 2007, 172 p.

19. Ivanov, V. B., Bystrova, E. I., Seregin, I. V. Comparative Impacts of Heavy Metals on Root Growth as Related to Their Specificity and Selectivity. Russian Journal of Plant Physiology. 2003, T. 50. № 3, 398–406.
20. Egorshina, A. A., Khairullin, R. M., Lukyantsev, M. A., Kuramshina, Z. M., Smirnova, Y. V. Phosphate-Mobilizing Activity of the Endophytic Bacillus subtilis Strains and their Effect on Wheat Roots Micorrhization Ratio. Scientific Journal of Siberian Federal University. Krasnoyarask, 2011, 1, 172–182.
21. Cho, U. H., Seo, N. H. Oxidative Stress in Arabidopsis thaliana Exposed to Cadmium is due to Hydrogen Peroxide Accumulation. Plant Sci. 2005, 168, 113–120.
22. Kuramshina, Z. M., Smirnov, Y. V., Khairullin, R. M. Protective effect of endophytic strain of bacteria at cadmium ions toxic effect on Sinapis alba plants. Herald of the Bashkir University. 2013, 18(3), 739–742.

CHAPTER 16

RESUSCITATING FACTORS FOR NONCULTURABLE CELLS

YU. D. PAKHOMOV, L. P. BLINKOVA, O. V. DMITRIEVA, O. S. BERDYUGINA, and N. N. SKORLUPKINA

FGBU "I.I. Mechnikov Research Institute for Vaccines and Sera" RAMS, Moscow 105064 Maliy Kazenniy per. 5a. Russia, Tel.: +7-495-916-11-52, Fax: +7-495-917-54-60; E-mail: labpitsred@yandex.ru

CONTENTS

ABSTRACT

In this study several resuscitating factors for nonculturable cells were tested. Objects of the research were nonculturable cells in populations of

Lactococcus lactis and in lyophilized probiotic preparations containing *Escherichia coli, Bifidobacterium bifidum* and *Lactobacillus acidophilus*. We found that some of the studied additives have potential for increasing numbers of CFU/mL of microorganisms in question.

16.1 INTRODUCTION

Existence of viable but nonculturable bacteria and their resuscitation into vegetative state are the major problems in modern microbiology. Nonculturable cells are formed as a response to a wide variety of stressful factors and their combinations. It is particularly important for food producers and manufacturers of probiotic preparations, since bacteria may loose culturability as a reaction to sanitation procedures, lyophilization or during fermentation processes in ripening foodstuffs [1–4]. Main task is to search for resuscitating agents that convert nonculturable cells into active forms. According to literary data such agents include fetal serum, vitamin K, yeast cells, live cells of higher organisms etc. [5–9]. Our aim was to study several resuscitating factors were tested on nonculturable cells of *Lactococcus lactis* and a number of lyophilized probiotics.

16.2 MATERIALS AND METHODS

16.2.1 STRAINS AND MEDIA

In this study we used following microorganisms: lyophilized probiotic cultures of *E. coli* M-17, *Bifidobacterium bifidum, Lactobacillus acidophilus* and three nisin producing strains of *Lactococcus lactis ssp. lactis* MSU, 729 and F-116 that were incubated in carbohydrate starvation conditions [2] and contained more than 99.9% nonculturable cells.

For growing microorganisms following media were used: tryptic soy medium for *E. coli*, Elliker medium for *L. lactis, B. bifidum* and *Lactobacillus acidophilus* and 0.1% fat milk for *L. lactis*. Tryptic soy and Elliker media were used in liquid, semisolid (0.4% agar) and solid (1.5% agar) forms.

16.2.2 RESUSCITATION FACTORS

In experiments with *L. lactis* we used the following factors: mixture of 7 amino acids (glutamine – 0.39 g/L, methionine – 0.12 g/L, leucine – 0.47 g/L, histidine – 0.1 g/L, arginine – 0.12 g/L, valine – 0.33 g/L, isoleucine – 0.21 g/L) [10]; concentrated inactivated biomass of homologous strain (0.1, 0.5 and 1%); inactivated culture of homologous strain (5%) [11]; oleic acid (0.1%, 0.5%, 1% and 5%). Blood substitute "Aminopeptidum" (0.5%, 1% and 10%) was applied to all species. Media without factors were used as controls.

16.2.3 VIABILITY AND CULTURABILITY ASSAYS

For visual detection of viable and dead bacteria in samples we used Live/Dead (BacLight™) kit and a luminescence microscope (Karl Zeiss). Culturability was assessed using plating, most probable number techniques and also by counting colonies in columns of semisolid agar. For total cell counts we used Goryaev or Thoma chambers. Number of viable but nonculturable cells were measured as a result of comparing total viable counts and CFU/mL.

16.3 RESULTS AND DISCUSSION

For *Lactococcus lactis* resuscitating effect was observed in the following cases (see Table 16.1): for strain MSU (after 3.5 months of incubation) addition 1% of inactivated, concentrated biomass was to solid Elliker's medium yielded 2.65 ($p<0.05$) times increase in CFU/mL. For the same strain (after 4.5 months of incubation) 3.75 ($p<0.05$) times increase was noted when 0.5% of biomass was added to liquid Elliker's medium. For strain F-116 we managed to increase culturability four times ($p<0.05$) by adding mixture of amino acids to semisolid Elliker's medium. Other resuscitating factors (additions to culture media) had no significant effect on nonculturable cells of *L. lactis* "Aminopeptidum" showed reactivating effect on lyophilized probiotic preparations. For *E. coli* addition of 10% of "Aminopeptidum" yielded 6.45 times ($p<0.05$) increase in

TABLE 16.1 Screening Different Factors for Resuscitation of Nonculturable Cells

Microorganism	Age of the culture	Resuscitating factor	Quantitative characteristics		
			Control, KOE/mL	Media with factors, KOE/mL	Increase in CFU/mL (times)/p value
	3.5 мес.	0.1% inactivated. concentrated homologous biomass in Elliker's broth	$0.85<4<18.7\times10^5$	$0.32<1.5<7.02\times10^6$	>0.05
L. lactis MSU		1% inactivated. concentrated homologous biomass on solid Elliker's medium	**$3.65\pm0.4\times10^5$**	**$9.68\pm1.06\times10^5$**	**2.65** < 0.05
	4.5 мес.	10% "Aminopeptidum" in Elliker's broth	$1.05<4<15.2\times10^5$	$1.84<7<26.6\times10^5$	>0.05
		0.5% inactivated. concentrated homologous biomass in Elliker's broth	**$1.05<4<15.2\times10^5$**	**$0.39<1.5<5.7\times10^6$**	**3.75** < 0.05
L. lactis 729			$0.37<1.4<5.32\times10^5$	$0.68<2.6<9.88\times10^5$	>0.05
L. lactis F-116	8 мес.	7 amino acids in semisolid Elliker's broth	**$1\pm0.11\times10^4$**	**$4\pm0.44\times10^4$**	**4** **< 0.05**
		10% "Aminopeptidum" in Elliker's broth	1×10^4	6×10^3	>0.05
		5% inactivated homologous culture in Elliker's broth	1×10^4	2×10^3	>0.05

TABLE 16.1 Continued

Microorganism	**Age of the culture**	**Resuscitating factor**	**Quantitative characteristics**		
			Control, KOE/mL	**Media with factors, KOE/mL**	**Increase in CFU/mL (times)/p value**
		7 amino acids in 0.1% fat milk	10^3	10^4	>0.05
		10% "Aminopeptidum" in 0.1% fat milk	10^3	10^3	>0.05
		5% inactivated homologous culture in 0.1% fat milk	10^3	10^3	>0.05
E. coli M–17 (Colibacterin, batch 40–3)		10% of "Aminopeptidum"	**6.2×10^7**	**4×10^8**	**6.45** **< 0.05**
Lactobacillus acidophilus (Acipol batch 11)			**$2.2\pm2.42\times10^5$**	**$1.1\pm1.21\times10^6$**	**5** **<0.05**
Bifidobacterium bifidum		"Aminopeptidum" 1%	**9×10^5**	**1.81×10^6**	**2** **< 0.05**
Bifidumbacterin batch 735		10%	**9×10^5**	**2.3×10^6**	**2.56** **< 0.05**

culturable counts. For *Lactobacillus acidophilus* culturability increased 5 times ($p<0.05$) in the medium with 10% of "Aminopeptidum" For *B. bifidum* (see Figure 16.1) when 1% of "Aminopeptidum" was added value of CFU/mL increased 2 times ($p<0.05$) and with addition of 10% of "Aminopeptidum" 2.56 times ($p<0.05$) increase in CFU/mL was observed. It should be noted that addition of "Aminopeptidum" markedly increased growth rate of *Lactobacillus acidophilus*. In the medium with the additive colonies formed within 48 hours while in controls 2/3 of colonies were appeared by 72 hours.

It is evident that all probiotic bacteria reacted on addition to "Aminopeptidum." This preparation is a blood substitute so its chemistry is close to the blood. Since probiotics are brought into human or animal intestine and adhere to its walls, they interact with body fluids and thus react to the additive. So we suggest that even if probiotics may contain significant portions of nonculturable cells (particularly ones containing lactobacilli), their effect may be increased when these cells resuscitate after

FIGURE 16.1 Stimulating effect of "Aminopeptidum" on *B. bifidum* from bifidumbacterin batch 735. Left – control, middle – 1% of "Aminopeptidum" right – 10% of "Aminopeptidum".

ingestion. For many bacterial species resuscitation via passage through host organism has already been shown [12].

Role of nonculturable bacteria is not yet sufficiently studied in many areas of microbiology. Research on this problem may significantly extend understanding of behavior of sanitary significant microorganisms, causes and manifestations of dysbiotic conditions, help better characterize pathogenic agents and improve viability assessment of probiotics used for correction of dysbiosis.

16.4 CONCLUSIONS

We conducted a search for factors that promote resuscitation from nonculturable back into actively growing state. For bifidobacteria, *E. coli* and lactobacilli the most efficient was "Aminopeptidum."

KEYWORDS

- ***Lactococcus lactis***
- **nonculturable cells**
- **probiotics**
- **resuscitation**

REFERENCES

1. Ganesan, B., Stuart, M. R., Weimer, B. C. (2007). Appl. Environ. Microb., 73(8), 2498.
2. Pakhomov, Yu. D., Blinkova, L. P., Dmitrieva, O. V., Berdyugina, O. S., Stoyanova, L. G. (2013). J. Bacteriol Parasitol, 5(1), doi: 10.4172/2155–9597.1000178.
3. Hoefman, S., Van Hoorde, K., Boon, N., Vandamme, P., De Vos, P., Heylen, K. (2012). PLoS ONE, 7(4), e34196. doi: 10.1371/journal.pone.0034196.
4. Blinkova, L., Martirosyan, D., Pakhomov, Y., Dmitrieva, O., Vaughan, R., Altshuler, M. (2014). Functional Foods in Health & Disease, 4(2), 66.
5. Oliver, J. D., Hite F, Mcdougald, D., Andon, N. L., Simpson, L. M. (1995). Appl. Environ. Microb., 61(7) 2624.

6. Steinert, M., Emody, L., Amann, R., Hacker, (1997). J. Appl. Environ. Microb., 63(5), 2047.
7. Senoh, M., Ghosh-Banerjee, J., Ramamurthy, T., Hamabata, T., Kurakawa, T., Takeda, M., Colwell, R. R., Nair, G. B. Takeda, Y. (2010). Microbiol. Immunol., 54(9), 502.
8. Senoh, M., Ghosh-Banerjee, J., Ramamurthy, T., Colwell, R. R., Miyoshi, S., Nair, G. B., Takeda, Y. (2012). Microbiol. Immunol., 56(5), 342.
9. Peneau, S., Chassaing, D., Carpentier, B. (2007). Appl. Environ. Microb., 73(7), 2839–2846.
10. Juillard, V., Le Bars, D., Kunji, E. R.S., Konings, W. N., Gripon, J.-C., Richard, J. (1995). Appl. Environ Microb., 61(8), 3024.
11. Miura. M., Seto, Y., Watanabe, M., Yoshioka, T. (2009). US Patent Application Publication, № US2009/0317892 A1.
12. Fakruddin, M. D., Bin Mannan, K. S., Andrews, S. (2013). ISRN Microbiology, Article ID 703813, 6 p.

CHAPTER 17

A CASE STUDY ON APPLICATION OF MICRO-SIZED PARTICLES FOR BIOLOGICALLY ACTIVE COMPOUNDS

LUBOV KH. KOMISSAROVA and VLADIMIR S. FEOFANOV

N.M. Emanuel Institute of Biochemical Physics of the Russian Academy of Sciences, Kosygin St. 4-117977 Moscow, Russia, Tel.: +8(495)9361745 (office), +8(906)7544974 (mobile); Fax: (495)1374101; E-mail: komissarova-lkh@mail.ru

CONTENTS

ABSTRACT

We have worked out new methods to modify the surface of the ferrocomposites microsized particles (iron, magnetite, and iron-carbon) by albumin, gelatin or dextran and have been studied immobilization hemoglobin and barbiturates (sodium phenobarbital and barbituric acid) and immobilization and dynamics of L-borophenilalanin (L-BPA) desorption. The optimal ferrocomposite types and the methods modifications their surface are suggested as sorbents for extracorporal detoxification of patients blood and purification of donor conserved blood and as carriers for magnetically-guided targeted delivery of L-BPA at Boron Neutron Capture of Tumor Therapy (BNCT).

17.1 AIMS AND BACKGROUND

Magnetic nano- and microsized particles can be used for various biomedical applications: cell separation, immobilization of enzymes and viruses, detoxification of biological liquids, magnetic drug targeting und others [1–5]. The most widespread for neutron capture therapy (NCT) have become compounds with ^{10}B (BNCT). The two boron containing compounds, one of them L-borophenilalanin (L-BPA) are used in clinical practice [6]. The aim of the research is to work out new methods to modify the surface of different chemical content microsized ferrocomposites particles by biocompatible materials for the following immobilization of biologically active compounds and to evaluate possibility to use them as sorbents for extracorporal detoxification of patients blood and conserved donor blood purification from free hemoglobin and barbiturates by the method of magnetic separation and as carriers for magnetically-guided targeted delivery of L-BPA at BNCT.

17.2 EXPERIMENTAL PART

We have studied composites: iron-silica ($FeSiO_2$) of content: 50%Fe, 50% SiO_2, iron-carbon (FeC) of content 44%Fe, 56%C, iron-carbon-silica ($FeCSiO_2$ of content 50%Fe, 40%C and 10% SiO_2, iron of content: 90% restored iron and 10%, Fe_3O_4, sized 0.02–0.1 mkm, obtained by plasmochemical method [5] magnetite (Fe_3O_4) sized 0.1–0.5 mkm, synthesized by chemical method [2]. Diameter of ferrocomposites microparticles in 0.6% of albumin solution 1–2 mkm [2]. The powders of ferromagnetics were treated (suspension in distillate water) by ultrasonic waves (frequency 22 kHz) in order to eliminate aggregation and to attain a homogeneous distribution of the particles in suspension. The particles' surface, besides the same of composites $FeSiO_2$ and $FeCSiO_2$, which is biocompatible, was covered by albumin, or gelatin, or dextran. Carboxilate-magnetic particles were obtained by bovine albumin or gelatin coating with the following aldehydes modification, aldehyde-magnetic particles – dextran coating with $NaJO_4$ activation. We coated the particles by mixing a suspension of particles and albumin or gelatin or dextran with Mm 67,000 Da (Sigma) using ultrasound, with the following 1 hour incubation at 20°C, separation of particles on Sm-Co magnet with inductance of 0.1–0.15 Tl. Thereafter the particles were incubated in the modification solutions: formaldehyde (Russia), or glutaraldehyde, or $NaJO_4$ (Sigma) and washed with distillate water. Surface-modified particles were kept at 10% concentration in physiological solution.

We have used bovine hemoglobin (Sigma), which contained to 75% methemoglobin and to 25% oxyhemoglobin. Immobilization of hemoglobin and barbiturates (Russia) was carried out by 30 sec. incubation with the suspension of particles in physiological solution and in a model biological liquid (0.6% albumin in physiological solution) at 20°C (pH 7.4) at different weight ratios of composite/substance: 10, 20, 50 and thereafter the particles were separated on Sm-Co magnet. We have chosen incubation time 30 sec according to the length of contact of biological liquids with suspension of magnetic microparticles in the device for extracorporal detoxification of blood by the method of magnetic separation [1]. Concentrations of compounds in the solutions were measured by differential visual and UV-spectroscopy. The sorption efficiency of ferrocomposites was evaluated as the ratio of the quantity of the adsorbed substance to its initial amount (w/w), expressed in % and in mg/g composite (absorptive capacity) for a certain weight ratio of composite/substance.

Immobolization L-BPA (Lachema) carried out by 10 min incubation with the suspension of particles in acidified water solutions at different weight ratios of composite/L-BPA. The dynamics of L-BPA desorption was studied by incubation of magnetic preparations with immobilized of L-BPA with fresh aliquots of 0.6% albumin at 37°C and by following registration of supernatant absorption UV-spectra. Concentration of desorbed BPA was evaluated on the calibri curve.

17.3 RESULTS AND DISCUSSION

17.3.1 IMMOBILIZATION OF HEMOGLOBIN AND BARBITURATES

The maximal sorption efficiency to hemoglobin on unmodified ferrocomposites particles showed for Fe_3O_4 and Fe-particles: 40.0 mg/g and 37.8 mg/g, respectively. The results of sorption efficiency of modified ferrocomposites particles to hemoglobin are presented in Table 17.1.

Table 17.1 shows that the sorption efficiency of magnetite to hemoglobin after covering the surface by gelatin is decreasing, glutar-modification leads to its further decrease. Sorption efficiency of magnetite does not exchange practically after formaldehyde modification by gelatin-covered particles. The same character of sorption efficiency of gelatin and albumin- modified particles to hemoglobin is discovered for Fe and FeC particles: sorption efficiency of is decreasing after covering of particles by proteins, the glutar-modification leads to its further decrease and does not change after formaldehyde modification. Immobilization of hemoglobin on aldehyde-modified particles, evidently is due to forming hydrogen connections between carboxilate groups of proteins and amino-groups of hemoglobin. Decrease of hemoglobin adsorption efficiency of particles with glutar-modified surface is predetermined, obviously, by a stereochemical factor.

The results on immobilization of hemoglobin on modified ferrocomposites particles are shown in and Table 17.2 and Figure 17.3.

Figure 17.2 and Table 17.2 demonstrate that iron particles covered by dextran and activated by $NaJO_4$ have shown maximal sorption efficiency to hemoglobin, which is, obviously, accounted for forming hydrogen connections between aldehydes groups of dextran and amino-groups of hemoglobin.

TABLE 17.1 Sorption Efficiency of Gelatin-Modified Fe_3O_4-Particles to Hemoglobin at Different Weight Ratios Fe_3O_4/Hb in Physiological Solution at pH 7.4

Types of composites, m	Sorption, average± SD (%)						
	Absorptive capacity, average± SD (mg/g)						
	Unmodified		Gelatin-covered		Gelatin+glutarald.–modified		Gelatin+formald. modified.
Fe_3O_4/Hb	20	50	20	50	20	50	20
	41.6 ±4.7	70.4 ±9.3	34.9±4.9	57.8 ±6.1	26.9±3.6	47.8 ±5.9	41.0± 4.2
	20.8 ±2.4	14.1 ±4.7	17.4±2.4	11.6 ±1.2	13.4±1.8	9.6±1.2	20.5±2.1

TABLE 17.2 Sorption Efficiency of Modified Ferrocomposites to Hemoglobin (Composite/Hb, w/w=20) in Physiological Solution at pH 7.4

	Sorption, average±SD (%)			
	Absorptive capacity, average±SD (mg/g)			
Types of composites	Fe + gelatin+ formald.	Fe_3O_4+ gelatin + formald.	Fe+dextran + $NaJO_4$	Fe+albumin + formald.
	38.0± 4.6	41.0±4.2	94.5±11.3	68.4 ±7.5
	19.0±2. 3	20.5±2.1	47.2 ± 5.6	34.2±3.8

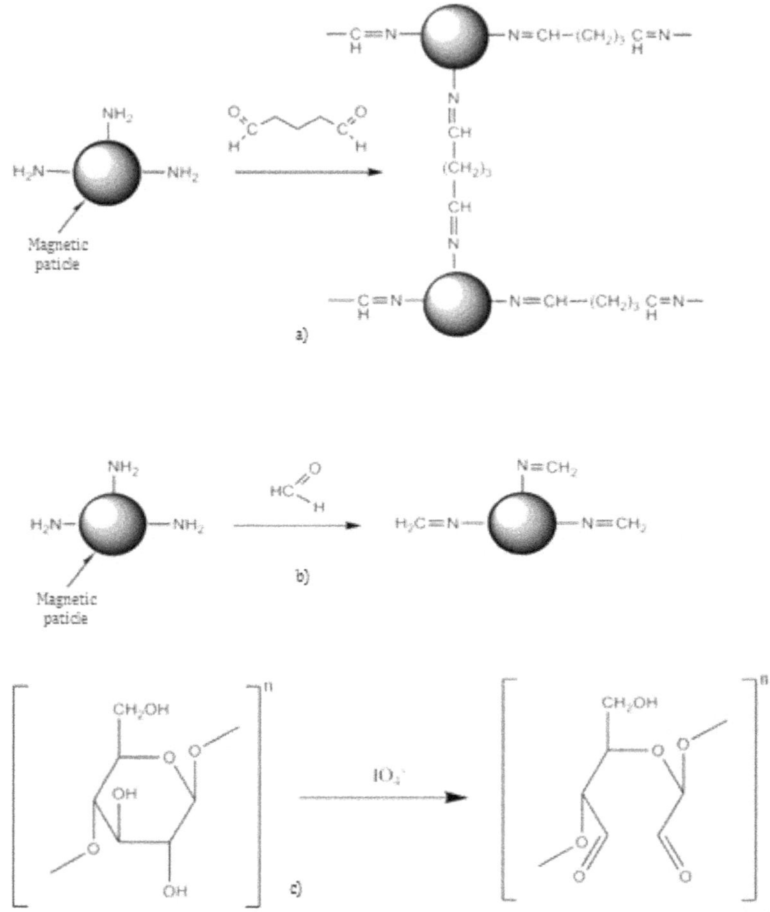

FIGURE 17.1 Modification of proteins (albumin or gelatin) NH_2 groups by glutar (a) and formaldehydes (b), activation of dextran OH-groups with $NaJO_4$.

Sorption efficiency of modified ferrocomposites to hemoglobin in a model biological liquid (0.6% albumin in physiological solution) is presented in Table 17.3.

Results on sorption efficiency of modified particles magnetite and iron to hemoglobin in model biological liquid (Table 17.3) showed, that maximal absorptive capacity manifested Fe-particles, modified by albumin (32.6 mg/g) and dextran (25.0 mg/g). These meanings are lower than those in physiological solution. This can be explained by decreasing of the sorption processes velocity due to increasing of solution viscosity. In fact,

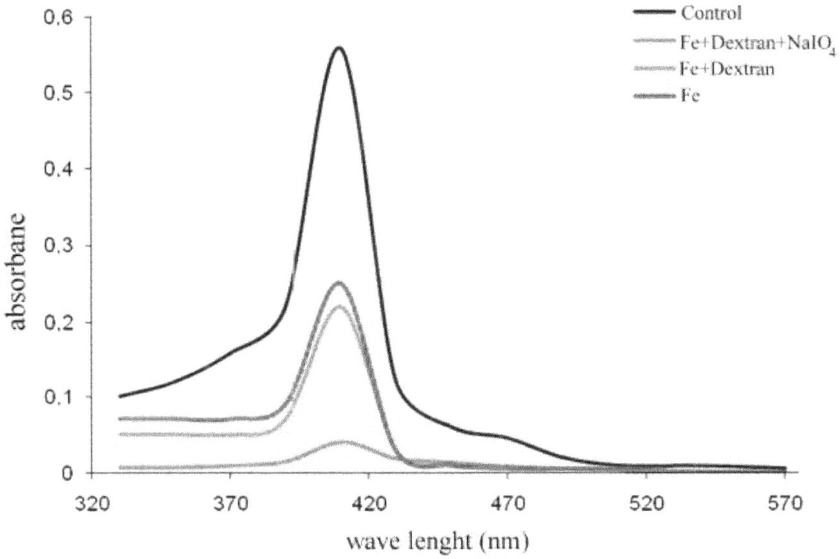

FIGURE 17.2 Absorption spectra of hemoglobin after immobilization on dextran-modified Fe-particles (Fe/Hb, w/w=20) in physiological solution at pH 7.4.

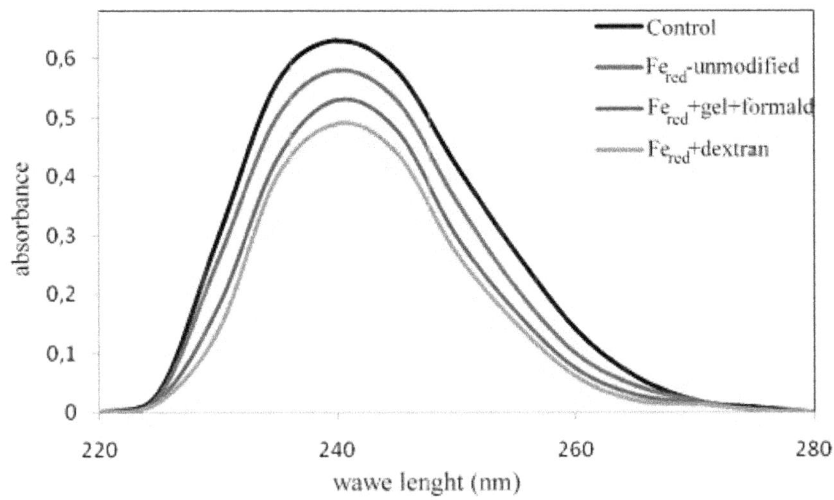

FIGURE 17.3 Absorption spectra of phenobarbital after immobilization on different chemical content Fe-particles (Fe/PhB, w/w=20) in physiological solution at pH 7.4.

the sorption efficiency increases at increasing the incubation time from 30 sec. to 60 sec. The interesting results on sorption efficiency of hemoglobin, carboxyhemoglobin and methemoglobin on gelatin-modified Fe-particles

TABLE 17.3 Sorption Efficiency of Modified Ferrocomposites to Hemoglobin (Composite/Hb, w/w=10) in 0.6% Albumin in Physiological Solution at pH 7.4

	Sorption, average±SD (%)			
	Absorptive capacity, average±SD (mg/g)			
Types of composites	**Fe_3O_4 + gelatin+ formald.**	**Fe+gelatin+ formald,**	**Fe+albumin+ formald.**	**Fe + dextran+$NaJO_4$**
	8.2±1.7	16.3 ±2.4	32.6 ±3.8	25.0 ±3.2
	8.2±1.7	16.3 ±2.4	32.6 ±3.8	25.0 ±3.2

have becalmed in donor blood. These meanings are equal: 60.7%, 52.9% and 22.5% accordingly.

It is important emphasize that adsorption of albumin on an modified particles reached to 40% for all composites types, after modification of composites surface adsorption of albumin was not more than 10%.

The sorption efficiency results of different chemical content modified ferrocomposites to phenobarbital in physiological solution are represented in Figures 17.3 and 17.4, and in Tables 17.4 and 17.5.

Table 17.4 shows that modification of Fe-microparticles surface by albumin led to considerable increase of phenobarbital immobilization: from 18.9% to 51.4%. The immobilization is realized, probably, by means of conjugation of phenobarbital with carboxilate-groups of albumin. The Figures 17.3 and 17.4 shows absorption spectra of phenobarbital after immobilization on different chemical content microparticles.

Maximal meanings of sorption efficiency of phenobarbital have demonstrated Fe-silica composites. Formation of hydrogen connections plays, apparently, a prevailing role in immobilization phenobarbital on Fe-silica composites (Figure 17.4 and Table 17.5).

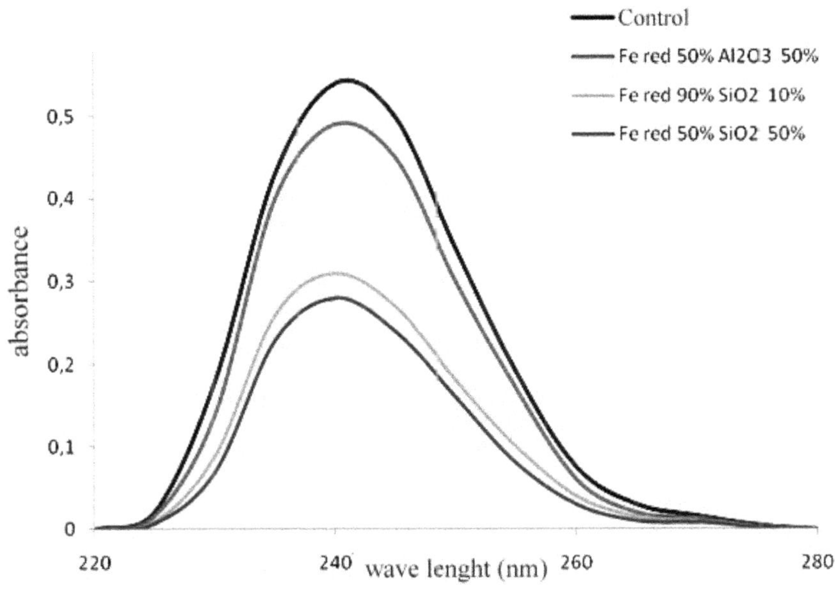

FIGURE 17.4 Absorption spectra of phenobarbital after immobilization on different chemical content Fe-particles (Fe/PhB, w/w=20) in physiological solution at pH 7.4.

TABLE 17.4 Sorption Efficiency of Different Chemical Content Modified Ferrocomposites to Phenobarbital (PhB), (Composite/PhB, w/w=20) in Physiological Solution at pH 7.4

Types of composites	Feunmodified	Fe+gel+formald.	Fe+albumin+formald.	Fe-Al_2O_3
Sorption, average±SD (%)	18.9 ±1.8	36.2 ±2.3	51.4 ±5.8	14.2±1.5
Absorptive capacity, average±SD (mg/g)	9.4±0.9	18.1 ±1.2	25.7 ±2.9	7.0±0.7

TABLE 17.5 Sorption Efficiency of Different Chemical Content Ferrocomposites to Phenobarbital (PhB), (composite/PhB, w/w=20) in Physiological Solution at pH 7.4

Types of composites	Fe unmodified	Fe+Dex	Fe-Silica (Fe90%, $SiO_2$10%)	Fe-Silica (Fe50%, $SiO_2$50%)
Sorption, average±SD (%)	18.9 ±1.8	15.9 ±1.4	42.6 ±4.2	48.1±6.4
Absorptive capacity, average±SD (mg/g)	9.4±0.9	7.9 ±0.7	21.3 ±2.1	24.1±3.2

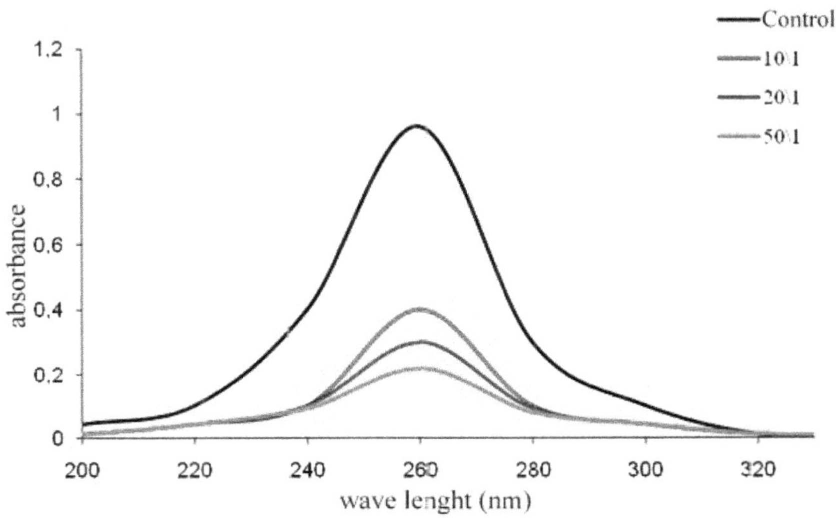

FIGURE 17.5 Absorption spectra of barbituric acid after immobilization on $FeCSiO_2$ particles at weight ratious composite/BA = 10, 20, 50.

The Figure 17.5 shows spectra of barbituric acid after immobilization on $FeCSiO_2$ microparticles at different weight ratios composite/BA. The maximal sorption efficiency of barbituric acid was found for FeC SiO_2 composite of content: 50% Fe, 40%C, 10%SiO_2. The meanings of sorption and absorptive capacity of barbituric acid for this composite: 75.0% and 58.0 mg/g at weight ratio $FeCSiO_2$/BA 50 and 10, accordingly. Apparently, it occurs physical adsorption of barbituric acid in microporous of composite.

17.4 IMMOBILIZATION AND DESORPTION OF L-BOROPHENILALANIN

The Figure 17.6 shows spectra of L-BPA after immobilization on FeC microparticles at different weight ratios composite/L-BPA. Apparently immobilization occurs by physical adsorption into porous of composite. The highest absorption capacity of L-BPA for this composite 78.0 mg/g was detected at weight ratio composite/L-BPA equal 5. The maximal adsorption capacity of L-BPA 160.0 mg/g was reached for dextran-modified iron-particles.

The desorption of L-BPA at λ225nm from FeC composite and dextran-modified Fe-particles are presented in Figures 17.7 and 17.8. Analysis of the results on desorption has shown that quantity of the desorbed L-BPA

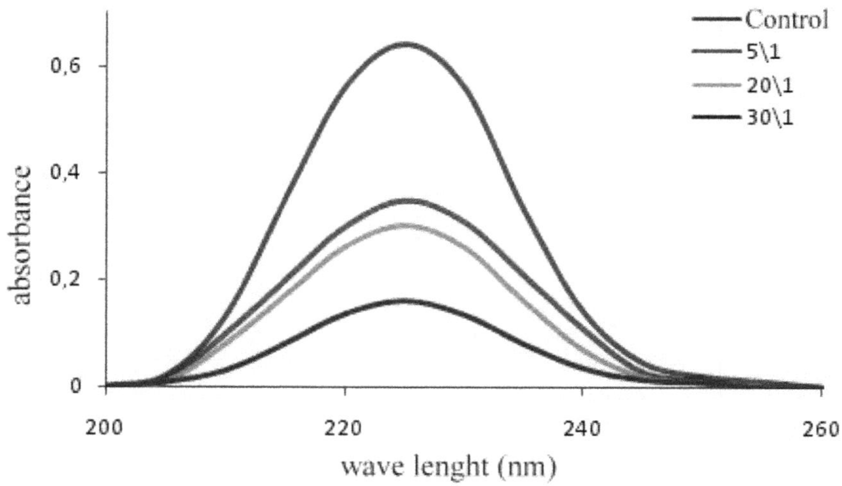

FIGURE 17.6 Absorption spectra of L-BPA after immobilization on FeC microparticles at different weight ratios composite/L-BPA.

from magnetic operated preparations is enough to create therapeutic concentration of boron atoms in tumor. Nevertheless it is required to continue investigations in order to chose the optimum ferrocomposites types with more longer of desorption time for working out magnetic operated preparations of L-BPA on their basis.

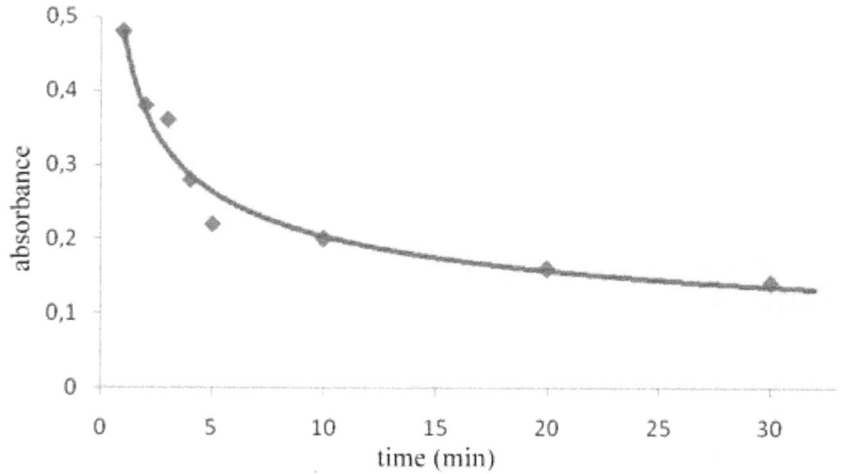

FIGURE 17.7 The Dynamics of L-BPA Desorption (λ225nm) from FeC Composite in 0.6% Albumin (T 37°C, pH 7.4).

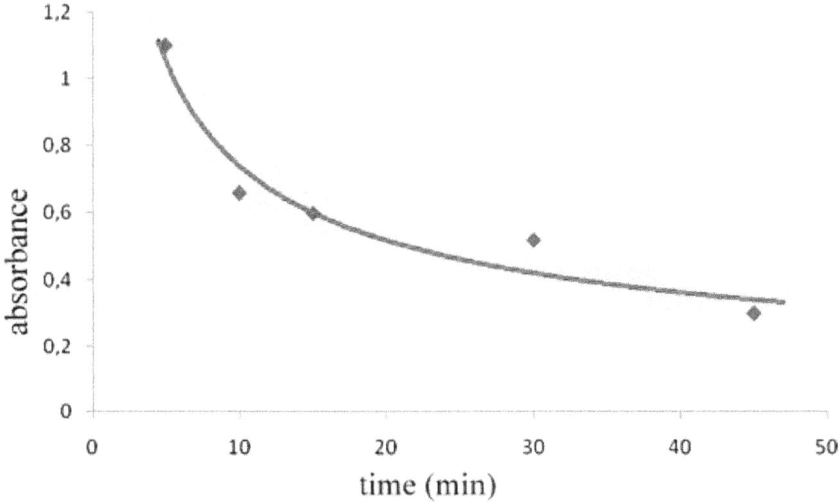

FIGURE 17.8 The dynamics of L-BPA desorption (λ225nm) from dextran-modified Fe-particles in 0.6% albumin (T 37°C, pH 7.4).

Spectrophotometric study of the reaction interaction L-BPA with dextran in the water solutions showed its conjugation with dextran (Figure 17.9).

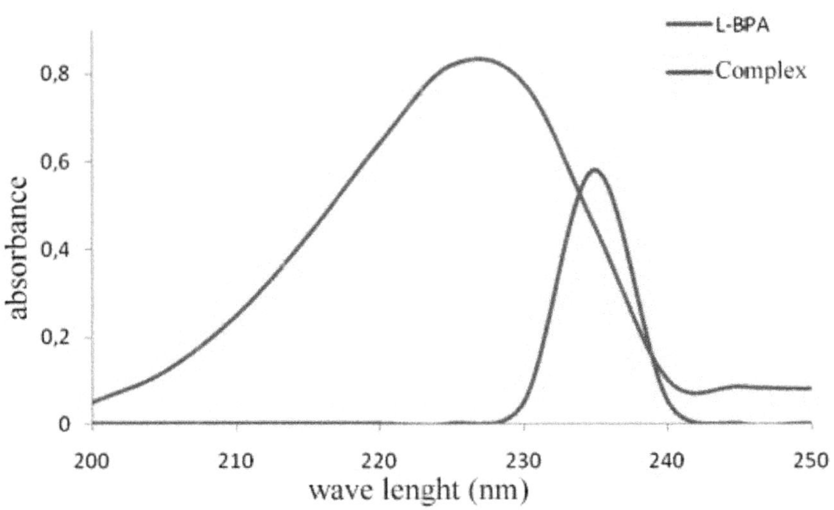

FIGURE 17.9 Conjugation of L-BPA with dextran: T 20°C, pH 7.0, incubation time 1 min.

17.5 SUMMARY

We have worked out new methods to modify the surface of the ferrocomposites microsized particles (iron, magnetite, iron-carbon) by albumin, gelatin or dextran and have been studied their sorption efficiency and the same for composite iron-silica and iron-carbon-silica to bovine hemoglobin and barbiturates: sodium phenobarbital and barbituric acid. Optimal Fe-composites for hemoglobin immobilization are by albumin and dextran modified microsized Fe-particles; for phenobarbital: albumin-modified Fe-particles and Fe-silica composite; for barbituric acid: FeC-silica composite. These ferrocomposites can be recommended for use as sorbents for extracorporal detoxification of patient's blood and purification of conserved donor blood from free hemoglobin and barbiturates by the method of magnetic separation. Dextran modified microsized Fe-particles are perspective as carriers for magnetically-guided targeted delivery of L-BPA at Boron Neutron Capture of Tumor Therapy.

KEYWORDS

- **barbiturates**
- **hemoglobin**
- **immobilization of biologically active compounds**
- **L-borophenilalanin**
- **microsized particles of ferrocomposites**
- **modified surface**
- **sorption efficiency**

REFERENCES

1. Komissarova, L. Kh., Filippov, V. I., Kuznetsov, A. A. In the Proceeding of the 1st Symposium on Application of biomagnetic carriers in medicine (in Rus.), Moscow, Science (Nauka) Publishing House, 2002, 68–76.

2. Brusentsov, N. A., Bayburtskiy, F. S., Komissarova, L.Kh. et al. Biocatalytic Technology and Nanotechnology. Nova science Publishers, New York. 2004, 59–66.
3. Yanovsky, Y. G., Komissarova, L.Kh., Danilin, A. N. et al. Solid State Phenomena. 2009, V. 152–153, 403–406.
4. Zhiwei Li, Chao Wang, Liang Cheng et al. Biomaterials. 2013, No 4. 9160–9170.
5. Kutushov, M. W., Komissarova, L. Kh., Gluchoedov, N.P. Russian patent. No 210952, 1998.
6. Bunis, R. J., Riley, K. J., Marling, O. K. In: Research and Development in Neutron Capture Therapy. Ed. Monduzzi. Bologna. 2002, 405.

CHAPTER 18

A STUDY ON ANTIOXIDANT SYSTEM OF THE BODY

N. N. SAZHINA,[1] I. N. POPOV,[2] and G. LEVIN[2]

[1]*Emanuel Institute of Biochemical Physics Russian Academy of Sciences, 4 Kosygin Street, 119334 Moscow, Russia, E-mail: Natnik48s@yandex.ru*

[2]*Research Institute for Antioxidant Therapy, 137c Invalidenstr., 10115, Berlin, Germany, E-mail: ip@antioxidant-research.com*

CONTENTS

ABSTRACT

An assessment of the total antioxidant activity (AOA) of blood serum of patients with the liver disease by two chemiluminescent methods having different models of free radical oxidation: "Hb-H_2O_2-luminol" and

"ABAP-luminol" is carried out. The comparative analysis showed not high correlation of results (r = 0.798), that is explained, mainly, by a distinction of free radical initiation mechanisms and influence of some blood serum components (proteins and bilirubins) on initiation process. More strongly it is shown in model with "Hb-H_2O_2". In this regard, more preferable in clinical practice for an AOA assessment it is necessary to consider model "ABAP-luminol." A comparison of antioxidant parameters of blood serum of patients with the affected liver with some general clinical blood characteristics, such as the content of uric acid, total and direct bilirubin, albumin, parameters of lipid metabolism is carried out. The aim and scope of this study is the comparative analysis of the total blood serum AOA of patients with liver pathology at parallel measurements by two chemiluminescence devices with various models of free radical oxidation for the aim of more suitable model choice in clinical practice. Comparison of antioxidant and some general clinical parameters of blood serum for patients with liver pathology are carried out to establish a correlation between these characteristics.

18.1 INTRODUCTION

Definition of the total antioxidant activity (AOA) of person blood serum is an important task for medico-biological researches as AOA determines the protection of person organism in fight against an oxidative stress. Blood is represented difficult substance for researches, antioxidant composition which is caused, first of all, by availability in it of amino acids, uric acid, vitamins E, C, hormones, enzymes, and also the intermediate and final metabolism products [1]. The total AOA is the integrated value characterizing possibility of combine antioxidant action of all blood plasma components taking into account their potential synergism. During 25–30 last years attempts of creation of techniques for total AOA definition at the same time all inhibitors of free radical reactions which are present in blood plasma are made. However, the role of each inhibitor can significantly differ at various ways of activation of oxidizing processes. Therefore, the choice of adequate systems of AOA assessment has paramount value for the correct interpretation of the received results within clinical laboratory diagnostics [1–3]. Definition of AOA assumes not only detection of one or several substances, and identification of their "functional" activity that can be made in suitable oxidizing system. The main components of any test system for blood serum AOA definition are: the system of radical generation and molecule target which being exposed to oxidation, changes the registered physical and chemical properties. Informational content of the received results depends on a choice of these objects. Now chemiluminescent (CL) methods of blood serum AOA definition are widely used [1, 2]. They are rather sensitive, operative and allow controlling oxidation kinetics directly. One of main distinctions of CL methods is the way of free radicals generation. It can be carried out on chemical or physical-chemical principles (e.g., at interaction of gem-containing derivatives with hydrogen peroxide, at thermo – destruction of azo-compounds and radiation of photosensitizes) [2–4].

One of main organs in antioxidative system of an organism is the liver, its synthetic and secretor activity: it belongs as to endogenous antioxidants, so to synthesis of components of inflammatory and anti-inflammatory syndromes [2, 5]. In the liver occur many vital metabolic processes, resulting in the formation and enter in blood substances necessary for the organism, including various endogenous antioxidants, which primarily include uric acid [6] and bilirubin and biliverdin [7]. The liver is the "repository" and some exogenous antioxidants, such as ascorbic acid, which is a synergist for many bioantioxidants and exerts its activity in important for the body's "moments" of oxidative stress [8]. Therefore, studying of blood serum antioxidant properties for patients with liver pathology is an important aspect for understanding of the occurring phenomena in organism protective system.

In the present work of the comparative analysis of the total blood serum antioxidant activity patients with liver pathology at parallel measurements by two chemiluminescence devices with various models of the free radical oxidation is carried out. The contribution to the total antioxidant activity of blood serum of its individual biochemical analytes was assessed also.

18.2 EXPERIMENTAL PART

Blood serum samples of 16 patients with liver pathology (atrophic cirrhosis, liver new growths, etc.) and 18 donors with necessary clinical blood indicators were transferred for research by Myasnikov Institute of clinical cardiology. Measurements of the total AOA in parallel by two CL devices were carried out in Emanuel Institute of biochemical physics RAS.

In the first model of free radical oxidation the system "hemoglobin (Hb) – hydrogen peroxide (H_2O_2) – luminol" in which generation of radicals by Hb and H_2O_2 interaction was used, and luminol plays a role of a *chemiluminogenic* oxidative substance. Distinctive feature of this model from other oxidation models is that the formed in it radicals can initiate free radical oxidation reactions in vivo as blood contains Hb and H_2O_2. The detailed measurement technique of this model for studying of blood serum AOA and its separate components is given in [4]. The kinetics and detailed scheme of reactions proceeding at "Hb and H_2O_2" interaction are rather difficult. The estimated scheme of the reactions leading to generation of luminol oxidation radical initiators is given in Figure 18.1.

As appears from this scheme, H_2O_2 and metHb (Hb-Fe^{+3}) – the oxidized Hb form, making the main part of the commercial Hb preparations, can induce luminol oxidation on two main mechanisms [9]. So, on the one hand, interaction of H_2O_2 with metHb is accompanied by gem destruction and exit from it of iron ions, which participate in education of $OH^{\bullet-}$radicals. Besides, as a result of this interaction active ferril-radicals (Hb($^{\bullet+}$)–Fe^{4+}=O) are formed. Being formed radicals initiate the luminol oxidation which sequence of reactions is well-known now [10]. In the oxidation process

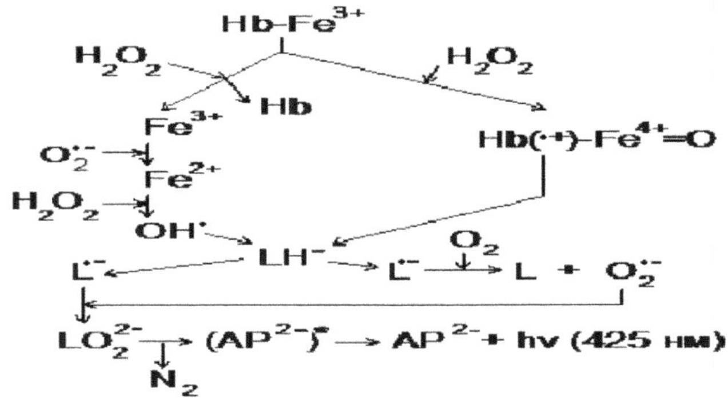

FIGURE 18.1 The estimated scheme of reactions in the system "Hb – H_2O_2 – luminol" [4] (LH^- – luminol anion, $L^{\bullet-}$ – luminol radical, $O2^{\bullet-}$– superoxide anion, LO_2^{2-}––luminol-endoperoxide, AP^{2-} and $(AP^{2-})^*$ – aminophthalate anion in the basic and excited states respectively).

$L^{\bullet -}$, $O2^{\bullet -}$-radicals are formed, a luminol-endoperoxide LO_2^{2-}, and further an aminophthalate anion in excited state $(AP^{2-})^*$ upon which transition to the main state light quantum hv is highlighted. Introduction of antioxidants in "metHb-H_2O_2-luminol" system leads to change of CH kinetics and increase in the latent period. Advantage of model: all reagents are available and aren't toxic. Restrictions: instability of H_2O_2 and need of frequent control of its concentration. For realization of this method in the present work the device "Lum-5373" (OOO"DISoft," Russia, www.chemilum.ru) was used according to detail technique [4].

In the second model of the thermo-initiated CL (TIC) luminol as an oxidation substance is used so [2, 3]. Initiation of free radicals happens at thermal decomposition of water-soluble

R-N=N-R azo-compound 2.2'-azo-bis (2-amidinopropane) dihydrochloride (ABAP). In the luminol oxidation processes in the presence of oxygen $ROO^{\bullet}$– radicals, $O_2^{\bullet -}$– radicals and further LO_2^{2-} and $(AP^{2-})^*$ are formed as in the first model. The corresponding scheme of reactions is given in Figure 18.2 [2]. An advantage of this model – the constant speed of peroxide radical initiation at a stable temperature during a long time. In the water environment at pH=7.4 and temperature 37°C, radical generation speed Ri, (mol/L)/s=1.36×10^{-6}·[ABAP], where [ABAP] – concentration of an ABAP [11]. A restriction – use of toxic chemical azo-compound.

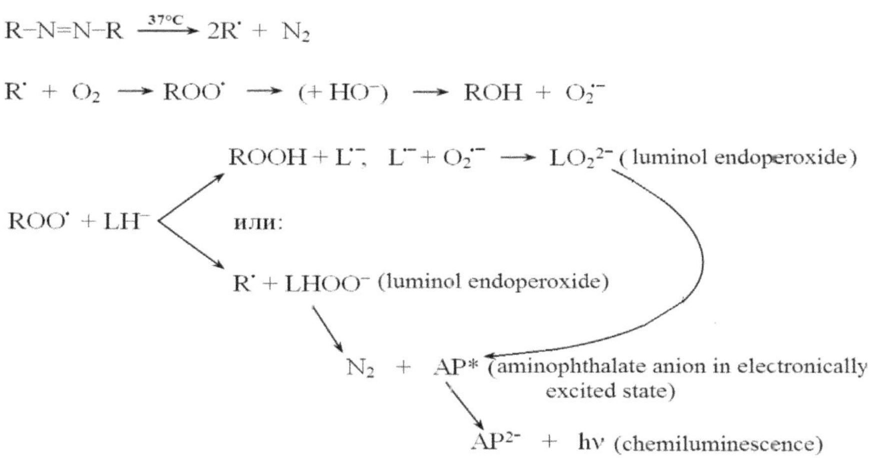

FIGURE 18.2 The estimated scheme of reactions in the model "ABAP – luminol" [2].

The TIC recordings were carried out with the device "minilum" (ABCD GmbH, Germany) at a temperature of 37±0.01°C (www.minilum.de).

In both CL systems the key measured parameter for determining of the total water-soluble blood serum component AOA (ACW – "integral antiradical capacity of water soluble compounds") is the latent period. It decides as time from the oxidation initiation moment to a point of intersection on an axis of time of the tangent attached to CL-curve in the point corresponding to a maximum of its first derivative dI/dt (Figure 18.7). Calibration of devices was carried out on ascorbic acid, and the total AOA of water-soluble components (ACW) was expressed in the equivalent ascorbic acid content in one liter of blood serum (μmol/L). The measurement error of this parameter for the first device made no more than 20%, for the second didn't exceed 5%.

18.3 RESULTS AND DISCUSSION

In Figure 18.3 kinetic CL-curves received when using both oxidation models for tests of patients are given: test №1-with the lowest value of total AOA (ACW), №2 – with an average, №3 – with the highest. In the first model (Figure 18.3a) for different tests it is characteristic not only change of the latent period, but also considerable change of luminescence intensity while in the second model (Figure 18.3b) only the latent period significantly changes. Especially it is characteristic for patients with the raised content of bilirubin (test №3). For the first oxidation model influence of protein components of blood serum is especially strongly expressed. As shown in Ref. [4], they generally and suppress luminescence intensity.

In Figure 18.4 comparison results of the total AOA (ACW) of water-soluble components of blood serum of studied patients received by both methods are presented. Results show a wide spacing of ACW values: from 300 to 2400 μmol/L for the first model, and from 15 to 940 μmol/L for the second (in this case at norm about 100–500 μmol/L). Abnormally high ACW values were observed for patients with the raised content in serum of the general bilirubin to 600 μmol/L above (at norm 1.7–20.5 μmol/L). Low correlation of ACW values for used models (correlation coefficient r=0.7977) is explained, mainly, by distinction of free radical initiation mechanisms and possible influence of some blood serum components

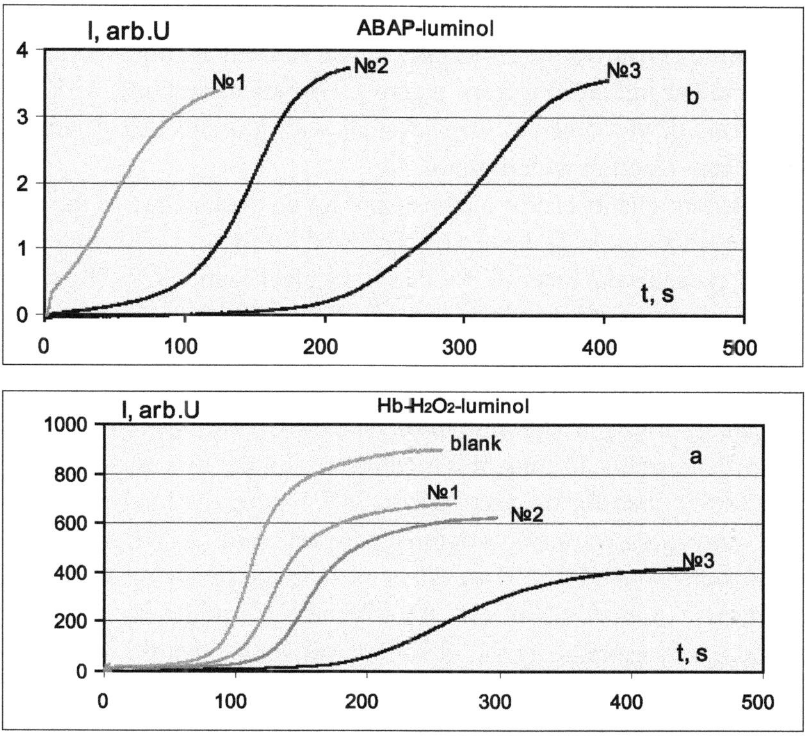

FIGURE 18.3 CL kinetics for "Hb – H_2O_2 – luminol" model (a) (blood serum volume v=2μL (№1), v=1.5 μL (№ 2,3)) and for the "ABAP-luminol" model (b) (v=2 μL (№1–3)). №1 – test with the lowest ACW value from the donor, test №2 and №3 – from recipients with an average and with the highest ACW. On ordinate axis – intensity of CL (I).

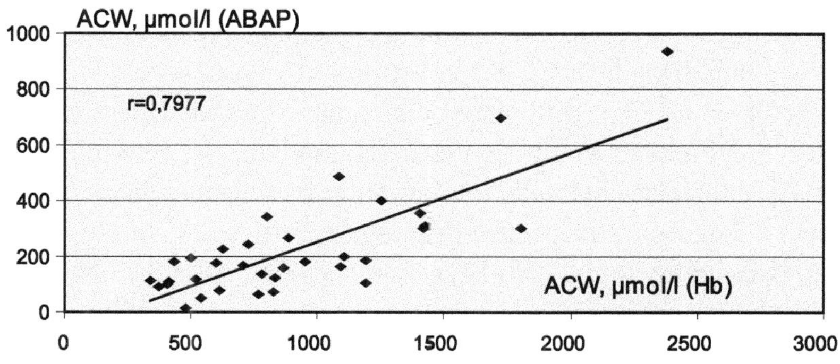

FIGURE 18.4 Correlation of ACW measurement results for donors and recipients (ACW – "integral antiradical capacity of water soluble compounds") for model with hemoglobin (ACW, μmol/L (Hb)) and with ABAP (ACW, μmol/L (ABAP)); r-correlate coefficient.

(especially proteins) on the initiation speed that can change the observed latent period considerably. Especially it concerns the first model in which one of radical initiators is very active OH•-radical reacting with many components of blood serum, in particular with proteins [2], which "distract" it from reaction with luminol.

Similar arguments can be adduced and for an explanation of the different dependences of ACW values on the blood serum uric acid content (the biochemical analysis) received for these models (Figure 18.5). The relative "stoichiometric" coefficient for uric acid in "ABAP-luminol" system makes about 2.0 [12], while in "Hb-H_2O_2-luminol" – less 1.0 [4]. Interacting with various radical intermediates, uric acid influences in the first model not only on the latent period, but substantially and on the CL intensity. Correlation of ACW with the uric acid content is much worse for the first model (r=0.583, Figure 18.5a), than for the second (r=0.745, Figure 18.5b). The values of similar correlation coefficients defined with uric acid on "ABAP-luminol" model in other works [12, 13] are rather close to coefficient received in ours experiments. In work [2] using "ABAP-luminol" model on the basis of blood serum measurements for 45 donors it is shown that the contribution of uric acid to ACW makes 64%, and proteins of 5%.

It was shown [14] that in systems of the photo-sensitized (PCL) and thermo-initiated (TIC) CL native amino acids and albumin don't possess anti-radical activity, but get it in processes of oxidizing modification. In "Hb-H_2O_2-luminol" system their share makes about 50% of ACW [4]. Therefore, measurement results are subject to influence of the serum albumin content which at patients with liver pathology is sharply underestimated owing to violation of its synthesizing ability (Figure 18.6). Average value of albumin made 25.21±5.33 g/L (min = 14.4 g/L, max = 37.1 g/L) at norm of 34–48 g/L. The difference of albumin values for different patients reaches 100% and above. It is an essential hindrance at determination of the ACW parameter and calls into question its informational content in respect of diagnostics or control of treatment efficiency. In our opinion, the only possibility to use "Hb-H_2O_2-luminol" method in the clinical purposes is the serum deproteinization. However, thus a possibility of quantitative assessment of oxidative stress degree is lost. It is shown in system of TIC by comparison of anti-oxidizing protection with extent of oxidizing damage of blood serum proteins [2].

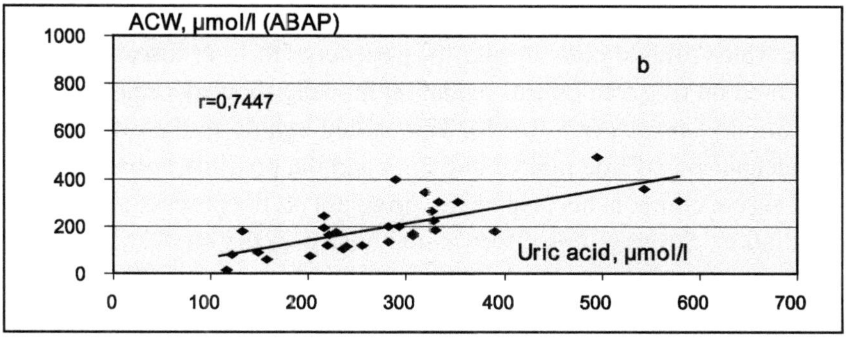

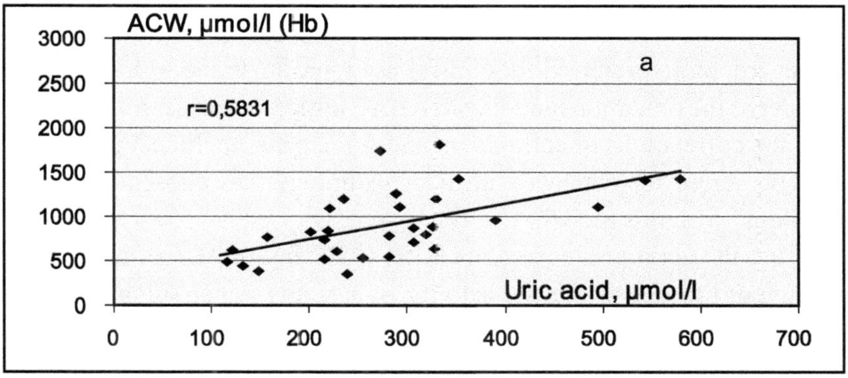

FIGURE 18.5 Dependence of blood serum ACW for donors and recipients on the uric acid content (the biochemical analysis) for oxidation model "Hb-H_2O_2-luminol" (a) and "ABAP-luminol" (b).

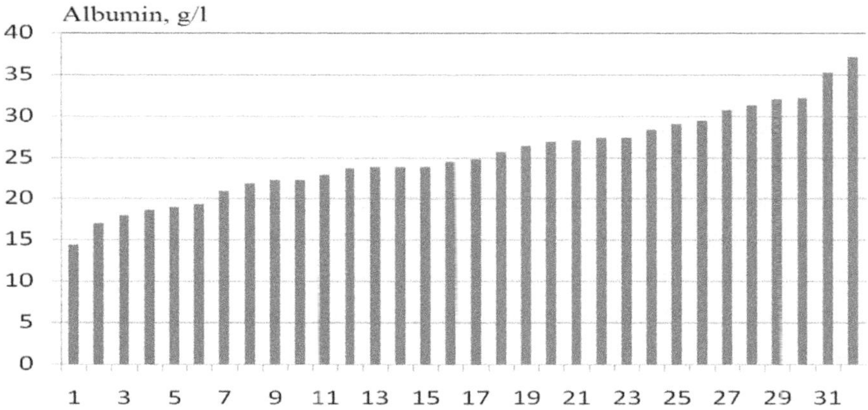

FIGURE 18.6 The albumin content in 32 studied blood serum tests (the biochemical analysis).

Thus, a comparative analysis of the total antioxidant activity of blood serum water-soluble components for patients with liver disease (ACW), performed on two free radical oxidation models, showed a relatively low correlation of results (r = 0.798). This is due mainly to the difference in the mechanisms of free radical initiation and the possible impact of some blood serum components (especially proteins) on the process and the rate of initiation. Stronger this effect is manifested in the model "Hb-H_2O_2," where an active OH•-radical-initiator reacts with a number of serum components. The discrepancy in measurement results significant for patients with abnormally high content of certain blood serum components which are differentially inhibit the luminol oxidation due to side reactions. In this regard, more preferred for clinical use to estimate the AOA should be considered the oxidation model with ABAP initiator. Therefore, for further study the correlations of antioxidant and some general clinical parameters of blood serum for patients with liver pathology was chosen the device "minilum" with this model.

Using the appropriate reagents (kits) (ABCD GmbH, Germany www.minilum.de) to determine the activity of fat- and water-soluble antioxidants, antioxidant parameters caused antiradical properties of uric (UA) and ascorbic (ASC) acids and high protein (ARAP – anti-radical ability of proteins) were measured.

Figure 18.7 shows kinetic TIC curves from the series of analyzes. The right curve corresponds to ACW for patient with hyperbilirubinemia, the left – the result of sample re-measurement after its preincubation with the urate oxidase enzyme. The difference of the latent periods corresponds to the contribution of antiradical capacity of uric acid (UA) in ACW. It should be noted that the normal contribution of UA in ACW are 40–80%. In this case only 9.6%. It indicates on the predominant contribution of bilirubin in ACW for this patient, which was confirmed by the results of the laboratory analysis: total bilirubin was 608.5 μmol/L (normal 1.7–20.5), direct – 387.2 μmol/L (normal 0–5.1). The middle curve in Figure 18.7 corresponds to the ACW, which lies within the normal range. Left curve was obtained by analyzing samples from a single donor. The extremely low value of the latent period, which coincides with the blank sample, indicates the complete absence of antioxidant protection. In this case it is necessary to control ACW of donors and take measures to protect organ

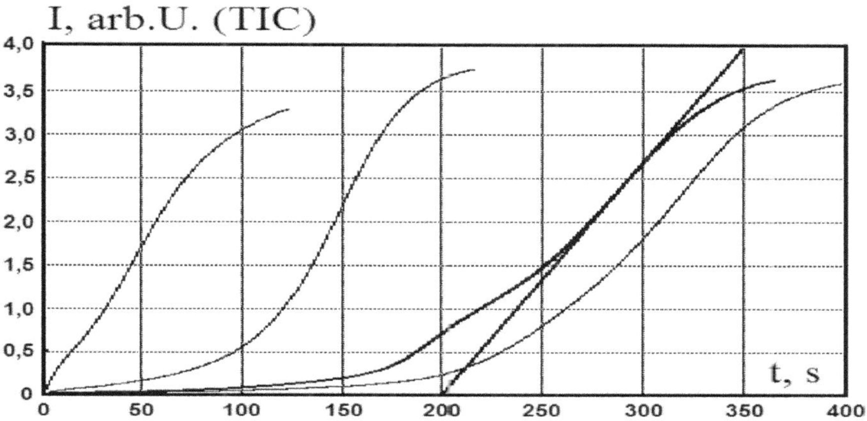

FIGURE 18.7 Kinetic TIC curves, obtained in the measurement of serum ACW for three patients and contribution UA (uric acid) in ACW. (Explanation in the text).

intended for transplantation. This is necessary to preserve its viability and preventing damage during storage. Efficient use of ascorbic acid for this purpose has been demonstrated previously [15] and repeatedly confirmed in recent years [16, 17].

During ACW measurements it was found that ascorbic acid is absent in blood serum of almost all recipients, since pre-incubation of samples with ascorbate oxidase did not change the character of TIC curves and latent period values. This fact can be explained by "homogeneous" group of seriously ill patients with similar pathology. Similar results were also observed in some cases for less severe diseases [2].

The data in Figure 18.8 show that for most patients the ACW values practically correspond to UA values. The small difference (2.5%) is explained by the contribution of ARAP parameter in ACW due to the action of thiol protein SH-groups. This indicates that the total antioxidant capacity of endogenous hydrophilic antioxidants (ACW) for these patients is mainly determined by the capacity of uric acid. For two patients with abnormally high levels of total and conjugated bilirubin ACW significantly exceeded the UA. Which forms of bilirubin: direct (bilirubin-glucuronide), indirect (unconjugated pigment in the bloodstream transporting albumin) or free is a major contributor to the ACW, not yet determined. There is evidence that all three forms can be antioxidant activity, but preference is given to

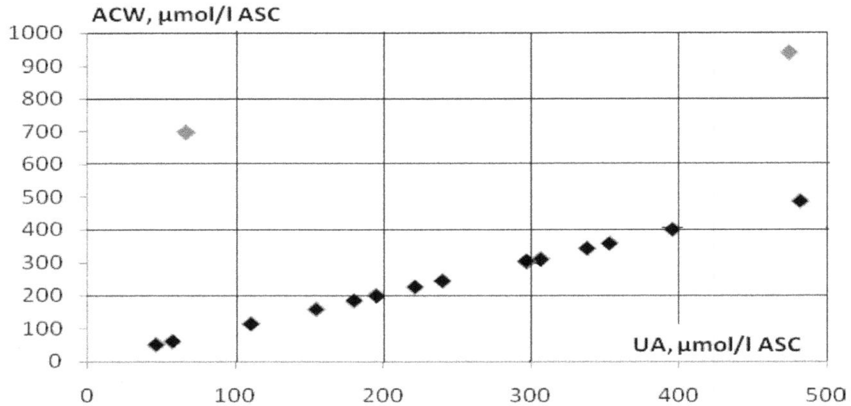

FIGURE 18.8 Comparison of the total blood serum antioxidant capacity (ACW) with antioxidant capacity (UA) of uric acid. Two deviated points – samples of recipients with pronounced hyperbilirubinemia.

indirect bilirubin. It is believed that it prevents oxidation of other ligands for albumin, especially fatty acids, in the complete absence of the reduced form of ascorbic or uric acids [18].

Earlier (Figure 18.5b) the compative results for serum ACW of patients with their uric acid content were presented. Significant variations in the ACW values are observed. This can be explained by the fact that the significance, which may not correspond to its content. Uric and ascorbic acids may inactivate sequentially two free radicals parameter UA does not reflect the uric acid content in the blood serum, but its antioxidant capacity, its functional remaining in the fully oxidized or semi-oxidized form.

One of the specific tests of liver pathology as parenchymatous organ is the content of the blood serum bilirubin – a bile pigment. Bilirubin is formed by hydrolysis in the spleen splenocytes of a tetrapirroll – heme of hemoglobin. The hydrolysis products are hydrophobic, and unconjugated bilirubin delivers the transport protein albumin in the liver. Attempts to find in our study the interconnection between the antioxidant parameters of water and lipid phases of blood serum and different forms of bilirubin, parameters of lipid metabolism, free and bound cholesterol, free fatty acids, triglycerides and phospholipids revealed no significant results. In respect of bilirubin, given its known role in extra- and intracellular lipid and protein protection from oxidation [19] it should be noted that a significant increase in its level in the blood forms the "protection" of blood serum proteins from damage, resulting in low values of ARAP. However, significant correlation between bilirubin in all its forms and parameters ARAP and ACW in the group of studied patients studied in this work not revealed.

Pathology of the liver in the acute stage is accompanied by the loss of his body many vital functions, such as, for example, the synthesis of serum albumin (it was in the normal range only for two patients, for rest patients it was significantly underestimated). The liver function in maintaining the antioxidant homeostasis is disturbed. This function consists, on the one hand, in the storing of ascorbic acid and its release into the bloodstream as required, and on the other hand – in regulating the synthesis of uric acid as an endogenous antioxidant complementary to ascorbic acid. If for any reason (enhanced load by xenobiotics, viral infection, etc.), the reaction of the liver is not sufficient to suppress the oxidative stress caused by these factors, then "wakes up" evolutionarily ancient mechanism of antioxidant protection, manifested in the activation of the heme oxygenase enzyme and the production of bilirubin. This phenomenon is also observed in the intact liver under strong oxidative stress, and with a lack of ascorbic acid in preterm infants, and for patients with severe inflammation [20, 21]. Thus, for an effective antioxidant defense may be sufficient even small (nanomolar) concentrations of bilirubin [22, 23]. Evolutionary reason for displacement of this antioxidant defense type is a toxicity of bilirubin at high concentrations [24]. It is noted however that recipients with high preoperative bilirubin levels in liver transplant are more favorable for postoperative period compared with patients with low content in the preoperative period [25].

18.4 CONCLUSIONS

1. The comparative analysis of the total antioxidant activity of blood serum water-soluble components for patients with liver disease (ACW), performed on two free radical oxidation models: "Hb-H_2O_{2-} luminol" and "ABAP- luminol" showed, that more preferred for clinical use should be considered the oxidation model with ABAP initiator.
2. Measurement results of the antiradical capacity of blood serum and its some individual components for recipients with liver disease

showed significant disturbance of liver function in maintaining the antioxidant homeostasis.

3. At the liver pathology, the absence of exogenous ascorbic acid and expressed human antioxidant defense, uric acid and bilirubin compensatory become in vivo the major hydrophilic antioxidants.
4. The increased hyperuricemia at different pathological processes can be considered as a activation test of the biological reaction of an inflammation and syndrome of anti-inflammatory compensatory protection.
5. It should be possible to lower hyperuricemia and compensatory function of uric acid in pathological processes by long-time use of optimal doses of ascorbic acid.

ACKNOWLEDGMENT

We thank professor V.N. Titov from Myasnikov Institute of Clinical Cardiology of Russian cardiologic R&D production complex of Russian Minzdrav, doctor S.A. Solonin and professor M.A. Godkov from Sklifosofskiy research Institute of emergency medical care for providing blood serum samples and biochemical analysis.

KEYWORDS

- **antioxidant activity**
- **blood serum**
- **chemiluminescence**
- **free radical oxidation**
- **liver**

REFERENCES

1. Bartosz, G. Total antioxidant capacity. *Adv. Clin. Chem.* 2003, 37, 219–292.
2. Popov, I., Lewin, G. Antioxidative homeostasis, its evaluation by means of chemiluminescent methods. *In: Handbook of chemiluminescent methods in oxidative stress assessment*. Transworld Research Network, Kerala, 2008, 361–391.

3. Popov, I., Lewin, G. Antioxidant system of the body and the method of the thermo-initiated chemiluminescence to quantify its condition. *Biofizika*. 2013, 58(5), 848–56. (in Russian).
4. Teselkin Yu. O., Babenkova, I. V., Lyubitsky, O. B., Klebanov, G. I., Vladimirov Yu.A. Inhibition of oxidation of luminol in the presence of hemoglobin and hydrogen peroxide by serum antioxidants. *Questions of medical chemistry.* 1997, 43(2), 87–93.
5. Titov, V. N. Endogenous oxidative stress system confrontation. The role of DHEA and oleic fatty acid. *Uspekhi sovremennoy biologii*. 2009, 129(1), 10–26. (in Russian).
6. Glantzounis, G. K., Tsimoyiannis, E. C., Kappas, A. M., Galaris, D. A. Uric acid and oxidative stress. *Curr Pharm Des*. 2005, 11(32), 4145–51.
7. Nakagami, T., Toyomura, K., Kinoshita, T., Morisawa, S. A beneficial role of bile pigments as an endogenous tissue protector: anticomplement effects of biliverdin and conjugated bilirubin. *Biochem Biophys Acta.* 1993, 1158(2), 189–93.
8. Popov, I., Lewin, G. Antioxidative homeostasis: characterization by means of chemiluminescent technique. In: Packer, L., ed. *Methods in enzymology*. New York. Academic Press; 1999, 300, 437–56.
9. Puppo A, Halliwell, B. Formation of hydroxyl radicals from hydrogen peroxide in the presence of iron. Is haemoglobin a biological Fenton reagent?. *Biochem, J.* 1988, 249(1), 185–190.
10. Faulkner, K., Fridovich, I. Luminol and lucigenin as detectors for $O_2^{\cdot-}$. *Free Radiic. Biol. Med.* 1993, 15(4), 447–451.
11. Niki, E. Free Radical Initiators as Source of Water- or Lipid-Soluble Peroxyl Radicals. *Methods in enzymology*. Eds. L. Packer & A. N. Glazer. New York. Academic Press; 1990, 186: 100–108.
12. Uotila, J. T., Kirkkola, A. L., Rorarius, M. et al. The total peroxyl radical-trapping ability of plasma and cerebrospinal fluid in normal and preeclamptic parturients. *Free Radic. Biol. Med.* 1994, 16(5), 581–590.
13. Lissi, E. A., Salim-Hanna, M., Pascual, C., Castillo, M. D. Evaluation of total antioxidant potential (TRAP) and total antioxidant reactivity from luminol-enhanced chemiluminescence measurement. *Free Radic. Biol. Med.* 1995, 18(2), 153–158.
14. Popov, I., Lewin, G. Photochemiluminescent detection of antiradical activity. VI. Antioxidant characteristics of human blood plasma, low-density lipoprotein, serum albumin and aminoacids during in vitro oxidation. *Luminescence:* 1999, 14: 169–174.
15. Popov, I., Gäbel, W., Lohse, W., Lewin, G., Richter, E., Baehr, R. V. Einfluss von Askorbinsäure in der Konservierungslösung auf das antioxidative Potential des Blutplasmas während der Lebertransplantation bei Minischweinen. *Z. Exp. Chirurgie.* 1989, 22, 22–26.
16. Wang, N. T., Lin, H. I., Yeh, D. Y., Chou, T. Y., Chen, C. F., Leu, F. C., Wang, D. and Hu, R. T. Effects of the Antioxidants Lycium Barbarum and Ascorbic Acid on Reperfusion Liver Injury in Rats. *Transplantation Proceedings*. 2009, 41, 4110–4113.
17. Adikwu, E., Deo, O. Hepatoprotective Effect of Vitamin C (Ascorbic Acid). *Pharmacology & Pharmacy*. 2013, 4, 84–92.
18. Hunt, S., Kronenberg, F., Eckfeldt, J., Hopkins, P., Heiss, G. Association of plasma bilirubin with coronary heart disease and segregation of bilirubin as a major gene trait: the NHLBI family heart study. *Atherosclerosis.* 2001, 154, 747–754.

19. MacLean, P., Drake, E. C., Ross, L., Barclay, E. Bilirubin as an antioxidant in micelles and lipid bilayers: Its contribution to the total antioxidant capacity of human blood plasma. *Free Rad. Biol. & Med.* 2007, 43, 600–609.
20. Fereshtehnejad, S. M., Bejeh Mir, K. P., Bejeh Mir, A. P., Mohagheghi, P. Evaluation of the Possible Antioxidative Role of Bilirubin Protecting from Free Radical Related Illnesses in Neonates. *Acta Medica Iranica;* 2012, 50(3), 153–163.
21. Patel, J. J., Taneja, A., Niccum, D., Kumar, G., Jacobs, E., Nanchal, R., The Association of Serum Bilirubin Levels on the Outcomes of Severe Sepsis. *J Intensive Care Med.* 2013, 28(3), 230–36.
22. Dore, S., Takahashi, M., Ferris, C. D. et al. Bilirubin, formed by activation of heme oxygenase-2, protects neurons against oxidative stress injury. *Proc. Natl. Acad. Sci.* 1999, 96(5), 2445–50.
23. Baranano, D. E., Rao, M., Ferris, C. D., Snyder, S. H. Biliverdin reductase: a major physiologic cytoprotectant; *Proc. Natl. Acad. Sci.* 2002, 99(25), 16093–98.
24. Ames, B. N., Cathcart, R., Schwiers, E., Hochstein, P. Uric acid provides an antioxidant defense in humans against oxidant- and radical-caused aging and cancer: a hypothesis. *Proc. Natl. Acad. Sci.* 1981, 78, 6858–62.
25. Igea, J., Nuno, J., Lopez-Hervas, P. et al. Evaluation of delta bilirubin in the follow-up of hepatic transplantation. *Transplant. Proc.* 1999, 31(6), 2469.

CHAPTER 19

EXTRACTION AND COMPARATIVE CHARACTERIZATION OF THERMOSTABLE PROTEIN COMPLEXES

D. DZIDZIGIRI, M. RUKHADZE, I. MODEBADZE, N. GIORGOBIANI, L. RUSISHVILI, G. MOSIDZE, E. TAVDISHVILI, and E. BAKURADZE

Department of Biology Faculty of Exact and Natural Sciences, Iv. Javakhishvili Tbilisi State University, Tbilisi, Georgia, E-mail: d_dzidziguri@yahoo.com

CONTENTS

19.1 INTRODUCTION

Identification of the functions of proteins and other polymeric complexes or cell proteomes, in which the achievements of proteomics contributes greatly, is the subject of intensive research [1]. Modern technology now allows us to investigate not only individual protein molecules in living cell, but also to understand their interaction with other macromolecules and reveal their previously unknown functions. Several facts are determined: participation of polyfunctional macromolecular protein complexes in the biosynthesis of fatty acids, involvement of erythrocyte membrane proteins macromolecular complexes in exchange of CO_2/O_2, biological effects of some growth factors (polyfunctional proteins), which sometimes is achieved by interactions of other protein complexes, etc. [2–4].

Earlier we have identified the protein complex with such properties in various cells of adult white rats [5–7]. The main feature of those complexes is the thermostability of the containing components. With gel electrophoreses and chromatography of hydrophobic interaction it was determined that components with high molecular weight (45–60 kD) are hydrophobic whereas components with low molecular weight (11–12 kD) are hydrophilic according to the column retention time. Through the inhibition of transcription, complex reduces the mitotic activity of homo- and heterotypic cells in growing animals [5, 7]. Components of complex are water-soluble and are not characterized by species specificity. Thus, we can assume that they are maintained in cells of phylogenetic distance organisms and in case of confirmation of this fact they may belong to the conservative family of proteins. In order to determine general regularities of effects of complex described by us, it is necessary to more detailed examination of components phylogenesis.

19.1.1 THE GOAL

Extraction and comparative characterization of thermostable protein complexes from phylogenetic distance organisms.

19.2 EXPERIMENTAL PART

19.2.1 MATERIAL AND METHODS

The thermostabile protein complexes obtained by alcohol extraction from normal cells of various organisms (bacteria, snail, lizard, guinea pig, rat, etc., also human postsurgical material and cell culture) were used for research.

Thermostable protein fractions were obtained by the method of alcohol precipitation of Balazs and Blazsek [8], with modification. Animals were decapitated under diethyl ether. Organs were removed quickly, separated from capsules of connective tissues and vessels, rinsed with the physiological solution, and crushed. Acueous homogenates were prepared in a tissue/distilled water ratio of 1:8. The homogenates were saturated step-wise with 96% ethanol to obtain 81% ethanol fraction, which was heated in a water bath (100°C) for 20 min, cooled and centrifuged (600 g, 15 min). The supernatant was frozen in liquid nitrogen and dried in an absorptive-condensate lyophilizer. As a result a residue was obtained of a thermostable protein complex (TSPC), a light-gray powder soluble in water. Samples were kept at 4°C. Protein concentration was determined by the method of Lowry et al. [9].

Hydrophobic interaction chromatography (HIC) was used for comparative analysis of TSPC [10]. A hydrophilic polymeric sorbent, HEMA BIO Phenyl-1000 (particle size 10 mm) modified by phenyl groups, served as the stable phase. The mobile phase was phosphate buffer (pH 7.4) with ammonium sulfate. Elution was performed with the mobile phase in molar concentration range from 2.0 M to 0.0 M (pure buffer) with respect to $(NH4)_2SO_4$. For coelution of hydrophilic and hydrophobic components of protein fractions, Brij-35 polyoxyethylene dodecyl ether with increasing concentration from 0% to 3% was added to the mobile phase. UV detection was usually set at 230 nm.

19.3 RESULTS AND DISCUSSION

Dynamic Interaction between the protein molecules (protein–protein) determines vital activity of the cells. In the last few years as a result of intensive research in this field some knowledge about formation and function of protein complex has been accumulated. Dynamic interactions of proteins are studied not only in individual species but also in different types of cells and tissues. Therefore, in the first stage the aim of the research was to obtain and compare the thermostable protein complex from organisms of different classes.

It was established that all the protein complex samples contain qualitatively different two groups of proteins. I group is hydrophilic and II group – hydrophobic proteins with a retention time of 5–6 min. and 20 min., respectively (Figure 19.1a). The same subfractions were revealed in case of protein complexes obtained from different tissues of white rat by using this method (Figure 19.1b).

It is known that dysfunctions of protein-protein interactions can lead to development of various diseases, including cancer, neurodegeneration, autoimmune diseases and more. Therefore, the analysis of protein networks based on the protein-protein interaction may be used in developing various therapeutic approaches [11]. On the next stage of the research we have performed comparative analyses of protein complexes obtained from rat kidney and normal and transformed renal tissue of human. The differences between normal and transformed cells were revealed in this

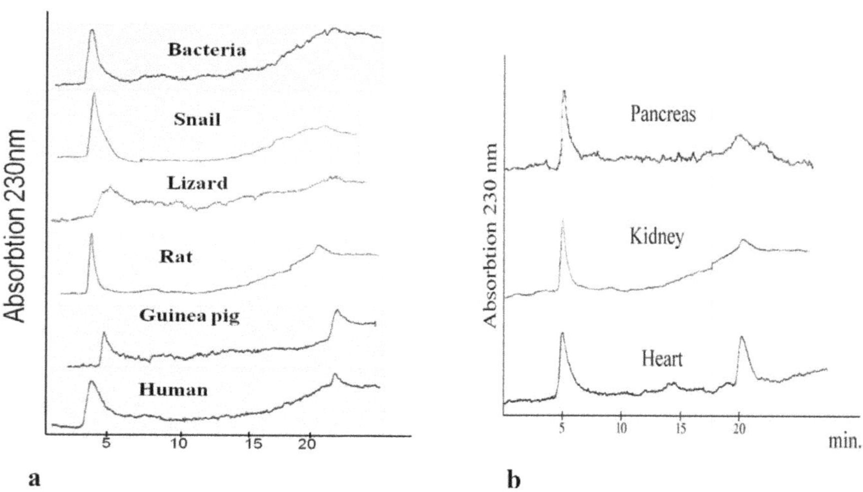

FIGURE 19.1 Chromatography of protein complex: a. protein complex from phylogeneticaly distance animals; b. protein complex form various tissues of white rat.

experiment. In particular, the components with low molecular weight was not observed in the protein complex obtained from human kidney cancer, which indicates on the changes of the complexes composition and their function during of cancer development (Figure 19.2).

The low molecular weight subfraction of protein complex has mitosis-inhibitory properties, reduces number of active and moderately active nucleoli, and decreases the activation of RNA synthesis in nuclei of cardiomiocytes of newborn rat [12, 13]. Usually, this subfraction is essential component of any thermostable protein complex obtained from various organs of adult rat. It is always seen as a major subfraction in native gel-electrophoresis (PAAG electrophoresis) (Figure 19.3).

Consequently, on the next stage of experiment the low molecular weight components of protein complexes obtained from intact (rat

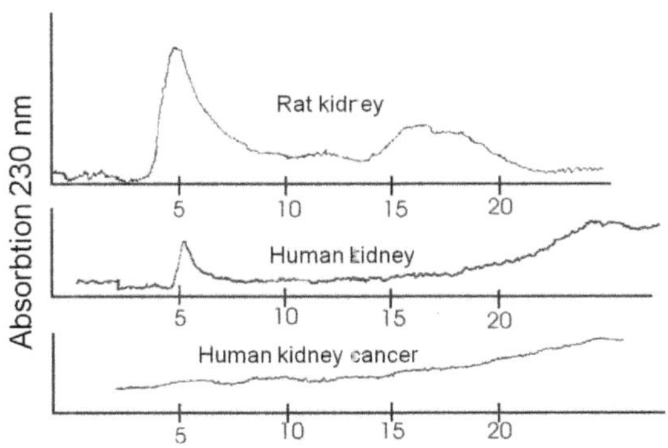

FIGURE 19.2 Chromatography of protein complexes obtained from rat kidney and normal and transformed renal tissue of human.

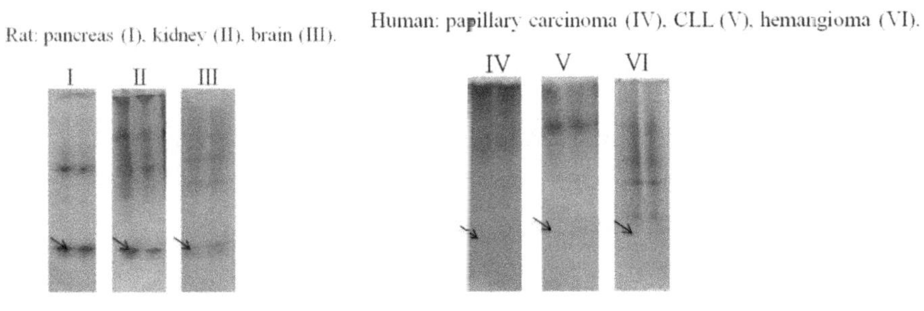

FIGURE 19.3 Electrophorogram of TPC from different organ of white rat and human (arrows indicate on low molecular weight subfraction of TPC).

pancreas, kidney, brain) and transformed cells (with different degrees) (human papillary carcinoma, CLL, hemangioma) were examined.

As it is shown at Figure 19.3 component with low molecular weight are well expressed in Protein complexes from various organs (kidney, pancreas and brain) of adult rat (Figure 19.3). Different picture is shown in complexes from transforming cells. In particular, electrophorogrames shows that protein components with low molecular weight are manifested in the samples obtained from benign as well as malignant cells. However, intensity of silver nitrate staining is much lower compared to the norm.

19.4 CONCLUSIONS

From these results it can conclude that pro-and eukaryotic cells contain a thermostable protein complex that inhibits cell proliferation. Quantitative content of protein components in the complex is changing with the growth of transformation degree of cells. Currently, development of the relative proteomics allows us to determine the identity of proteins within these complexes to reliably identify the set of proteins which is responsible for and participates in the regulation of proliferation are constantly presented in the cell. With the help of comparative proteomics it was identified Nilaparvata lugens proteins that are involved in the process of proliferation and their expression changes in response to insecticide treatment [14].

KEYWORDS

- **alcohol extraction**
- **comparative characterization**
- **extraction**
- **polymeric complexes**
- **protein complexes**
- **thermostable**

REFERENCES

1. Shaojun Dai, Taotao Chen, Kang Chong, Yongbiao Xue, Siqi Liu, and Tai Wang. Proteomics Identification of Differentially Expressed Proteins Associated with Pollen Germination and Tube Growth Reveals Characteristics of Germinated, *Oryza sativa.* Pollen *Mol Cell Proteomics*, 6, 207–230, 2007.
2. Ge, L. Q., Cheng, Y., Wu, J. C., Jahn, G. C. Proteomic analysis of insecticide triazophos-induced mating-responsive proteins of *Nilaparvata lugens* Stal (Hemiptera: Delphacidae), *J. Proteome Res.*, 10 (10), 4597–4612, 2011.
3. http://belki.com.ua/belki-struktura.html
4. Dijke, P. T., Iwata, K. K. Growth factors for wound healing. Biotechnology, 7(8), 793–798, 1992.
5. Giorgobiani, N., Dzidziguri, D., Rukhadze, M., Rusishvili, L., Tumanishvili, G. Possible role of endogenous growth inhibitors in regeneration of organs: Searching for new approaches. Cell Biology International, 29, 1047–1049, 2005.
6. Dzidziguri, D., Iobadze, M., Aslamazishvili, T., Tumanishvili, G., Bakhutashvili, V., Chigogidze, T., Managadze, L. The influence kidney protein factors on the proliferative activity of MDSK cells. Tsitologiya, 46(10), 913–914, 2004.
7. Amano, O., Iseki, S. Expression and localization of cell growth factors in the salivary gland. Kaibogaku Zasshi, 76(2), 201–212, 2001.
8. Balazs, A., Blazsek, I. Control of cell proliferation by endogenous inhibitors. Akademia Kiado (Budapest), 302, 1979.
9. Lowry, D. H., Rosebrough, N. J., Farr, A. L., Randell, R. J. Protein measurement with the folin phenol reagent. J Biol Chem, 193, 265–275, 1951.
10. Queiroz, J. A., Tomaz, C. T., Cabral, J. M. S. Hydrophobic interaction chromatography of proteins. J Chromatogr, 87, 143–59, 2001.
11. Terentiev, A. A., Moldogazieva, N. T., Shaitan, K. V. The dynamic proteomics in the modeling of living cell. Protein-protein interactions. Success of biological chemistry (Advance in Biological Chemistry), 49, 429–480, 2009 (article in Russian)
12. Rusishvili, L. Giorgobiani, N., Dzidziguri, D., Tumanishvili, G. Comparative analysis of cardiomiocyte growth-inhibitory factor in animals of different classes. Proc. Georgian Acad. Sci., Biol. Ser. B, 1(1–2), 42–45, 2003.
13. Modebadze, I., Rukhadze, M., Bakuradze, E., Dzidziguri, D. Pancreatic Cell Proteome – Qualitative Characterization and Function. Georgian Medical News, 7–8(220–221), 71–77, 2013.
14. Lin-Quan Ge, Yao Cheng, Jin-Cai Wu, and Gary, C. Jahn. Proteomic Analysis of Insecticide Triazophos-Induced Mating-Responsive Proteins of Nilaparvata lugens Stål (Hemiptera: Delphacidae). School of Plant Protection, Yangzhou University, Yangzhou 225009, P. R. China, J. Proteome Res., 10(10), 4597–4612, 2011.

CHAPTER 20

INFLUENCE OF BIOLOGICAL FACTORS IN DIFFERENT AGROECOSYSTEMS

RAFAIL A. AFANASEV and GENRIETTA YE. MERZLAYA

Pryanishnikov All-Russian Scientific Research Institute of Agrochemistry, d. 31A, Pryanishnikov St., Moscow, 127550, Russia, Tel: +7-499-976-25-01; E-mail: rafail-afanasev@mail.ru; lab.organic@mail.ru

CONTENTS

ABSTRACT

The authors revealed new regularities of mobile phosphorus dynamics in the soils of different agroecosystems during prolonged systematic fertilizer application. They had shown that mobile phosphorus content in different soils increased only at the first rotations of field crop rotations. Later, in spite of positive phosphorus balance the content of its mobile forms didn't increase and even tended to decrease as a result of the transition of phosphorus in stiff state. In case of negative balance mobile phosphorus content in soils was recompensed through slow-moving phosphorus supply. Ecological balance at the agroecosystems was maintained due to these processes of phosphorus transformation. This equilibrium prevented losses of phosphorus due to surface and subsurface flows of the element for environment; so risks of eutrophication decreased. Also influence of phosphorous fertilizer on the biodiversity of soil microflora was ascertained. The mobile phosphorus dynamics in different soils, which was revealed during the prolonged field experiments, could be the model of phosphorous fertilizer transformation in condition of agricultural production.

20.1 INTRODUCTION

Mobile phosphorus content in soils at the time of intensive chemicalization is simpliciter connected with the level of artificial and organic fertilizer application. In Russia in 1965 before intensive chemicalization rate of artificial fertilizer (NPK – Nitrogen-phosphorus-kalium) usage per 1 hectare of arable land per year was on average 20 kg of primary nutrient; while by five-year periods its application increases: in 1966–1970 – up to 28 kg/ha; in 1976–1980 – up to 65 kg/ha; in 1986–1990 – up to 99 kg/ha. While fertilizer (especially phosphorous) application in the country increased the soil fertility improved (particularly mobile phosphorus supply increased). So in period 1971–1990 the tillage share with low content of mobile phosphorus decreases from 52 to 22% while increasing the share of soils that had high and medium phosphorus content [1].

At the same time the increase of mobile phosphorus content in soils is but one effect of phosphorous fertilizer intensive application. A considerable portion of the phosphorus above its carry-over with crop yields transformed into not mobile forms creating the reserve of phosphorus plant nutrition [2–4]. According to Ref. [5] during 25 years about 300 kg/ha of phosphorus were applied above its carry-over; the entire amount was stayed in soil. The amount was sufficient to harvest 2 t/ha of grain yield during 25–30 years unless the phosphorous fertilizer application.

Since 1995 the balance of phosphorus in the agriculture of our country tended the pattern: carry-over became exceed apply. That resulted in trend of decrease of mobile phosphorus in arable lands, although the trend was not so evident as was expected earlier. At the same time the regularities of transformation of applied phosphorous fertilizers were studied insufficiently. Outstanding scientist K.E. Ginsburg has drawn attention to this fact: "The absorbing capacity of soils in the case of phosphorus confuses our calculations of increase of mobile phosphorus content in soil because while applying mobile phosphorus unknown but considerable part of these transforms into poorly soluble and hard-to-reach for plants forms" [4, p. 124]. This chapter allows spy out longstanding dynamics and character of transition of mobile soil phosphorus into poorly soluble forms. Our study uses results of prolonged trials.

20.2 EXPERIMENTAL PART

Experimental technique was based on the materials of prolonged trials and our proper studies. We analyzed the effect of systematic organic and mineral (including phosphorous) fertilizer application in increased rates on change of mobile phosphorus content. Different soils were studied: the soddy gleyic heavy textured loamy soil (Lithuania), the sod-podzol heavy textured loamy soil (Moscow Region), the soddy-podzolic easy-loamy soil (Smolensk Region), the sod-podzol loamy sandy soil (Belarus) and ordinary chernozem (Stavropol).

We calculated the economic balance of phosphorus by principal variants of field experiments taking into account the applied phosphorus content with rotation of field rotations. We detected the influence of systematic long-termed phosphorous fertilizer application on mobile phosphorus content in soils.

The content of mobile forms of phosphorus in different soils was determined by the methods accepted in agrochemistry [6]. The main methods of determination of soluble phosphorus content in different soils were: the method of Egner-Rim – 0.04 normal solution of $(CH_3CHOHCOO)_2Ca{\cdot}5H_2O$ soil extract at a pH of 3.5–3.7; the method of Kirsanov 0.2 normal solution of HCl $5H_2O$ soil extract; the method of Machigin – 1% solution of $(NH_4)_2CO_3$ extract. The methods of determinations are indicated when depicting the results of studies.

20.3 RESULTS AND DISCUSSION

K.I. Plesyavichius who studied the efficiency of long-termed mineral fertilizer application on the soil [7]; these experiments were realized in Lithuanian Agriculture Research Institute. In the variant $N_{225}P_{324}K_{350}$ applied during the rotation with annual phosphorous fertilizer application on an average in seven fields deployed in nature when the economic balance of phosphorus was 68 kg/ha in the first rotation and 73 kg/ha in the second – the mobile P_2O_5 content (according to Egner-Rim) to the end of rotations amounted to 42 and 43 mg/kg soil (Figure 20.1).

The change of the index compared with their beginning equaled respectively 12 and 1 mg/kg when productivity of crop rotation was for the first period (1961–1968) on an average 44.8 centner of grain units/ha and for the second period (1969–1974) – 46.5 C/ha. So almost around 140 P_2O_5 centner/ha applied above its carry-over during two rotations of field crop rotations to the end of the second rotation transformed in soil in the forms not extractable by according to Egner-Rim.

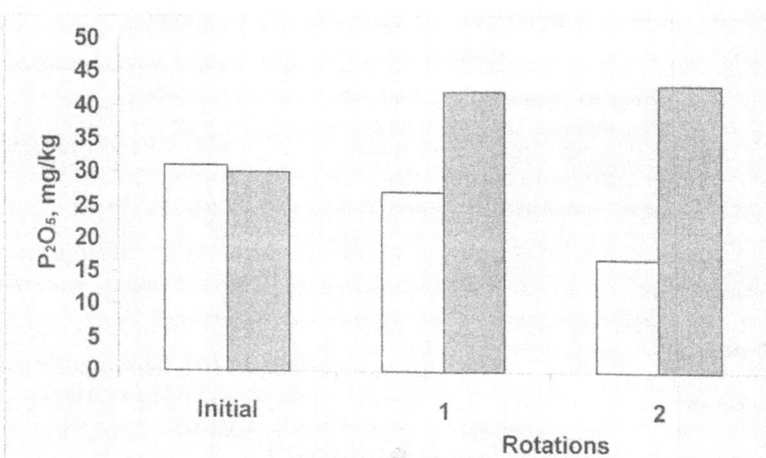

FIGURE 20.1 Dynamics of distribution of phosphorus in soddy-gleyic heavy-loamy textured dried along rotation of field crop rotations (Lithuania) Bright columns – control, dark columns – rotation of field rotations.

On our calculation the expenditures of fertilizer phosphorus for increase of mobile phosphorus content in surface soil by 10 mg/kg at the first rotation were 57 kg/ha P_2O_5, at the second rotation – the mathematical procedure above.

Also the change of mobile phosphorus content in the soddy-podzolic heavy textured loamy soil of another experiment was enough expressed. This long-termed experiment was carried out at the Central Experimental Station (Pryanishnikov All-Russian Agrochemistry Research Institute, Moscow Region), [8]. During seven rotations of a four course rotation at 28-year-long regular application of fertilizers maximum mobile phosphorus content (according to Kirsanov) was determined at the fourth rotation. Then the content was decreased (Figure 20.2), though the positive balance of phosphorus for 28-year-long experiments was 2700 kg/ha.

The expenditure of fertilizer phosphorus applied above its carry-over for increase of mobile phosphorus content in surface soil by 10mg/kg per 1 hectare on an average equaled 119 kg P_2O_5.

In this variant for 12-year-long aftereffect of fertilizers winter wheat yield (on the average for tree rotations) was 17.2 centner per hectare as compared with 12.0 centner per hectare in the control variant. So due to

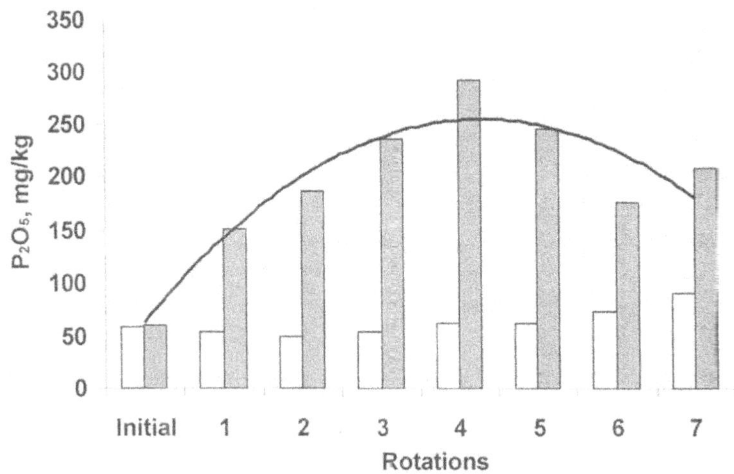

FIGURE 20.2 Dynamics of mobile phosphorus in soddy-podzolic heavy textured-loamy soil along rotation of field crop rotations (Moscow Region) (bright columns – control, dark columns – rotation of field rotations).

formerly applied fertilizers preceding increment was 5 centner of winter wheat per hectare.

During 30-year-long regular application of mineral fertilizers for all the crops except perennials to the soddy-podzolic easy loamy soil (Smolensk Region) in the case of the mineral system of fertilization $N_{990}P_{990}K_{990}$ was applied at the first rotation of crop rotation (1979–1989), $N_{450}P_{450}K_{450}$ was applied both at the second (1990–1995) and the third (1996–2001) rotations and $N_{405}P_{405}K_{405}$ was applied at the fourth rotation (2002–2008). For the first two rotations of crop rotation with balance of phosphorus of 943 kg/ha, the content of mobile phosphorus in soil increased from 149 to 210 mg/kg, hence, the value of increment was 61 mg/kg. By the end of the fourth rotation, with balance of phosphorus of 523 kg/ha in sum for the third and the fourth rotations, even a decrease in the mobile phosphorus content in the arable layer was observed: from 210 to 174 mg/kg, for example, by 36 mg/kg (Table 20.1, Figure 20.3).

TABLE 20.1 Influence of Fertilizers on Productivity of Crop Rotation and Mobile Phosphorus Content in the Soddy-Podzolicl Easy Loamy Soil (Smolensk Region)

Index	Control without fertilizers	NPK	Manure	NPK + manure
1–2 rotations (17years)				
Productivity in average year, centners g.u. per hectare	24.0	34.4	28.6	34.6
Applied P_2O_5, kg/ha	–	1440	340	1780
Carry-over P_2O_5, kg/ha	369	497	422	500
Balance P_2O_5, kg/ha	–369	+943	–82	+1280
P_2O_5 content in the soil in the beginning of the rotations, mg/kg	170	149	143	166
P_2O_5 content in the soil in the end of the rotations, mg/kg	65	210	85	185
Variation of P_2O_5 content in soil, mg/kg	–105	+61	–58	+19
3–4 rotations (30 years)				
Productivity in average year, centner g.u. per hectare	20.1	29.6	26.0	26.0
Applied P_2O_5, kg/ha	–	855	252	1107

TABLE 20.1 Continued

Index	Control without fertilizers	NPK	Manure	NPK + manure
Carry-over P_2O_5, kg/ha	222	332	287	300
Balance P_2O_5, kg/ha	–222	+523	–35	+807
P_2O_5 content in the soil in the beginning of the rotations, mg/kg	65	210	85	185
P_2O_5 content in the soil in the end of the rotations, mg/kg	56	174	160	213
Variation of P_2O_5 content in soil, mg/kg	–9	–36	+75	+28

Note: g.u. – grain unit, equivalent 1 kg of wheaten grains.

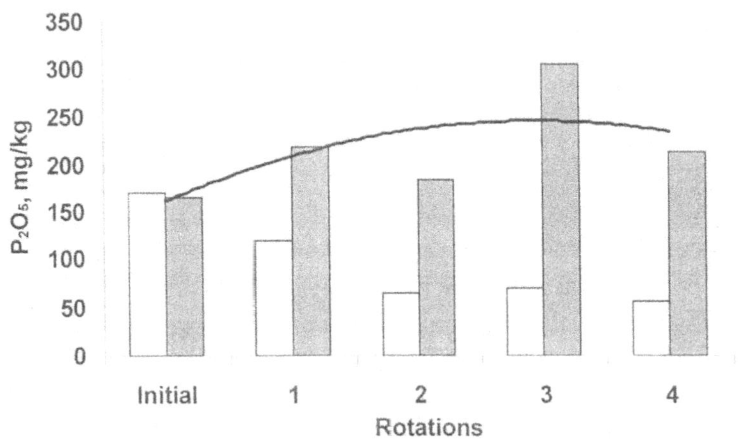

FIGURE 20.3 Mobile phosphorus content in the soddy-podzolic easy loamy soil along rotation of field crop rotations (Smolensk Region) (bright columns – control, dark columns – rotation of field rotations).

In the subsurface soil mobile phosphorus content (P_2O_5) in the control variant of the experiment in the end of the fourth rotation (2008) compared with middle of the first rotation (1983) decreased from 1.8 to 1.2 t/ha at the expense of element carry-over by yield (Figure 20.4).

In the variant with application of organic-mineral system for this period mobile phosphorus supply in the subsurface soil (20–100 cm) remained at the level 1.9 t/ha, for example, it was impossible to notice appreciable change the mobile phosphorus content in the subsurface soil. This records

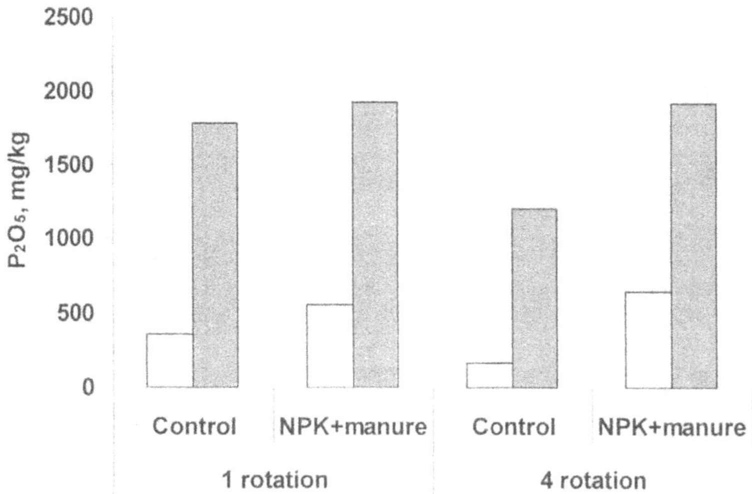

FIGURE 20.4 Mobile phosphorus supply content in the soddy-podzolic easy loamy soil along layer of profile (Smolensk Region) (bright columns – 0–20 cm, dark columns – 20–100 cm).

an exhaustive transformation of their migratory phosphorus into slow-moving forms if not to take into slow-moving forms its migration outside the limits of controlled soil layer.

Studies of the Stavropol Scientific Research Institute of Agriculture [9] showed that in conditions of ordinary loamy chernozem annual fertilizer phosphorus application for crops of a six coarse rotation rate of $N_{120}P_{90}K_{120}$ first rotation of crop rotation ensured positive phosphorus balance (the content of phosphorus was determined by the method of Machigin) 270 kg/ha (Table 20.2, Figure 20.5).

Mobile phosphorus content in subsurface soil in the end of the rotation increased to 26 mg/kg or 13 mg/kg compared with the beginning of the rotation. During the second and third rotations at the same rate and the same total phosphorus balance 542 kg/ha mobile phosphorus content in the soil of the variant increased to 52 mg/kg, for example, 26 mg/kg compared with the beginning of the second rotation. But to the end of fifth rotation of crop rotation, for example, 12 years after the end of the third rotation in spite of systematic mineral fertilizer application at the same rate with phosphorus balance for these years 250-kg/ha mobile phosphorus content in the surface soil increased by only 2 mg/kg.

TABLE 20.2 Influence of Mineral Fertilizers (With Increasing Rates of Phosphorus) on Productivity of Crop Rotation and Mobile Phosphorus Content in Ordinary Chernozem (Stavropol Territory)

Index	Control without fertilizers	$N_{120}P_{30}K_{120}$	$N_{120}P_{90}K_{120}$	$N_{120}P_{150}K_{120}$
	1 rotation (6 years)			
Productivity in average year, centners g.u. per hectare	27.7	34.3	35.2	35.7
Applied P_2O_5, kg/ha	–	150	450	750
Carry-over P_2O_5, kg/ha	139	177	180	182
Balance P_2O_5, kg/ha	–139	–27	+270	+568
P_2O_5 content in the soil in the beginning of the rotations, mg/kg	13	13	13	13
P_2O_5 content in the soil in the end of the rotations, mg/kg	13	21	26	58
Variation of P_2O_5 content in soil, mg/kg	0	+8	+ 13	+45
	2–3 rotations (18years)			
Productivity in average year, centners g.u. per hectare	26.5	32.7	34.5	33.3
Applied P_2O_5, kg/ha	–	300	900	1500
Carry-over P_2O_5, kg/ha	277	342	358	350
Balance P_2O_5, kg/ha	–277	-42	+542	+1150
P_2O_5 content in the soil in the beginning of the rotations, mg/kg		21	26	58
P_2O_5 content in the soil in the end of the rotations, mg/kg	12	28	52	76
Variation of P_2O_5 content in soil, mg/kg	–1	+7	+26	+18
	4–5 rotations (30years)			
Productivity in average year, centners g.u. per hectare	19.0	23.2	24.5	24.6

TABLE 20.2 Continued

Index	Control without fertilizers	$N_{120}P_{30}K_{120}$	$N_{120}P_{90}K_{120}$	$N_{120}P_{150}K_{120}$
Applied P_2O_5, kg/ha	–	180	540	900
Carry-over P_2O_5, kg/ha	226	281	290	291
Balance P_2O_5, kg/ha	–226	–101	+250	+609
P_2O_5 content in the soil in the beginning of the rotations, mg/kg	12	28	52	76
P_2O_5 content in the soil in the end of the rotations, mg/kg	20	30	54	70
Variation of P_2O_5 content in soil, mg/kg	+8	+2	+2	–6

Note: g.u. – grain unit, equivalent 1 kg of wheaten grains.

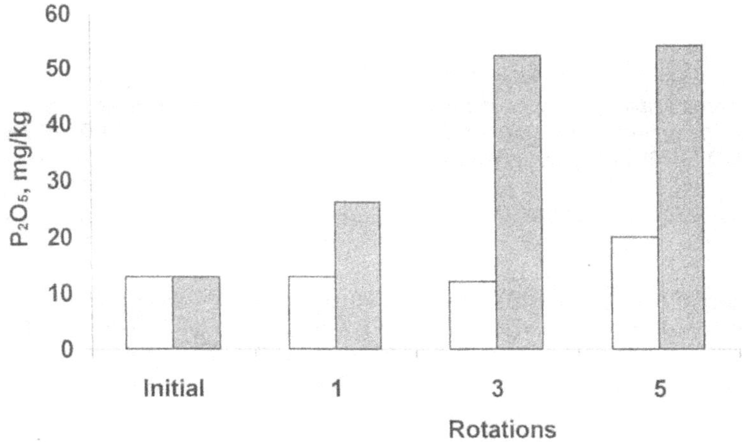

FIGURE 20.5 Mobile phosphorus content in ordinary chernozem (Stavropol Region) (bright columns – control, dark columns – $N_{120}P_{90}K_{120}$).

The application of a higher dose of phosphorus, such as $N_{120}P_{150}K_{120}$, at the balance of phosphorus for the first rotation of 568 kg/ha, with total balance for first three rotations of 1718 kg/ha, and with balance for the fourth and fifth rotations of 609 kg/ha, resulted in the content of mobile phosphorus in the surface layer of 58, 76 and 70 mg/kg, correspondingly.

Hence, for the last two rotations of the field crop rotation, the application of 609 kg/ha of phosphorous fertilizers not only has not increased the content of mobile phosphorus in soil, but also even had lowered the content by 6 mg/kg as compared to the end of the third rotation.

The analysis of mobile phosphorus content not only in the surface layer but in the subsurface soil of soil profile of ordinary chernozem revealed the close dependence of the intensity of its transformation in slow-moving forms on the value of positive balance of phosphorus in the agroecosystem. As can be seen from Figure 20.6 with the increase of balance P_2O_5 in the variant and $N_{120}P_{90}K_{120}$ and content of mobile phosphorus transformed in slow-moving state increased. On the whole during five rotations of crop rotation more than 1000 kg/ha P_2O_5 transformed in slow-moving forms.

According to the research of Pryanishnikov All-Russian Scientific Research Institute of Agrochemistry together with Lomonosov Moscow state University unilateral use of mineral fertilizer decreased the total number of microorganisms in soil and Shannon index biodiversity (Table 20.3).

Manure application increased the total number of microorganisms, but decreased Index biodiversity. With use of organic-mineral fertilizer system increase of the total number of microorganisms almost

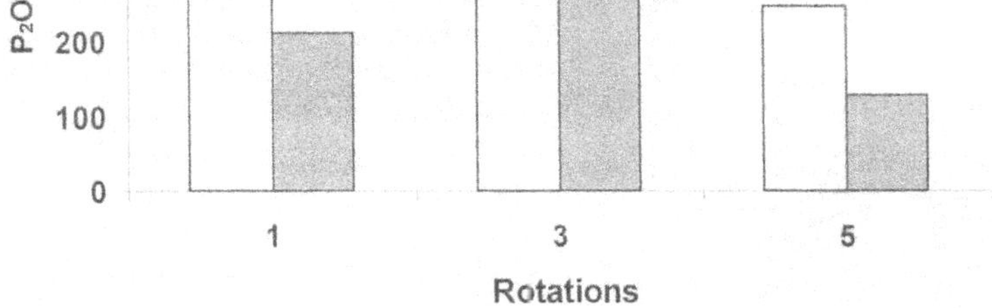

FIGURE 20.6 Dependence of the mobile phosphorus transformation in slow-moving forms (losses) in 1-meter-long layer of ordinary chernozem (the variant $N_{120}P_{90}K_{120}$) on the value of balance P_2O_5 in rotations of crop rotation (Stavropol Territory) (bright columns – balance, dark columns – losses of mobile phosphorus).

TABLE 20.3 Influence of Fertilizers on the Content of Microorganisms in the Sod-Podzol Easy Loamy Soil (Smolensk Region)

Index	Control without fertilizers	NPK	Manure	NPK+manure
Proteobacteria, cells/gram $\times 10^6$	13.6	17.1	22.0	20.3
Actinobacteria, cells/gram $\times 10^6$	19.3	13.4	18.1	13.3
Firmicutes, cells/gram $\times 10^6$	11.9	5.1	20.4	22.5
Bacteroidetes, cells/gram $\times 10^6$	2.0	2.5	5.9	3.7
The total number of microorganisms, cells/gram $\times 10^6$	46.8	38.1	66.4	59.8
Index biodiversity	5.0	4.7	4.6	4.9

without its biodiversity reducing was to be observed. Reductions in the size of microorganisms, which were marked in the table for the most part, occurred at the expense of Actinobacteria, and increase – at the expense of Proteobacteria and Bacteroidetes. In conditions of fertilizer aftereffect the number of proteolytic bacteria increased in 1.5–2 times. These bacteria (e.g., *Pseudomonas fluorescens*, *Pseudomonas putida, Brevundimonas vesicularis*) actively transformed organic phosphorus [10]. Generally soil microbial community was represented by more than 40 species belonging to 34 genera. The high number of microorganisms and their diversity indicate a sufficient degree of cultivation of investigated soddy-podzolic soil. In the long field experiences conducted in Belarus by Institute of Soil Science and Agrochemistry [11] in grain-tilled rotation on highly cultivated soddy-podzolic sandy loam soil in 1986–2009, in the table of contents rolling phosphate was observed dynamics similar to the transformation of phosphates in loamy soils (Figure 20.7). In the variant with an annual application of N78P72K119 at the original content of mobile phosphorus in the soil 137 mg/kg to 2009, it increased to 260 mg/kg or by 370 kg/ha. A significant increase of mobile phosphates was observed in the first years of fertilization with subsequent going out curve on the plateau, and reducing it at the end of the experience. A positive balance of phosphorus during the years of the experiment amounted to 1500 kg/ha, while the actual increase in the content of mobile phosphorus in the soil was not more than 370 kg/ha. It follows that on one ha more than 1 ton P_2O_5 passed into the forms that are not covered by Kirsanov method.

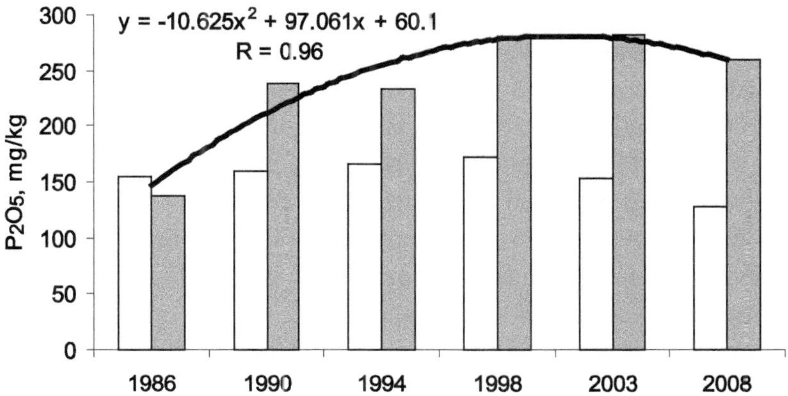

FIGURE 20.7 Mobile phosphorus dynamics in soddy-podzolic loamy sandy soil on years of research (Belarus) (1st row – control, 2nd row – N78P72119).

The analysis of the long-term dynamics of the mobile forms of phosphorus in long field experiments with clay and loam soils showed the regularity its transition after a time in slow-moving forms with a decrease in concentration of mobile forms. Depending on agrochemical qualities of soils time constraints of the transition matured in different periods: in the soddy gleyic heavy textured loamy soil (Lithuania) – after 7 years (in the second rotation of crop rotation), in the soddy-podzoic easy-loamy soil (Smolensk Region) – after 17 years (in the third-fourth rotations of crop rotation), and in the ordinary chernozem after 18 years (in the fourth-fifth rotations of crop rotation.

Inasmuch as soil is bio inert system which has formed under the influence of biological factors it is inherent the function of conservation and transformation of substances [12, 13] and it reacts to soluble phosphorous fertilizer application according to Le Chatelier principle: If a system at equilibrium experiences a change then the equilibrium shifts to partially counter-act the imposed change [14]. In different soils these functions are manifested in accordance with root natural causes. In soddy-podzolic soils with increased iron and aluminum compounds content applied phosphorous fertilizers transform in phosphate sesquioxides while in carbonate-enriched chernozems and chestnut soils the function of phosphorus conservation manifests itself in emergence of phosphates with different basicity including sparing soluble compounds, for example, apatite. Obviously the specific reasons determined a small increase or even a decrease of moving phosphorus content in the last rotations of the indicated crop rotations were due to increase of intensity of phosphorous fertilizer transformation in sparing soluble forms from the first rotations to following rotations of crop rotations owing to strengthening reaction of soil as a system to the redundant water-soluble phosphorus fertilizer application which was increased summarized over rotations.

From the above is follows that traditional soil test, for example, moving phosphorus determination as index of its effective fertility didn't create an adequate representation of its natural fertility.

There are reports [15] that content of 600–800 kg residual phosphates in the soil per 1 hectare ensures almost all of the soils its optimal phosphate status. So for an objective assessment of the phosphate level it is expedient to take into account not only data on the mobile phosphate content in the soil but also supplies of slow-moving forms which would be used by crops in future. This is important for economic estimation of arable lands. The estimation allows take into account its natural fertility.

20.4 CONCLUSION

1. Prolonged systematic fertilizer application with rates of phosphorus, which exceed carry-over of the element by crop yields increases mobile phosphorus content in soils of agroecosystem in the beginning of intensive fertilizer application only. In the further due to phosphorus transition in sedimentary forms mobile phosphorus content in soils increases very slightly or even has a tendency to decrease. Intensity of mobile phosphorus transition in slow-moving forms largely depends on soil type, also depends on the duration of the interaction of mobile phosphates with soil and on the value of positive phosphorus balance.
2. Transition of applied phosphorus in sedimentary phosphates of soil as a result of its immobilization ensures phosphorus plant nutrition during for a number of years even upon termination of phosphorus fertilizer application. The processes in many respects predetermine crop production level including the grain production.

3. Fertilizer use in general increases soil fertility, enriches their with soil microflora. Unilateral fertilizer application decreases diversity of microorganisms, while use of combined organic-mineral fertilizer contributes to its conservation. For all that many species of protolithic bacteria influence on processes of phosphorus transformation in conditions of aftereffect fertilizer in agroecosystems.

KEYWORDS

- **agroecosystems**
- **fertilizers**
- **mobile phosphorus dynamics**

REFERENCES

1. Sychev, V. G., Mineev, V. G. Role of Pryanishnikov All-Russian Scientific Research Institute of Agrochemistry in solving complex problems of agriculture chemicalization."Fertility" (in Rus.), 2011, №3, 2–4. (in Russian).
2. Cook, J. W. Soil fertility regulation. Moscow: "Kolos" ("Ear," in Rus.) Publishing House, 1970, 520 p. (in Russian).
3. Black, K. A. Plant and soil. Moscow: "Kolos" ("Ear," in Rus.) Publishing House, 1973, 501 p. (in Russian).
4. Ginsburg, K. E. Phosphorus of general soil types in, U.S.S.R. Moscow: "Kolos" ("Ear," in Rus.) Publishing House, 1981 542 p. (in Russian).
5. Sychev, V. G., Shafran, S. A. Agrochemical properties of soils and mineral fertilizer efficiency. Moscow: Pryanishnikov All-Russian Scientific Research Institute of Agrochemistry, 2013, 296 p. (in Russian).
6. Agrochemical methods of investigation of soils. Moscow: "Nauka" ("Science" in Rus.) Publishing House, 1975, 656 p. (in Russian).
7. Plesyavichius, K. I. Fertilizer application systems comparison in heavy soils. The results of long-termed researchers with fertilizers in regions of the country. Proceedings of Pryanishnikov All-Russian Scientific Research Institute of Agrochemistry. 1982, Issue 12, 4–82 (in Russian).
8. Efremov, V. F. Study of the role of organic manure organic matter in raising sod-podzol soil fertility. The results of long-termed researchers in the system of geographical network of experiments with fertilizers in Russian Federation. Moscow:

Pryanishnikov All-Russian Scientific Research Institute of Agrochemistry, 2011, 47–71. (in Russian).
9. Shustikova, E. P., Shapovalova, N. N. Productivity of ordinary chernozems while long-termed systematic mineral fertilizer applying. The results of long-termed researchers in the system of geographical network of experiments with fertilizers in Russian Federation. Moscow: Pryanishnikov All-Russian Scientific Research Institute of Agrochemistry, 2011, Issue 1, 331–351. (in Russian).
10. Merzlaya, G. Ye., Verkhovtseva, N. V., Seliverstova, O. M., Makshakova, O. V., Voloshin, S. P. Interconnection of the microbiological indices of soddy-podzolic soil on application of fertilizer over long period of time. Problems of agrochemistry and ecology. 2012, № 2, 18–25 (in Russian).
11. Lapa, V. V., Ivahnenko, N. N., Productivity of crop rotations, nutrient balance and soil fertility changes on loamy sandy soddy-podzolic soil under long-term fertilization //"Fertility" (in Rus.), 2014, №5, 5–8. (in Russian).
12. Williams, V. R. The collected works. Moscow: "Sel'hozgiz" ("Agricultural Edition," in Rus.) Publishing House, 1951, V.6. 576 p. (in Russian).
13. Shein, E. V., Milanovsky, E.Yu. Spatial heterogeneity of properties on different hierarchic levels as a basis of soils structure and functions. Scale effects in the study of soils. Moscow: Publishing house of Moscow state University, 2001, P. 47–61. (in Russian).
14. Glinka, N. L. General chemistry. Leningrad: "Khimiya" ("Chemistry," in Rus.) Publishing House, 18th edition. 1976, 728 p. (in Russian).
15. Fertilizers, their properties and how to use them [Ed. Korenkov]. Moscow: "Kolos" ("Ear," in Rus.) Publishing House, 1982, 415 p. (in Russian).

CHAPTER 21

GENETICS OF PLANT DEVELOPMENT

OLGA A. OPALKO[1] and ANATOLY IV. OPALKO[1,2]

[1]*National Dendrological Park "Sofiyivka" of NAS of Ukraine, 12-a Kyivska Str., Uman, Cherkassy Region, 20300, Ukraine*

[2]*Uman National University of Horticulture, Instytutska Str., Uman, Cherkassy Region, Ukraine 20305; E-mail: opalko_a@ukr.net*

CONTENTS

ABSTRACT

The data of literary sources concerning the bloom of angiosperm plants and deviation in the development of a flower and inflorescence, in particular untimely flowering, was generalized; our observation results of some peculiarities of re-bloom of woody plants in the National dendrological park "Sofiyivka" of NAS Ukraine were discussed. It has been shown that re-bloom is mostly associated with abrupt fluctuations of meteorological conditions both in the year of re-bloom and during flower bud formation in the year which precedes flowering. Most cases of untimely flowering registered in NDP "Sofiyivka" represent a group of early summer-fall flowering, some late spring flowering occurred though. An adaptive nature of re-bloom takes place provided seeds with full germination are developed.

21.1 INTRODUCTION

To understand the processes of sexual reproduction, namely the manifestation of sex, and first of all sexual di- and polymorphism of the plants, has always been of interest for tillers since Ancient Egypt times, where along with many androgynous plants such dioecious plants as date palm (*Phoenix dactylifera* L.), fig (*Ficus carica* L.), ricin (*Ricinus communis* L.), asparagus (*Asparagus officinalis* L.) and others were grown. In the course of a century-long research of the peculiarities connected with sexual differentiation of higher plants at various organization stages (a flower, an individual, a population, a species), numerous mechanisms, which ensure a timely beginning of a sexual process, the organ of which in angiosperm plants is a flower, were identified [1–3]. Among the most outstanding achievements made by the scientists from different countries from the times of Ch. Darwin [4], it is worth mentioning the description of the course of sexual differentiation and flowering itself, the identification of geographic localization and systematic belonging of the plants of various sexual forms [5–9], the explanation of flowering hormone—florigen—made by M.Kh. Chailakhian in 1936 [10, 11], and the identification of material carriers and the conditions of gene expression (21st century), under their control this hormonal complex is synthesized and transported [12–16].

Very complicated and multivariate flowering process was formed during the evolution of a propagation system of angiosperm plants as a basis of fertilization and further fruit and seed development, as a basis of sexual plant propagation. As a result of vernalization and photoperiodism reactions, which were also developed during a long-term evolution, flowering (under regular conditions) occurs in the most favorable period for pollination and fertilization of every plant [10, 11, 17–20]. It concerns not only a certain period of the season, but also certain hours of the day. However, various deviations, malformation of a flower and inflorescence, fasciation, the formation of additional petals, heterostyly, untimely (most frequently double, sometimes three- and four-fold) flowering occur in this perfect process of generative organ formation of angiosperm plants.

The phenomenon of untimely flowering was described by many researchers [16, 21–28]. The connection between untimely flowering and stressful environmental conditions in which plants developed was discussed, namely, re-growth and re-bloom of the plants in the ashes [24], re-bloom of early fruitful walnuts—*Juglans regia* [29], cytomorphological features of re-bloom flowers *Cerasus vulgaris*, in particular, sour cherry, sour cherry-sweet cherry and sour cherry-chokecherry hybrids and perspectives of their use in sour cherry breeding [30]. Other cases of re-bloom were described: white mulberry—*Morus alba* L. [31], species of micro sour cherry—*Microcerasus* Webb. Emend Spach. [32], cotoneasters—*Cotoneaster* Medik. [33] and various exotic species [25], features of germ seeds of cherry plum—*Prunus cerasifera* Ehrh., received from re-bloom [34].

Alongside with this, an increased number of reports about re-bloom (at the end of summer-at the beginning of fall) of the representatives of various woody plant species whose flowers usually blossom in May–June prompts the analysis of the available information concerning the mechanisms of flowering and the causes which lead to deviation of flowering processes.

Flowering of the woody plant representatives of a collection fund of the National Dendrological Park "Sofiyivka" of Ukraine's NAS (NDP "Sofiyivka") was studied; statistics about re-bloom in different cities of Ukraine was monitored. While summing up the available information, special attention was paid to the data received from observation analysis, made in different years by the researchers from different scientific schools, concerning induction mechanisms of generative organ development and a feasibility to regulate higher plant flowering [6, 7, 13, 16, 21–25, 27–36]. The classification of re-bloom facts was carried out according to V.L. Vitkovskyi [35].

The angiosperm diversity, variation in floral structure, as well as micro- and macro-sporogenesis processes, pollination, fertilization, embryo and fruit formation, were shaped under age-long evolutional pressing [17, 37].

According to the ability to be fertilized with own pollen and that of another genotype autogamous (self-pollinated) and allogamous plants (cross-pollinated) are classified. Self-pollinated plants are considered to be originated from cross-pollinated wild species resulting from multiple subconscious limitations of cross-pollination and subconscious artificial selection. Which is why, a number of self-pollinated plants are smaller comparing to that of cross-pollinated ones [3, 38]. It was Ch. Darwin (1876) who paid attention to numerous diversity of biological mechanisms, which facilitated cross-pollination [4]. At present, many morpho-physiological and genetic mechanisms which favor allogamy (cross-pollination) are known. They are heterostyly and dichogamy (protandry and protogyny) in hermaphrodite plants, and metagyny and metandry in dioecious plants, various forms of self-incompatibility, selective fertilization, etc.

The most reliable mechanism of blocking self-pollination and favoring cross-pollination is separation of female and male generative organs and their formation on various plants (on separate male (staminate) or female (pistillate) individuals) or at a certain distance on the same plant. Such plants as date palm, dioecious hemp, spinach, hop, buckthorn, some nuciferous plants, etc. develop female and male flowers on different individuals. This is the reason why they are called dioecious plants. Male and female flowers of monoecious plants are located on the same plant but at some distance; corn, watermelon, cucumbers, coco palm, walnut, hazelnut and other crops are the examples. The number of monoecious plants is large: 10% of all mono- and 4% of all dicotyledonous plants.

Most of the plants have bisexual flowers. However, they have many other possibilities to prevent self-fertilization. It is protandry, when pollen matures faster than stigma pistil is able to accept it, and also protogyny, when anthers mature with delay after fertilization is done with another plant's pollen. Protandry occurs quite frequently, including beets, sunflower, lettuce, parsley, celery. Protogyny is characteristic of cabbage plants, many grain cereals and other species. Protogyny is meaningless for many fruit crops—apples, pears, plums, etc.—because self-fertilization is unlikely to happen due to numerous genetic and physiological mechanisms of self-incompatibility.

In view of the development of Darwin's ideas, contemporary researchers study peculiar aspects of a flower structure, quality, quantity and ways of pollen transfer and fertilization, a degree of self- and cross-incompatibility, the effect of inbreeding, a share of cross-fertilization for self-pollinated plants and that of self-fertilization for cross-pollinated plants, etc. [38]. Flowers, pollinated by insects, attract them by means of nectar and pollen (sources of ambrosia), smell and a wide diversity of colors. Flowers of many plants gain the strongest aroma exactly at the time when insects, which pollinate them fly very actively. It is a known fact that flowers of petunia, honeysuckle, pelargonium and other plants, pollinated by night butterflies, begin to smell most of all in the evening, whereas the flowers, pollinated by bees and other day insects, almost stop releasing aroma at sunset. The color of some flowers attracts certain insect species. Bees fly to blue and violet flowers first of all, whereas night butterflies – to white and straw-colored ones. A flower shape of some orchids resembles a female of insect-pollinators; this attracts male-pollinators (pollination of such orchids occurs before female-insects, which could "compete" with flowers for males' attention, appear).

In 1936, after a set of experiments on photoperiodic regulation of flowering had been carried out, M. Kh. Chailakhian suggested the existence of a natural complex of hormones which stimulated flowering of the plants and called it florigen. Physiologically active substances, which stimulate flowering are formed in leaves under favorable conditions (optimal air temperature and proper duration of daylight hours). They come through plant tissues from leaves into buds and initiate flower formation. Plant grafting trials prove that florigen moves from a flowering plant-donor to a vegetating plant-receptor facilitating the flowering of the latter. The same trials show that florigen does not have species specificity: it initiates flowering of various plant species and photo periodical groups [11].

It has been found out that gibberellin treatment initiates flowering of long-day plants, which experience lack of this phytohormone, however it does not affect that of short-day ones. Instead, grafting/inoculation of plants with different need in photoperiod facilitates flowering of both [11]. To speed up fruiting resulted from grafting of cuttings with juvenile seedlings of perennial trees has always been a common practice among fruit breeders aimed at faster fruit quality evaluation of newly developed hybrids [38]. Other ways of reducing a juvenile development stage of hybrid seedlings were found with help of empiric methods [39], but molecular mechanisms of flowering time regulation were determined/revealed after long-term fundamental researches of Swedish scientists headed by professor Ove Nilsson [36] on model genetic object *Arabidopsis thaliana* (L.) Heynh., only in the year of 2005 [14]. In 2007 Ove Nilsson received prestigious Marcus Wallenberg award for the confirmation of the feasibility to use principles, discovered on *Arabidopsis,* for flower time regulation of perennial woody plants. Numerous manipulations of transgenes are of FT gene from *A. thaliana* to *Populus* L. Gene were made for this purpose. FT gene functioned ectopically, which made it possible to speed up the course of plant juvenile development stage, and in turn they're fruiting period.

In natural conditions flowering of most temperate climate pants begins at air temperature 10–12°C. Its beginning as well as budding takes place due to reserve nutritive substances in a trunk, branches, leading roots, and it unlikely depends on root system functioning itself. Most frequently at 13°C flowering lasts for 10–12 days (day and night) and in dry weather at 20–25°C—only 5–6 days. In rainy weather at temperature, lower than 12°C it may last up to 15 days and longer, whereas in sunny weather at 30–32°C flowering lasts only 3–4 days [40]. Generative buds, which originate inflorescence and flowers, are developed at various terms. Some plants start budding in spring of the same vegetative period when a plant blossoms, others—in the year, which precedes flowering, usually in the second half of summer [18]. Woody plants start budding (according to plant species, cultivation zone and farm practices) in June-August of a previous year [20, 35]. The beginning of flower bud differentiation is very close to the moment when shoots cease their growth. This period is characterized by fast increase of tissue carbohydrate concentration, dry and hot weather accelerating this process [41]. The beginning of bud formation can change for a ten-day period or more depending on meteorological conditions, species/cultivar structure of orchards, the age and physiological condition of a plant [40]. Excessive irrigation may cause a delay in fruit bud differentiation terms [41]. Trees with abundant yield budding starts later compared with unfruitful trees: cherry and plum trees start budding later than apple trees; budding starts later in wet years versus dry ones [40].

At first vegetative buds start to develop—germination of a stem and leaves develops from meristem tissues on a cone top of outgrowth. Then generative meristem appears, swelling on a cone top of outgrowth, later flower protrusions distinguish—first conception of flowers. Further, in the process of differentiation, receptacle develops by winter, and perianth forms on it—sepals and petals, anthers and carpels develop, fruit receptacles and seminal beginnings, pollen and germinal vesicles, and in spring gametes develop [40].

In natural conditions of a temperate zone, usually a process of flower formation of fruit plants lasts up to 250–300 days, covering summer-fall and winter-early spring periods [35]. For apices of vegetative buds to differentiate intro generative ones, certain internal conditions, both in meristem tissues of cone outgrowth and in the branches they are located on, are required. First of all, such conditions are created when shoot growth weakens which is due to a considerable increase of the concentration and activity of growth inhibitors. Development or intensification of specific enzymes and phytohormones takes place during this process, in particular flowering hormone, which activate generative development genes, which in turn, results in initiation and differentiation of flower buds. However, a proper course of metabolism is necessary for this process—active synthesis of specific DNA, RNA, protein, carbohydrates and due supply with nitrogen, phosphorus, microelements, water, etc. These special conditions are required for the first stage of flower development (flower budding), conditions similar to the growth of vegetative organs are necessary for the second stage (differentiation). Most frequently, differentiation of flower buds in pommes fruit plants (apple-, pear trees) begins in the period of intensive fruit growth. Flower buds develop poorly or do not develop at all on the trees with heavy flowering and abundant yield, which explains irregularity (periodicity) of fruiting. As to stone fruit plants, except for late-yielding cultivars of plum and peach, processes of generative budding and active fruit growth do not coincide in time – flower bud differentiation often begins when fruit growth is almost over. Hence, necessary internal physiological-biochemical conditions are created for generative budding and for regular fruiting as well [40].

All above-mentioned and other processes of plant growth and development take place under the control of due genes, but manifestation rate of every gene in phenotype is due to many exogenous and endogenous factors. In the process of making genetic potential of individual a features of environmental condition, the development of any feature or property can be modified very much, namely, the formation of generative organs. Endogenous effects, consisting in the interaction of a gene studied with other genes of core and cytoplasm, which can hinder or enhance its manifestation in phenotype, are of great importance. Complementarities and epistasis can inhibit expression of even dominant genes, and for recessive gene to develop, besides releasing a complex of endogenous interactions, it should be transformed into homozygote condition [38]. All deviations concerning flower gene expression in phenotype, as well as any gene, take place within an evolutionally formed principle of response and have an adaptive meaning. However, if environmental fluctuations go beyond phylo-genetic bounds, for example, they acquire stressful nature, some inadequate changes may occur which make a genotype adaptive potential worse. Fasciation and other abnormalities of generative organ development, in particular untimely flowering, belong to such non-standard, mostly non-adaptive, changes. Most of the described cases of untimely flowering are unproductive, as very often fruits which may set during this time [17, 32, 35] are small as a rule [29], they neither ripen nor form good seeds; even if they do, to stimulate embryo development and seedling survival, scientists (in breeding-genetic trials) use special techniques [30, 34], including in vitro methods.

First news from European botanists about untimely flowering of plants dates back to fifteenth century [21]. Later, one could find more and more information about the appearance of flowers on fruit plants in late spring, summer, fall and even in winter time, when the term for regular spring flowering passed long ago (or it did not begin). According to V.L. Vytkovskyi, this interesting phenomenon was observed in many areas of a temperate zone. For instance, the facts of berry plant re-bloom on the territory of former Russia were described in the mid-nineteenth century, but multiple plant flowering aroused more interest in twentieth century [35].

Because of cumulative effect of irregular natural and anthropogenic factors in Ukraine, as well as in the whole world, abnormal climate changes take place; so, fluctuations of environmental parameters, especially an urbanized one, in which plants develop, go beyond evolutionally formed standards of response; the latter is one of the factors of re-bloom.

In October 2004, horse chestnut tree blossomed for the second time in Lviv, Kyiv, Ivano-Frankivsk, Uzhhorod, Vinnytsia and Ternopil.

In July 2005 apple-trees in Kirovohrad showed re-bloom, at the beginning of September 2006 horse chestnut blossomed in the downtown of Donetsk. In the year of 2007 re-bloom was registered on lilac—near Yalta, linden trees—in Odesa, apple-, cherry-trees, horse chestnuts—in Rivne area. In 2008 horse chestnuts started to blossom again in the mid-September in Odesa and Kyiv, at the beginning of October, 2009—in the Carpathian area, at the beginning of September, 2012—in Cherkassy and Poltava, in the first decade of October the same year—in Lviv and Ternopil. Along with horse chestnut moth spreading, September-October bloom of horse chestnuts became typical for damaged plants. At the end of June and the beginning of July, 2013 rowan trees *Sorbus aucuparia* L. and its form *S. aucuparia* 'Pendula' blossomed on Uman streets and at the end of July and the beginning of August, 2014 elderberry *Sambucus nigra* L. and black locust *Robinia pseudoacacia* L. showed the same phenomenon. In 2014 re-bloom was observed on cherry trees which damaged by cocomycosis (*Coccomyces hiemales* Higg.) lost their leaves in the mid-summer, but after partial leaf appearance they managed to develop few flowers.

As far as this phenomenon is concerned, plants of NDP "Sofiyivka" were no exception. At the beginning of September, 2007 re-bloom was seen on blackthorn bushes—*Prunus spinosa* L. (Figure 21.1), on two wild apple species: *Malus baccata* (L.) Borkh. (Figure 21.2) and *M. pallasiana* Juz. (Figure 21.3), and in the third decade of September pyramidal inflorescences of horse chestnuts—*Aesculus hippocastanum* L. appeared on branches with nuts and dried leaves damaged by horse chestnut leaf miner—*Cameraria ohridella* Deschka and Dimić (Figure 21.4). Besides, in September flowers of young seedlings of *M. spectabilis* (Ait.) Borkh. burst into blossom, and it was its first bloom. Cases of re-bloom occurred rarely in the following years.

However, in the fall of 2007 some plants—*Cerasus vulgaris* Mill. had re-bloom. Besides plants of this species experienced re-bloomed in the fall in 2009–2011; *Sorbus torminalis* (L.) Crantz—in the years of 2010–2012. Some trees—*Aesculus hippocastanum* L. showed re-bloom in 2010–2012. A young tree—*Sorbus aucuparia* L. 'Fastigiata' first blossomed in May, 2009, and in September the same year flowers burst into blossom (Figure 21.5). Double flowering of this young tree was seen in 2010–2012.

FIGURE 21.1 Early summer-fall flowering of blackthorn (NDP "Sofiyivka," September 4, 2007).

FIGURE 21.2 Early summer-fall flowering of *M. baccata* (L.) Borkh. (NDP "Sofiyivka," September 6, 2007).

FIGURE 21.3 Early summer-fall flowering of *M. pallasiana* Juz. (NDP "Sofiyivka," September 6, 2007).

FIGURE 21.4 Early summer-fall flowering of *Aesculus hippocastanum* L. (NDP "Sofiyivka," September 26, 2007).

FIGURE 21.5 Early summer-fall flowering of *Sorbus aucuparia* L. 'Fastigiata' (NDP "Sofiyivka," September 17, 2009).

Based on the classification of V.L. Vytkovskyi [35], all cases of re-bloom can be classified into four groups:

- **late spring flowering,** which occurs 10–15 days after main spring flowering. In these cases flowers of re-bloom develop from buds, set in a summer-fall period of the previous year along with standard flower buds, but for some reason their development slowed down and they burst later than other flowers of the same tree;
- **early spring-early summer flowering** is observed at the ends of new shoots about 3–4 weeks after first (main) flowering. Re-bloom flowers of the plants that belong to this group develop rapidly due to the differentiation of outgrowth axillary cones, which set in the previous year and had enough time to accomplish their development cycle during fall-winter-spring period. Such abnormalities can be caused by excessive nutrition in a previous year as well as

different damages of shoots, namely winter ones, and delay in apical bud growth. As a result, only some flowers, which avoided damage, burst into blossom during main (spring) flowering. By contrast, re-bloom may be quite plentiful. The intercalary shoot growth from flower buds with apical flower location on each of them precedes in some genotypes;

- **early summer-fall flowering** results from accelerated flower formation in summer-fall period of a current year. This is the most common abnormality. Similar to regular development flowers develop from apical, for example, former axillary, cones of outgrowth which have already finished their development in the previous year. So there is no need in lower temperatures. On the contrary, the development process of such flowers goes more actively than at higher (not high) temperatures;
- **early winter flowering** starts at sharp increase of the temperature in winter when natural bud dormancy is over. Flowers formed in summer-fall period burst into blossom, they normally differentiate from the cells of outgrowth apical cones, which finish their development cycle earlier. Under regular conditions flowers have to start blossoming in spring, but abnormal warming (spring imitation) provokes untimely flowering.

Although the classification suggested by V.L. Vitkovsky [35] has mostly a stated nature, it was good enough when being formulated, and, with certain conditions, it can be applied nowadays. Accordingly, using this classification, abnormal cases can include facts of early summer-fall flowering and early winter flowering. And late spring flowering can be adaptive response of damaged plants to exogenous stresses, due to which the feasibility of sexual propagation remains. Early spring-early summer flowering is a dual-natured category, which, under certain conditions, may have adaptive meaning and facilitate seminal propagation.

Early winter flowering of entomophilous plants is abnormal in most cases, whereas it may be productive for wind-pollinated plants, for example, representatives of *Corylus* L., in particular in the years with mild winters and/or in the regions with meteorological conditions similar to subtropical ones.

Apples, pears, plums, currants and many other plants in the years with early spring-early summer flowering from outgrowth axillary cones, resulted from intercalary growth for substitution of damaged apical flower buds, can develop shoots, and flowers of re-bloom are located on their apices. Intercalary shoot growth from flower buds is considered to be abnormal for the plants mentioned, but for raspberries, quince and some other plants it is intercalary meristem that ensures the growth of peduncles with inflorescences or separate flowers. This type of flowering and fruiting is described as normal for All-Saints' cherry. It is also considered to be typical for rowan, hawthorn and some other crops [35].

Secondary flowering is usually connected with meteorological peculiarities of the year. Most frequently it is seen in very dry years, first of all after hot and dry spring. Drought during massive flowering of the trees reduces intensity of their flowering, and some nutritive substances mobilized in the previous season remain unused. Serious deficit of moisture leads to partial yellowing and falling of the leaves in summer and can cause early bud awakening on leafless shoots, which is a known fact. Similar effects are observed on horse chestnuts, damaged by chestnut moth, which lost most of their leaves in the mid-summer. However, heavy and long rains may also cause re-bloom, the procedure of re-bloom after heavy rains corresponding to the sequence of regular flowering of these plants in spring.

Plants which do not develop inflorescences and flowers for the next year are more predisposed to re-bloom. Contrary to this, the plants, whose flower buds of the next year set in June and July, blossom again, despite the fact that their flowers fully developed in restoration buds, including not only perianth, but also stamen and pistil [19].

Re-bloom of the trees which blossom before leaf burst (hazel, maple, etc.) occurs rather rarely, but cherry-, apple- and other fruit trees experience re-bloom quite often [19]. Early yielding walnuts also have re-bloom [29].

Under abnormal cases of untimely flowering, when regular seeds are not guaranteed, the germination efficiency of inadequate seeds received from such flowering can be enhanced by cultivating immature germs in vitro.

Most cases of untimely flowering registered in NDP "Sofiyivka" belong to a group of early summer-fall flowering, some facts of late spring flowering occur though.

Explaining the phenomenon of untimely flowering in view of a plant response to exogenous stressful factors, the mentioned facts of late spring flowering can be considered as an adequate genotype response to unfavorable conditions, which slow down a regular flower bud development. New flowers open soon after petal fall from the flowers of a normal flowering term (Figure 21.6). Provided proper pollination takes place, such flowers develop full seeds, which can be a source of natural propagation and can be used in breeding practice.

Early summer-fall flowering is connected with accelerated flower development as a response to stressful conditions which lead to shoot growth inhibition and differentiation enhancement of flower elements in buds. Quite often one can see flowers of re-bloom and fruits of more or less maturity on a plant (see Figures 21.1–21.5).

Summer-fall flowering can have certain breeding importance, as flowers develop fully, they blossom well, have viable pollen and set fruits which ripen perfectly (especially in the south). Plants grown from such fruit seeds turn to be earlier maturing than initial cultivars.

FIGURE 21.6 Late spring flowering of *M. prunifolia* (Willd.) Borkh. *var. Rinki* (Koidz.) Rehd. *f. fastigiatabifera* (Dieck) Al. Teod. (NDP "Sofiyivka," May 19, 2009).

21.2 CONCLUSION

Thus, the analysis of literary sources and our observation results give every ground to associate the phenomenon of numerous reports about the facts of untimely flowering of many woody plants with abnormal fluctuations of exogenous, in particular meteorological stressful conditions, which have been observed recently not only in Ukraine and are of global nature. Researches of untimely flowering in the plants of species *Aesculus*, *Cerasus*, *Malus*, *Prunus*, *Sorbus,* carried out in NDP “Sofiyivka,” prove the occurrence of re-bloom which is classified as early summer-fall flowering and also late spring flowering of *M. prunifolia.* Re-bloom may have an adaptive nature provided seeds with full germination are developed, but immature germs of underdeveloped seeds, formed from the flowers of early summer-fall flowering, can be used in breeding programs.

ACKNOWLEDGEMENT

This material is partly based on the work supported by the National dendrological park “Sofiyivka” of NAS of Ukraine (№ 0112U002032) in compliance with thematic plan of the research work. We thank corresponding member of NAS of Ukraine DSc of Biology Ivan Kosenko for consultations and discussion.

KEYWORDS

- **dichogamy**
- **fasciation**
- **florigen**
- **hermaphrodite plants**
- **heterostyly**
- **intercalary meristem**
- **protandry**
- **protogyny**
- **sexual polymorphism**

REFERENCES

1. Bilokin, I. P. Growth and development of plants. Kyiv: Higher School, 1975, 432 p. (in Ukrainian).
2. Godin, V. N. Sexual polymorphism as the factor of adaptation *Pentaphylloides fruticosa* (L.) O. Schwarz in the Altai-Sayan Mountain: The thesis for a doctor degree of biological sciences in specialty 06.01.05. Novosibirsk, 2009, 447 p. (in Russian).
3. Sidorsky, A. G. Evolution of sexual systems in flowering plants. Nizhny Novgorod: Volga-Vyatka Publishing House, 1991, 210 p. (in Russian).
4. Darwin, C. R. The effects of cross and self-fertilization in the vegetable kingdom. London: John Murray, 1876, 482 p.
5. Case, A. L., Barrett, S. C. H. Environmental stress and the evolution of dioecy: *Wurmbea dioica* (Colchicaceae) in Western Australia. Evolutionary ecology. 2004, Vol. 18, № 2. P. 145–164.
6. Delph, L. F. Sexual dimorphism in gender plasticity and its consequences for breeding system evolution. Evolution and development. 2003, Vol. 5, № 1. P. 34–39.
7. Golonka, A. M., Sakai, A. K. and Weller, S. G. Wind pollination, sexual dimorphism, and changes in floral traits of *Schiedea* (Caryophyllaceae). American journal of botany. 2005, Vol. 92, № 9. P. 1492–1502.
8. Ramula, S., Mutikainen, P. Sex allocation of females and hermaphrodites in the gynodioecious *Geranium sylvaticum.* Annals of botany. 2003, Vol. 92, № 2. P. 207–213.
9. Webb, C. J. Empirical studies: evolution and maintenance of dimorphic breeding systems. Gender and sexual dimorphism in flowering plants [Eds.: M. A. Geber, T. E. Dawson and, L. F. Delph]. N. Y.: Springer, 1999, P. 61–95.
10. Romanov, G. A. Mikhail Khristoforovich Chailakhyan: The fate of the scientist under the sign of florigen. Russian Journal of Plant Physiology. 2012, Vol 59, № 4. P. 443–450. (in Russian).
11. Chailakhyan, M. Kh. Regulation of flowering in higher plants. Moscow: Science, 1988, 560 p. (in Russian).
12. Lutova, L. A. Initiation of flowering, In: Genetics of plant development, L. A. Lutova, G. L. Ezhova, I. E. Dodueva and MA Osipova [Ed.: S. G. Inge-Vechtomov]. St. Petersburg: Publishing house N-L, 2010, Part III. Genetic control of plant morphogenesis, Ch., 12. P. 323–346. (in Russian).
13. Abe, M., Kobayashi, Y., Yamamoto, S. et al. FD, a bZIP protein mediating signals from the floral pathway integrator FT at the shoot apex. Science. 2005, Vol. 309. № 5737, 1052–1056.
14. Huang, T., Bohlenius, H., Eriksson, S. et al. The mRNA of the arabidopsis gene FT moves from leaf to shoot apex and induces flowering. Science. 2005, Vol. 309, № 5741, 1694–1696.
15. Tamaki, S. Matsuo, S. and Wong, H. L. Hd3a protein is a mobile flowering signal in rice. Science. 2007, Vol. 316, № 5827, 1033–1036.
16. Wilkie, J. D., Sedgley, M. and Trevor, O. Regulation of floral initiation in horticultural trees. Journal of experimental botany. 2008, Vol. 59, № 12, 3215–3228.
17. Plant life. In 6 vol. [Ch. Ed. Al. A. Fedorov]. Vol. 5. Part 1. Flowering plants [Ed. AL Takhtadzhyan]. Moscow: Enlightenment, 1980, 430 p. (in Russian).

18. Zlobin, U. A. Plant Physiology and Biochemistry: textbook. Sumy: VTD "University Book", 2004, 464 p. (in Ukrainian).
19. Kozhevnikov, A. V. The spring and autumn in the plant life. Moscow: Publishing House of the Moscow Society of Naturalists, 1950, 240 p. (in Russian).
20. Kosenko, I. S., Opalko, A. I. Adaptive significance of specific elements of the egg apparatus of the *Corylus* L. ssp. Book of abstracts of The International Scientific Conference "Plant cover evolution in the natural and cultigenic environment" dedicated to Charlz Darvin birth 200-th anniversary (Uman, 27–30 October, 2009). Uman: National dendrological park "Sofiyivka" NAS of Ukraine, 2009, 86–88. (in Ukrainian).
21. Galakhov, N. N. Re-bloom of plants. Russian Journal of the Nature. 1937, № 1, 25–28. (in Russian).
22. Galushko, R. V. On the secondary flowering woody plants of the Mediterranean on the southern coast of Crimea. Bulletin of The state Nikitsky botanical gardens. 1980, № 1 (41), 19–25. (in Russian).
23. Zhmylyov, P. Y., Karpuhina, E. A. and Zhmylyova, A. P. Secondary flowering: the induction and development abnormalities. Journal of General Biology. 2009, Vol. 70, № 3, 262–273. (in Russian).
24. Ivashin, D. S. On the regrowth and secondary flowering of plants in the burned areas of the Southern Urals. Botanical Journal. 1958, Vol. 43, № 10, 929–935. (in Russian).
25. Kuznetsova VM The secondary flowering of introducents in the state Nikitsky botanical gardens. Botanical Journal. 1979, Vol. 64, №1, 72–75. (in Russian).
26. Opalko, O. A., Opalko, A. I. Features re-bloom angiosperm woody plants. Autochthonous and alien plants. The collection of proceedings of the National dendrological park "Sofiyivka" of NAS of Ukraine. 2013, Vol. 9. P. 51–60. (in Ukrainian).
27. Nesterov, Y. S. The secondary flowering of fruit crops. Botanical Journal. 1961, Vol. 46, № 2. P. 266–270. (in Russian).
28. Bassi, D., Monet, R. Botany and taxonomy, In: The peach: botany, production and uses [Eds.: Desmond, R. Layne, Daniele Bassi]. Wallingford: CABI, 2008, P. 1–36.
29. Mamajanov DK On the early appearance of fruit forms of walnut. The Herald of Kyrgyz National Agrarian University. 2011, № 2 (20). P. 9–13. (in Russian).
30. Yandovka, L. F. Cyto-morphological features of secondary flowers bloom of *Cerasus vulgaris* (*Rosaceae*) and the prospects for their use in breeding. The successes of modern science. 2005, № 11. C. 52. (in Russian).
31. Bouyanov, M. F. The secondary (autumnal) flowering and fruiting of white mulberry. Botanical Journal. 1956, Vol. 41, №10. P. 1102–1108, (in Russian).
32. Malosieva, G. V. Features of growth and development of the representatives of the genus *Microcerasus* WEBB. Emend Spach in a Botanical Garden after name, E. Gareeva National Academy of Sciences of Kyrgyz Republic. The herald of Kyrgyz national agrarian university. 2011, № 2 (20). P. 48–52. (in Russian).
33. Abdzhunusheva, T. B. Assess the diversity of cotoneaster in a Botanical Garden after name, E. Gareeva National Academy of Sciences of Kyrgyz Republic. The herald of Kyrgyz national agrarian university. 2011, № 2 (20). P. 46–48. (in Russian).
34. Zdruykovskaya, A. I. Nurture of plum seed embryos derived from secondary (autumnal) flowering Agrobiology. 1954, № 4. P. 39–45. (in Russian).

35. Vitkovsky, V. L. Morphogenesis of fruit species. Leningrad: Kolos, 1984, 207 p. (in Russian).
36. Nilsson, O., Weigel, D. Modulating the timing of flowering. Current opinion in biotechnology. 1997, Vol. 8, № 2, 195–199.
37. Shishkova, S. O. Genetics of flower development, In: Genetics of plant development, L. A. Lutova, N. A. Provorov, O. N. Tikhodeev et al.; [Ed.: S. G. Inge-Vechtomov]. St. Petersburg: Science, 2000, Ch. 2.5. P. 201–255. (in Russian).
38. Opalko, A. I., Zaplichko, F. O. Fruit and vegetable crops breeding. Kyiv: Higher School, 2000, 440 p. (in Ukrainian).
39. Visser, T. Juvenile phase and growth of apple and pear sidling. Euphytica. 1964, Vol. 13, № 2. P. 119–129.
40. Kuyan, V. G. Fruit growing. Kyiv: Agricultural Science, 1998, 472 p. (in Ukrainian).
41. Chandler. W. H. Fruit growing. Boston: Houghton Mifflin, 1925, 794 p.

GLOSSARY

allogamy

Cross-fertilization in plants. This promotes genetic variation in the population especially in plants that are dioecious and monoecious. Most plants are hermaphrodite, but have mechanisms such as self-incompatibility that promote allogamy.

autogamy (autogamous)

Obligatory self-fertilization in flowering plants (fertilization of a flower by its own pollen). This restricts genetic variation but allows isolated individuals to reproduce. It is found particularly in pioneer weed species and in ecosystems such as the tundra, where insect vectors are rare. Transitional stages exist between autogamy and allogamy. Very few plants rely exclusively on autogamy. The autogamy as reproduction modes is observed more often among of crop plants, for example, pea, peanut, flax, barley, oats, lupine, rice, soybean, or wheat, etc.

dichogamy

The condition in which the anthers and stigmas mature at different times thus helping prevent self-pollination.

dioecious

Denoting a plant species in which male and female reproductive organs are borne on separate individuals.

fasciation

An abnormal flattening of stems due to failure of the lateral branches to separate from the main stem, as well as it can adhere itself to many parts of a plant including the stem, root, flower head and even the fruits in some rare instances.

florigen

The universal hormone that supposedly causes plants to change from the vegetative to the reproductive state; in 2005, after 70 years of hypothesis O. Nielson et al. (*Science* 2005) identified mRNA of the *FT* gene of thale cress (*Arabidopsis thaliana*) that is produced in leaves and induces flowering when transported to apex tissue.

hermaphrodite plants

A plant having both female and male reproductive organs in the same flower, for example, cotton, sugarbeet, alfalfa, rapeseed, rye, or sunflower etc.

heterogamy

The change in functions of male and female flowers or in their position on the plant (as an anomaly).

heterostyly

A polymorphism among flowers of the same species that ensures cross-fertilization through pollination by visiting insects; flowers have anthers and styles of different length.

intercalary meristem

An internodal meristem, situated between differentiated tissues; it produces cells perpendicular to the growth axis and causing internode elongation.

metagyny

The maturation of male flower before female flower (protandry in dioecious plants).

metandry

The maturation of female flower before male flower (protogyny in dioecious plants).

monoecious

Denoting plants species in which the male and female reproductive organs are borne on separate structures on the same individual plant. Monoecious flowering plants bear separate unisexual male and female flowers, for example, maize, cucumber and many temperate trees such as birch, hazel, oak, pine, spruce, walnut etc. Many monoecious plants are wind pollinated.

photoperiod
The relative length of the periods of light and darkness associated with day and night; in many species, floral induction occurs in response to day-length; species have been categorized according to their day-length requirements as short-day, long-day, intermediate-day, or day-neutral.

photoperiodism
The response of a plant to periodic, often rhythmic, changes in either the intensity of light or to the relative length of day.

protandry
The maturation of anthers before carpels.

protogyny
A condition in which the female parts develop first.

re-bloom
In ornamental plants, a valuable characteristic in which a plant blooms at its normal period and then, after a period of rest, produces a second set of flowers.

remotant
Reblooming or repeat bloomer; a plant that blooms continuously or more than once yearly.

sex dimorphism (sexual dimorphism)
The different morphology of individuals within a species caused by their sexual constitution (e.g., in *Melandrium alba,* hops, hemp, etc.)

vernalization
The treatment of germinating seeds with low temperatures to induce flowering at a particular preferred time. Certain plants will germinate and other

flower only if exposed to low temperatures (1–2°C) at an early period of growth, for example, they have a chilling requirement. Thus, winter cultivars of cereals will only flower in summer if sown the previous autumn. Spring-sown winter cultivars remain vegetative throughout the season unless they have been vernalized.

INDEX

E

H

I